Thermodynamics of Hydrocarbon Reservoirs

Thermodynamics of Hydrocarbon Reservoirs

Abbas Firoozabadi

McGraw-Hill

New York   San Francisco   Washington, D.C.   Auckland   Bogotá
Caracas   Lisbon   London   Madrid   Mexico City   Milan
Montreal   New Dehli   San Juan   Singapore
Sydney   Tokyo   Toronto

Library of Congress Cataloging-in-Publication Data

Firoozabadi, Abbas.
    Thermodynamics of hydrocarbon reservoirs / Abbas Firoozabadi.
        p.   cm.
    Includes index.
    ISBN 0-07-022071-9
    1. Hydrocarbon reservoirs.   2. Thermodynamics.   I. Title.
    TN870.57.F57   1999
    553.2—dc21   98-48667
                                                                        CIP

McGraw-Hill

A Division of The McGraw-Hill Companies

1 2 3 4 5 6 7 8 9 0   AGM/AGM   9 0 3 2 1 0 9 8

ISBN 0-07-022071-9

The sponsoring editor for this book was Robert Esposito and the production supervisor was Pamela Pelton. It was set in Century Schoolbook by Techset Composition Limited.

Printed and bound by Quebecor Martinsburg.

McGraw-Hill books are available at special quantity discounts to use as premiums and sales promotions, or for use in corporate training programs. For more information, please write to the Director of Special Sales, McGraw-Hill, 11 West 19th Street, New York, NY 10011. Or contact your local bookstore.

 This book is printed on recycled, acid-free paper containing a minimum of 50% recycled, de-inked fiber.

Dedicated to my wife Ghashang
and to my mother Azra

Contents

2   General Theory of Phase Equilibria and Irreversible Phenomena in Hydrocarbon Reservoirs   55

3   Equation of State Representation of Reservoir Fluids Phase Behavior and Properties   129

4 Equilibrium, Stability, and Criticality   209

Index   351

Preface

A sun, hidden in a single particle,
opens its mouth,
Suddenly creating heavens and earths
as they spin out of its tiny corner

Rumi, Persian philosopher and poet of the 13th century

Thermodynamics is a fascinating subject. A very broad range of problems are studied through the entropy maximum principle at constant total internal energy, total volume, and total amount of species or the equivalent principle of Gibbs free energy minimum at constant temperature, pressure, and total amount of species. Examples include the interaction between stars (Milne, 1966), the clouds and rainfall (Dufour and Defay, 1963), and the aggregation of molecules (see Chapter 5). No other subject in physical sciences enjoys such a broad applicability at different scales, and with more elegance.

In the description and exploitation of hydrocarbon reservoirs, there are a large number of issues that can only be studied using the rigor of thermodynamics. Since early in the 1900s, through proper application of equilibrium thermodynamics, it has been shown that in some hydrocarbon mixtures, when pressure is increased at constant temperature, the liquid phase may vaporize. This phenomenon is known as retrograde vaporization, which is contrary to intuition where one expects condensation upon pressure increase. There are a large number of other problems that are not as well known, some even unknown to those who work on the subject. For example, when a liquid mixture is held in a capillary tube, its vapor pressure (that is, its saturation pressure) may increase even when gas is the nonwetting phase and liquid is the wetting phase. The widely known Kelvin equation is synonymous with vapor-pressure lowering, which is valid for pure components. For certain fluid mixtures, the saturation pressure increases in a capillary tube. This book is intended to present a step-by-step derivation of basic expressions and to provide the basic concepts for a large number of problems, such as condensation and vaporization in a capillary tube using the Second Law of thermodynamics. The book is also intended to be a graduate-level text. The examples and problems at the end of each chapter are designed to make the book a textbook as well as a reference work.

Chapter 1 reviews basic expressions and simple mathematics for thermodynamic applications. The other chapters provide the solutions to a number of problems and the understanding of these problems. The following discusses the type of issues and problems that can be solved from the material presented in each chapter.

Chapter 1 presents the basic expressions of equilibrium thermodynamics. There is one major problem that is posed in this chapter; at a given temperature, and pressure, when one adds say 4 cm^3 of pure component 1 to 20 cm^3 of component 2, the volume of the mixture can be less than 20 cm^3 at the same temperature and pressure. Reservoir fluids in the critical region may show this behavior when some of the heavier components are mixed with them.

Chapter 2 is one of the two key chapters of the book. The presentation is general for application to a variety of problems in engineering and science. This chapter can be divided into two parts. In the first part, the general theory of equilibrium in the presence of gravity and curved interfaces is presented, and two main problems are covered. The first problem relates to the effect of gravity on equilibrium. Given pressure and composition of a fluid mixture at one point, it is desired to calculate both pressure and composition at any other point at a different height. The temperature is assumed constant. The second problem relates to the effect of interface curvature on equilibrium. Suppose a natural-gas mixture is placed in a network of nonuniform diameter interconnected capillary tubes at a pressure well above the upper dewpoint pressure in a wide container. The dewpoint pressure observed in a large-diameter container would correspond to the upper dewpoint of a flat gas–liquid interface. The pressure in the interconnected network is lowered from the initial pressure. Two items are of interest (1) where will the liquid form first; that is, whether it will form in the smaller or larger tubes, and (2) will the upper dewpoint pressure in the capillary-tube network be higher or lower than the dewpoint in the large-diameter container. In this problem, it is assumed that the process is isothermal, and the pressure lowering is carried out very slowly. In the second part of this chapter, irreversibility and entropy production are presented on a quantitative basis. It is shown that whenever there is a temperature gradient in a system, one may not invoke the criteria of equilibrium. The main problem in the second part is the calculation of composition variation in a nonisothermal system taking into account the effects of thermal diffusion and natural convection.

Chapter 3 contains a simple presentation of cubic equations of state and their strengths for pure components and mixtures. It is shown that cubic equations can be used to calculate (1) volumetric properties, (2) gas and liquid phase compositions, (3) thermal properties, and (4) sonic velocities. The last two items, thermal properties and sonic velocity, are related. Three types of problems are solved in this chapter: (1) two-phase compressibility, (2) two-phase sonic velocity, and (3) heating and cooling due to expansion for multicomponent mixtures. For the two-phase compressibility and two-phase sonic velocity, it is shown that the two-phase gas–liquid compressibility can be greater than the gas compressibility and the two-phase sonic velocity can be less than the gas-phase sonic velocity. The calculation procedures for these two problems are provided in detail. The problem of heating and cooling from expansion is well known, and is much simpler than the first two problems.

Chapter 4 is the second key chapter of the book. The physical meaning and the criteria for stability and criticality are provided from first principles in a step-by-step manner. No such derivations are available in existing books. Three problems are covered in this chapter. The first problem is, given overall composition at a fixed temperature and pressure, how many phases will be in equilibrium and what are the composition and amount of each phase. The second problem relates to the calculation of the stability limit of a given mixture. The third problem is the direct calculation of the critical state of a mixture with many components.

Chapter 5 presents the thermodynamics of wax and asphaltene precipitation. Both solid-solution and the multisolid models are presented for wax precipitation. It is shown that pressure has different effects on wax precipitation in a gas condensate than in a crude oil. The main problem posed for asphaltene precipitation is the pressure and composition effect. Chemicals to inhibit asphaltene precipitation are also discussed. Asphaltene polarity adds great complexity because of its aggregation in the crudes. A thermodynamic micellization model is selected to offer both predictive and explanatory value.

The concepts and theories discussed throughout the book exclude certain behavior of solids and some fluids. For systems that exhibit phase transitions of other than order one, the thermodynamic features are somewhat complicated. There are two kinds of phase transitions: transitions of the first order for which energy, volume, and crystal structure change discontinuously, and transitions of second and higher order or continuous transitions, in which energy and volume change continuously and the temperature derivatives of these quantities have singularities. The phase transitions of the second kind mark the onset of superconductivity, superfluidity, the establishment of various types of molecular order in alloys, etc. Gibbsian thermodynamics have also been applied to include continuous transitions by Tisza (1951), who provides a general theory of phase transitions and develops criticality criteria for solids from the fundamental principles of thermodynamics.

References

1. Dufour, L., and R. Defay, *Thermodynamics of Clouds*, Academic Press, New York, 1963.
2. Milne, E. A., "Transfer of Radiation," *Thermodynamics of Stars*. Menzel, D. H., (editor), Dover Publications, Inc., New York, 1966.
3. Tisza, L. *On the General Theory of Phase Transitions, Phase Transformation in Solids*, Smoluchowski, R., Mayer, J. E., and Weyl, W. A. (editors). John Wiley & Sons, New York, p. 1, 1951.

Acknowledgements

Much of this book draws upon my teachings of two graduate courses: 1) advanced gas engineering at Stanford University in 1985 and 1987, and 2) advanced thermodynamics of hydrocarbon reservoirs at the University of Texas—Austin in 1996. Gary Pope and the late Henry Ramey Jr. suggested the idea that I teach these courses at the University of Texas—Austin and Stanford, respectively. Without their initiative, I would not have been able to write this book.

Many friends and colleagues have furthered my understanding of thermodynamics and have ably assisted me in the preparation of this book. I would like to thank Huanquan Pan of RERI, Eric Chang of Arco Oil and Gas, Kramer Luks of the University of Tulsa, Keshawa Shukla of Rice University, Dimitri Gidaspow of the Illinois Institute of Technology, Terry Wong of Landmark Graphics Corporation, John Prausnitz of the University of California—Berkeley, Kassem Ghorayeb of RERI, and Peder Tyvand of Agricultural University of Norway for fueling my investigations with their collaboration.

I have been fortunate enough to develop friendships and colleagial relations with people from the oil industry, both in the United States and abroad, who have supported my work. The support and technical discussions with L. Kent Thomas of Phillips Petroleum Company have provided a unique opportunity to carry out my research in the last ten years. Norsk Hydro ASA of Norway has provided me ample opportunity to study various oil and gas reservoirs of the North Sea from which I have abstracted some fundamental problems. My thanks goes out to Theodore Van Golf-Racht who initiated the collaboration with Norsk Hydro ASA. Discussions with the many outstanding engineers and researchers of Norsk Hydro ASA have helped me to clarify certain ideas. Bret Beckner of Mobil has also been instrumental in support of my research activities at RERI. In addition to close working relationships with Norsk Hydro ASA, Phillips Petroleum Company, and Mobil

Oil Corporation, the following companies have supported my research in the past 10 years: Abu Dhabi National Oil Company, Amoco Production Company, ARCO Exploration and Production Technology, British Petroleum, Chevron Petroleum Technology Company, Elf Aquitaine Production, Exxon Production Research Company, Intevep S.A., Japan National Oil Corporation, Maersk Oil and Gas A.S., Marathon Oil Company, Petrobras S.A., Petrofina S.C.A., Petronas Research and Scientific Services SDN BHD, SAGA Petroleum a.s., Shell Exploration and Production Company, Statoil, Texaco Inc., Total, Unocal, and Union Pacific Resources. The US Department of Energy also provided funding.

I am indebted to the late Donald Katz of the University of Michigan from whose critical mind and clarity of thought I have learned and benefited. Likewise with Hassan Sadat, formerly of the National Iranian Oil Company, who was a source of moral support. In many more ways than can be mentioned here, these two individuals assisted in the undertaking of this project.

Felix Wong and Debbie Rowland, both students from Stanford University in mechanical engineering and environmental science, respectively, performed the seemingly infinite tasks of manuscript and graph preparation—with endless revisions. Their pleasant manner and willingness to help at all hours of the day or night kept my spirits up and allowed me to focus on the task at hand. Kathleen Simoudis, our office manager at RERI, took care of administrative tasks beyond her duties and kindled an atmosphere that allowed for uninterrupted hours of work. Finally, my son, Reza, has been immensely helpful and supportive in the preparation of this book.

To everyone, my thanks and gratitude goes out. I am lucky to enjoy the friendship and support of so many people.

Abbas Firoozabadi

Notation

All the symbols used in the book are listed here except for those that appear infrequently or only in one section.

A = Helmholtz free energy
A = (aP/R^2T^2)
$\mathcal{A}$ = surface area
a = attractive parameter of the EOS
a = sonic velocity
a = unit surface area of micellar core given by $(4\pi R^2/n_2)$, where R is micellar core radius and n_2 is the number of resin molecules absorbed onto the micellar core
a_p = effective cross-sectional area of a resin molecule
a_o = contact area of a resin molecule onto the surface of the micellar core
B = (bP/RT)
b = covolume in the EOS
c = number of components
c = volume-translation constant
C_T = isothermal compressibility
C_S = isentropic compressibility
c_S = number of solid phases
c_P = molar heat capacity at constant pressure
c_V = molar heat capacity at constant volume
C^P = coefficient of pressure diffusion term
C^T = coefficient of thermal diffusion term
C^x = coefficient of molecular diffusion term
D = molecular diffusion coefficient
D = shell thickness in a micelle
d = differential
d_- = infinitesimal change

$\bar{E}_i$	=	partial molar property of component i
e	=	thermal expansivity
F	=	feed moles
F	=	degree of freedom
f	=	fugacity
G	=	gas gravity; $(M/29)$
G	=	Gibbs free energy
g	=	acceleration of gravity
g	=	molar Gibbs free energy
$\bar{G}_i$	=	partial molar Gibbs free energy
H	=	enthalpy
H	=	height
$\bar{H}_i$	=	partial molar enthalpy of component i
h	=	molar enthalpy
h^*	=	ideal-gas molar enthalpy
i	=	component index in a mixture
J	=	curvature
J	=	Jacobian
$\vec{J}_i$	=	diffusion flux of component i
$\vec{J}_i^s$	=	entropy flux of component i
K	=	association constant
K	=	thermal conductivity
K_i	=	equilibrium ratio of component i
k	=	permeability
k	=	Boltzmann's constant, $(1.3805 \times 10^{-23}\ \mathrm{J/K})$
k_T	=	thermal diffusion ratio
L	=	characteristic length of a molecule
L	=	mole fraction of liquid phase
M	=	molecular weight
$\bar{M}$	=	average molecular weight
m	=	mass
m_a	=	molecular volume ratio of asphaltene to asphalt-free oil
m_r	=	molecular volume ratio of resin to asphalt-free oil
N_A	=	Avogadro's number (6.02217×10^{23})
N	=	number of moles
n	=	total number of moles of a mixture
n_T	=	total number of moles of a mixture
n_i	=	number of moles of component i
$\underline{\boldsymbol{n}}$	=	$(n_1, n_2, \ldots, n_c)$
$\boldsymbol{n}_i$	=	$(n_1, n_2, \ldots, n_{i-1}, n_{i+1}, \ldots, n_c)$
n_1	=	number of asphaltene molecules in the micellar core
n_2	=	number of resin molecules in the solvated shell
P	=	pressure
P	=	parachor
P_c	=	capillary pressure

$$
\begin{array}{rcl}
P^f &=& \text{pressure at the melting point} \\
p &=& \text{number of phases, number of subsystems} \\
Q &=& \text{heat quantity} \\
\vec{q}^{\,\text{cond}} &=& \text{energy flux by conduction} \\
R &=& \text{gas constant, } [8.3144\,\text{J}/(\text{gmole.K}) = 82.0567 \\
&& (\text{atm.cm}^3)/(\text{mole.K}) = 10.73(\text{psia.ft}^3)/(\text{lbmole.R})] \\
R &=& \text{radius of micellar core} \\
r &=& \text{radius of bubble or droplet} \\
S &=& \text{saturation} \\
S &=& \text{entropy} \\
\bar{S}_i &=& \text{partial molar entropy} \\
s &=& \text{molar entropy} \\
T &=& \text{temperature} \\
T^f &=& \text{fusion-point temperature} \\
U &=& \text{internal energy} \\
\bar{U}_i &=& \text{partial molar internal energy} \\
u &=& \text{molar internal energy} \\
V &=& \text{volume} \\
V &=& \text{volume fraction of gas phase} \\
V &=& \text{molecular volume} \\
\bar{V}_i &=& \text{partial molar volume of component } i \\
v &=& \text{specific volume} \\
v &=& \text{molar volume} \\
\vec{v} &=& \text{velocity vector} \\
v_x &=& \text{velocity in } x\text{-direction} \\
W &=& \text{width} \\
W &=& \text{work} \\
w &=& \text{mass fraction} \\
x &=& x\text{-axis} \\
x &=& \text{mole fraction (often in the liquid phase)} \\
\underline{x} &=& (x_1, x_2, \ldots, x_c) \\
\boldsymbol{x} &=& (x_1, x_2, \ldots, x_{c-1}) \\
y &=& \text{mole fraction (often of the gas phase)} \\
\underline{y} &=& (y_1, y_2, \ldots, y_c) \\
\boldsymbol{y} &=& (y_1, y_2, \ldots, y_{c-1}) \\
Z &=& \text{compressibility factor} \\
z &=& \text{vertical direction, positive upwards} \\
z &=& \text{overall mole fraction} \\
\underline{z} &=& (z_1, z_2, \ldots, z_c) \\
\boldsymbol{z} &=& (z_1, z_2, \ldots, z_{c-1}) \\
\alpha &=& \text{mole fraction of vapor phase} \\
\alpha &=& \text{temperature-dependent parameter of } a \text{ in the EOS} \\
\alpha &=& \text{thermal diffusion factor [for a binary mixture} \\
&& \alpha = k_T/(x_1 x_2)] \\
\gamma &=& \text{activity coefficient, symmetrical}
\end{array}
$$

γ^*	$=$	activity coefficient, asymmetrical
δ	$=$	interaction coefficient
δ	$=$	solubility parameter
δ_{jk}	$=$	dirac delta function, $+1$ if $j = k$, 0 if $j \neq k$
Δ	$=$	change in the variable from one state to another state,
ΔG^{oo}	$=$	standard Gibbs free energy of micellization
Δh^o	$=$	enthalpy of dimer formation
Δs^o	$=$	entropy of dimer formation
Δh^f	$=$	enthalpy of fusion
η	$=$	molar density
θ	$=$	contact angle
λ	$=$	positive multiplier
μ	$=$	chemical potential, per mole, or per molecule
$\tilde{\mu}$	$=$	μ/M, per mole
μ	$=$	viscosity
μ	$=$	Joule-Thomson coefficient
π	$=$	$3.14159\ldots$
π	$=$	osmotic pressure
ρ	$=$	fluid mass density
σ	$=$	entropy production per unit time per unit volume
σ	$=$	interfacial tension
$\vec{\vec{\tau}}$	$=$	stress tensor
τ_{ij}	$=$	stress in the ij plane
Φ	$=$	volume fraction
φ	$=$	fugacity coefficient
ϕ	$=$	porosity, often fraction
ψ	$=$	stream function, $v_x = -\partial\psi/\partial z$
ψ_m	$=$	modified stream function, $\rho v_x = -\partial\psi_m/\partial z$
Ω_a^o	$=$	constant of a parameter of the EOS
Ω_b^o	$=$	constant of b parameter of the EOS
ω	$=$	acentric factor

Superscripts

I	phase I
II	phase II
$'$	phase index
$''$	phase index
o	standard state
o	reference space
∞	flat interface
∞	infinite dilution
$*$	infinite dilution standard state
b	bubble

f	melting point, also fusion
G	gas
GL	gas–liquid
GS	gas–solid
(k)	kth Legendre transformation
LS	liquid–solid
L_1	liquid phase
L_2	second liquid phase
L	liquid
S	solid
tp	two-phase
V	vapor
$\rightharpoonup$	vector
$\rightrightarrows$	tensor

Subscripts

$'$	monomeric state
a	asphaltene
b	bulk phase
c	critical
i	axis index: as an example, x-axis
i	component index
j	axis index, as an example y-axis
j	phase index
k	derivation with respect to variable number k
kk	second derivative with respect to variable number k
r	reduced, relative to critical value
r	resin
x	derivative with respect to variable x; x can be V, T, P or any other variable

∇	gradient
$\nabla \cdot$	divergence
$\nabla \times$	curl

Chapter 1

Review of basic concepts and equations

Thermodynamics can be divided into subjects which deal with (1) equilibrium, (2) nonequilibrium, and (3) irreversible processes. All three of these subdivisions are important in hydrocarbon reservoirs and in the interpretation of laboratory experiments for the understanding of hydrocarbon reservoirs. However, equilibrium thermodynamics is by far the most important and the best understood subject. According to Tisza (1966), the subdivision of equilibrium thermodynamics can be carried out further into Gibbsian thermodynamics and the early thermodynamics of Clausius and Kelvin. The latter considered the thermodynamic system as a black box, and all the relevant information was then derived from the energy absorbed and the work done by the system. The concepts of internal energy, U, and entropy, S, from the observable quantities are then established. In Gibbsian thermodynamics, the concepts of internal energy and entropy are assumed to be known and are used to provide a detailed description of the subsystems in equilibrium (we will soon define some of the terms used above).

In this chapter, we will mainly consider the Gibbsian thermodynamics of phase equilibria relevant to problems in hydrocarbon reservoirs and use its concepts in the other chapters to solve practical problems. The thermodynamics of equilibrium processes also provide the framework for nonequilibrium and irreversible thermodynamics. It is our intention that the material covered in this book should be self-contained. The postulational approach introduced by Callen (1985), and Tisza (1966) is, therefore, adopted to make brief the presentation of basic concepts and equations.

Thermodynamics consists essentially of the first, second, and third laws. However, to obtain useful results, these laws should be combined with an equation of state (EOS) to provide a knowledge of the fluid properties at any point in the system. Since gravity and capillary forces are important in most hydrocarbon reservoirs, the concepts of phase-equilibria thermodynamics developed in Chapter 2 will include these forces. However, we assume the absence of elastic, electric, and magnetic effects. We also exclude chemical reaction.

Prior to addressing various concepts, a limited number of terms often used in thermodynamics will be defined.

Mole number $\equiv$ actual number of moles of each type of species (i.e., number of molecules, if there is no association between molecules divided by Avogadro's number, $N_A = 6.02217 \times 10^{23}$). Alternatively, mole number is the amount of mass of each type of species divided by its molecular mass. The mole number of species is denoted by n_i or N_i.

Mole fraction $\equiv n_i/n_T$, where $n_T = \sum_{i=1}^{c} n_i$ is the total mole number and c is the number of different species in the mixture; the subscript T is often dropped from n_T.

Molar quantity $\equiv$ the ratio of a particular property of a phase divided by the total number of moles, n_T. As an example, the molar volume is defined by $v \equiv V/n_T$, where V is the total volume.

Wall, enclosure, partition $\equiv$ a physical system idealized as a surface forming the common boundary of two different systems. The walls that completely enclose a system are called enclosures. Walls separating the subsystems of a composite system are called partitions.

Impermeable wall $\equiv$ a wall that is restrictive of matter.

Semipermeable wall $\equiv$ a wall that is restrictive to only certain chemical species and nonrestrictive to others.

Permeable wall $\equiv$ a wall that is nonrestrictive of matter and energy.

Adiabatic wall $\equiv$ a wall that is restrictive of matter and energy.

Rigid wall $\equiv$ a wall that does not move or deform.

Diathermal wall $\equiv$ a wall that is restrictive of matter but nonrestrictive of energy.

Isolated system $\equiv$ a system enclosed in a completely restrictive enclosure.

Closed system $\equiv$ a system in an impermeable wall.

Open system $\equiv$ system with a wall that is nonrestrictive to matter.

Heat $\equiv$ a transfer quantity associated with energy transferred across a rigid diathermic wall, $Q = \Delta U$.

Composite system ≡ a conjunction of spatially disjoint simple systems (subsystems or phases).

Subsystems or phases ≡ simple systems or phases that form a composite system. A phase may be homogeneous throughout or may have a continuous variation of properties.

Extensive variables ≡ variables that depend on the total quantity of matter in the system or subsystem. Examples are volume, V, and mole numbers, n_i.

Intensive variables ≡ variables that have point values and are independent of the size of the system or subsystem. Examples are T and P.

Equilibrium state ≡ the state in which the properties are determined by intrinsic properties that are time independent. A more precise definition is deferred to later.

Next we present the five postulates of thermodynamics. These postulates are mainly adopted from Callen (1985).

Postulate I.   The conservation of the internal energy, U, of a system, is postulated in the first law of thermodynamics. The internal energy of a closed system is defined in the form as

$$dU = d_-Q + d_-W, \tag{1.1}$$

where dU is the differential of the internal energy, U, and d_-Q and d_-W are the infinitesimal amounts of heat absorbed by the system, and of the work done on the system across the system boundary, respectively (see Fig. 1.1). In Eq. (1.1), dU is the differential of the state variable U. Q and W are not functions of state and do not have, in general, differentials dQ and dW in the state-space. Alternatively, it is said that U has an exact differential, and Q and W do not. (We will soon define an exact differential.) Note that Q and W depend on the interactions between the system and its surrounding. For the above system, if the enclosure is rigid and the system is not displaced in the vertical direction,

$$\Delta U = Q, \tag{1.2}$$

where Q is the heat added to the system at constant volume and ΔU is the change in the internal energy as a result of heat absorbed by the system.

On the other hand if the enclosure is adiabatic, but not rigid, then

$$\Delta U = W, \tag{1.3}$$

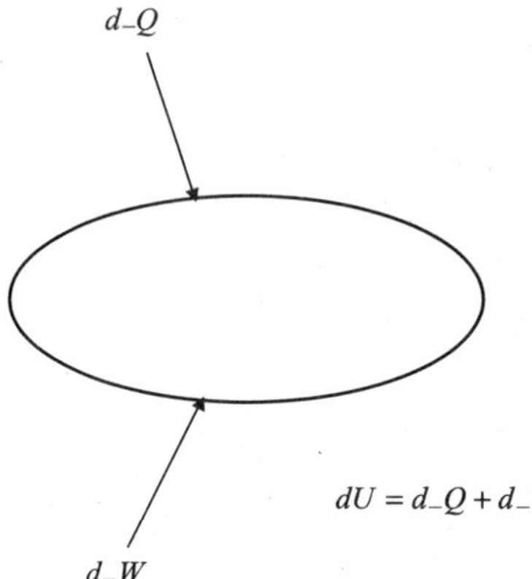

Figure 1.1 Sketch defining the first law of thermodynamics (the system is closed).

where W is the work done on the system. As an example, if the volume of the system increases by dV and the system is subjected to a uniform external pressure P, we have

$$d_-W = -PdV \tag{1.4}$$

or at constant pressure

$$W = -P\Delta V, \tag{1.5}$$

provided the work of expansion is the only kind of work.

Unlike Q and W, which have absolute values, U does not have an absolute value; Q and W are measurable quantities, U is measured only with respect to a reference state. Internal energy, U, is also an extensive variable.

Postulate II.   The equilibrium states of the subsystems of a composite system are fully defined by the internal energy, U, the volume V, and the mole numbers $n_1, n_2, \ldots, n_c$ of the composite system. Note that $U, V, n_1, n_2, \ldots, n_c$ are extensive variables and are, therefore, additive properties. ($U = \sum_{j=1}^{p} U_j$, $V = \sum_{j=1}^{p} V_j$, and $n_i = \sum_{j=1}^{p} n_{ij}$, where p is the total number of subsystems and n_{ij} is the number of moles of component i in subsystem j.)

Postulate III.   There exists an additive variable S (called entropy) that is a function of extensive variables $U, V, n_1, n_2, \ldots, n_c$ of the composite system and has the characteristic that at equilibrium it is a maximum with respect to all of the extensive variables $U, V, n_1, n_2, \ldots, n_c$ of the subsystems.

From Postulates II and III, the entropy S of the composite system consisting of primed and double-primed subsystems, is a maximum at equilibrium with respect to $U', V', n_1', n_2', \ldots, n_c'$, and $U'', V'', n_1'',$ $n_2'', \ldots, n_c''$.

Postulate IV.   The entropy, S, is continuous and differentiable and is a monotonically increasing function of the internal energy, U, of the composite system.

From postulates II and III, we can write the following expressions:

$$S = S' + S'' \tag{1.6}$$

$$S' = S'(U', V', n_1', \ldots, n_c') \tag{1.7}$$

$$S'' = S''(U'', V'', n_1'', \ldots, n_c''), \tag{1.8}$$

where the composite system is assumed to be composed of subsystems primed and double-primed. Equations (1.6) to (1.8) are based on the additivity property of the entropy. S' and S'' are first-order homogeneous functions of the extensive properties $U', V', n_1', \ldots, n_c'$ and $U'', V'', n_1'', n_2'', \ldots, n_c''$, respectively; i.e.,

$$S'(\lambda U', \lambda V', \lambda n_1', \ldots, \lambda n_c') = \lambda S'(U', V', n_1', \ldots, n_c'), \tag{1.9}$$

where extensive variables are multiplied by a positive constant λ. The definition of a homogeneous function will be given soon.

The mathematical expression for the monotonically increasing property of S with respect to U can be written as

$$(\partial S/\partial U)_{V, n_1, \ldots, n_c} > 0, \tag{1.10}$$

where all the variables are for the composite system. The thermodynamic temperature T is defined as

$$T = (\partial U/\partial S)_{V, n_1, \ldots, n_c}. \tag{1.11}$$

Equation (1.11) implies that T cannot be negative. However, as we will see in Chapter 3, there is not such a restriction on pressure; it can be negative.

The continuity, differentiability, and monotonic property imply that

$$S = S(U, V, n_1, \ldots, n_c) \tag{1.12}$$

can be uniquely solved for U given by

$$U = U(S, V, n_1, n_2, \ldots, n_c). \tag{1.13}$$

Postulates II, III, and IV define the second law. Finally the third law of thermodynamics is also defined on the postulational basis.

Postulate V.   The entropy, S, vanishes as $T \to 0$.

Therefore, similarly to V, and n_i, and unlike U, S has a uniquely defined zero.

Prior to the presentation on equilibrium, we will discuss exact differentials and homogeneous functions.

Exact differentials.   Let $F(x_1, x_2, x_3, \ldots)$ be a continuous function of independent variables $x_1, x_2, x_3, \ldots$ except at certain isolated singular points. The total differential of F is

$$dF = (\partial F/\partial x_1)_{x_2, x_3, \ldots} dx_1 + (\partial F/\partial x_2)_{x_1, x_3, \ldots} dx_2 + \cdots, \qquad (1.14)$$

which is called an exact differential. Since F is continuous, it possesses continuous higher-order derivatives except at isolated singular points. If F is integrated over some path connecting two points A and B in the $x_1, x_2, \ldots$ space, then $F(\text{at } B) - F(\text{at } A)$ depends on the points A and B and not on the path connecting A and B.

Now consider the differential

$$d\varphi = X_1 \, dx_1 + X_2 \, dx_2 + \cdots \qquad (1.15)$$

where $X_1, X_2, \ldots$ are functions of $x_1, x_2, \ldots$ The above differential is an exact differential if $X_i = (\partial \varphi/\partial x_i)_{x_1, \ldots, x_{i-1}, x_{i+1}, \ldots}$.

Homogeneous functions.   A function of several variables is said to be homogeneous of degree l if multiplying each variable by a positive constant, k, is the same as multiplying the original function by k^l. As an example, consider $F(x,y)$. This function is said to be a homogeneous function of degree l if $F(kx, ky) = k^l F(x, y)$. If $l = 1$, then F is said to be a first-order homogeneous function of x and y.

Conditions for equilibrium

The fundamental equation for U in an open system is

$$U = U(S, V, n_1, n_2, \ldots, n_c) \qquad (1.16)$$

which in the differential form can be written as

$$dU = (\partial U/\partial S)_{V,\underline{n}} dS + (\partial U/\partial V)_{S,\underline{n}} dV + \sum_{i=1}^{c} (\partial U/\partial n_i)_{S, V, n_i} dn_i, \qquad (1.17)$$

where $\underline{n} \equiv (n_1, n_2, \ldots, n_c)$ and $\boldsymbol{n}_i \equiv (n_1, n_2, \ldots, n_{i-1}, n_{i+1}, \ldots, n_c)$. Therefore, the $\boldsymbol{n}_i$ symbol denotes that all the variables $n_1, n_2, \ldots, n_c$ are held constant except for n_i.

The first and second partial derivatives on the left side are given by

$$(\partial U/\partial S)_{V,\underline{n}} = T \tag{1.18}$$

$$(\partial U/\partial V)_{S,\underline{n}} = -P. \tag{1.19}$$

Equation (1.18) was established before; Eq. (1.19) can be implied from $dU = TdS + d_-W$ and $d_-W = -PdV$ (for a closed system). The coefficient derivatives of the last term are defined by

$$\mu_i = (\partial U/\partial n_i)_{S,V,\underline{n}_i}, \tag{1.20}$$

and are called chemical potentials. The chemical potential has a function analogous to temperature and pressure. The temperature difference causes heat flow, and pressure difference in the absence of gravity may result in bulk phase displacement. The chemical potential difference results in diffusion from the region of higher to the region of lower chemical potential in the absence of gravity effect (see Example 1.1). The chemical potential concept was introduced by Gibbs (1957). This concept, as we will see later, facilitates the description of the phase behavior of open systems. Denbigh (1971) presents the interpretation that the last term in Eq. (1.17), i.e., $\sum_{i=1}^{c} \mu_i dn_i$, is a form of work which can be done by a system at constant volume due to its change of composition.

Combining Eqs. (1.17) to (1.20), the expression for dU is given by

$$dU = TdS - PdV + \sum_{i=1}^{c} \mu_i dn_i. \tag{1.21}$$

Note that the coefficients of the independent extensive variables in Eq. (1.21)—T, V, and μ_i—are intensive variables. Upon rearrangement,

$$dS = \frac{dU}{T} + \frac{P}{T}dV - \sum_{i=1}^{c} \frac{\mu_i}{T} dn_i. \tag{1.22}$$

In the following, we will use Eq. (1.22), $S = S(U, V, n_1, \ldots, n_c)$, and the entropy maximum principle from postulate III to derive the conditions for various types of equilibria.

Thermal equilibrium.   Consider an isolated composite system consisting of two simple subsystems separated by a rigid diathermic partition. Therefore, $dV' = 0, dV'' = 0, dn_i' = 0, dn_i'' = 0, i = 1, \ldots, c$, and $d(U' + U'') = 0$ (see Fig. 1.2).

Mathematically, the entropy maximum principle can be written as

$$dS = 0, \tag{1.23}$$

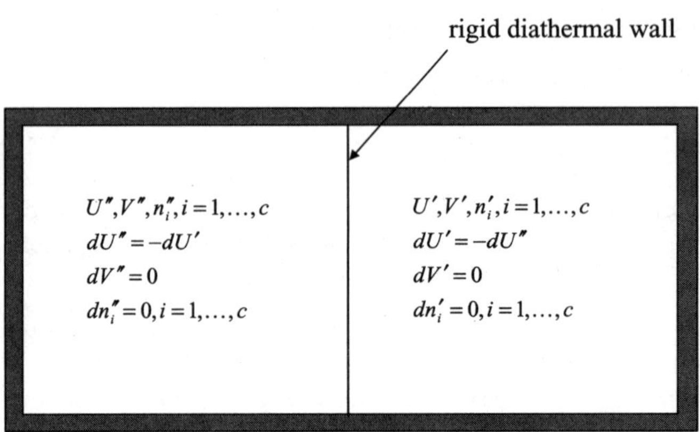

Figure 1.2 Isolated composite system with rigid diathermal wall.

where $S = S' + S''$. (As we will see in detail in Chapter 4, to ensure the maxima, another requirement has to be met.) Writing Eq. (1.23) for the two subsystems,

$$dS' = \frac{dU'}{T'} + \frac{P'}{T'}dV' - \sum_{i=1}^{c}\left(\frac{\mu_i'}{T'}\right)dn_i' \tag{1.24}$$

$$dS'' = \frac{dU''}{T''} + \frac{P''}{T''}dV'' - \sum_{i=1}^{c}\left(\frac{\mu_i''}{T''}\right)dn_i'', \tag{1.25}$$

and from $dV' = dV'' = dn_i' = dn_i'' = 0$,

$$dS = dS' + dS'' = \frac{dU'}{T'} + \frac{dU''}{T''}, \tag{1.26}$$

and since $dU' = -dU''$, then

$$dS = \left(\frac{1}{T'} - \frac{1}{T''}\right)dU'. \tag{1.27}$$

From Eqs. (1.23) and (1.27),

$$T' = T'', \tag{1.28}$$

which states that the temperatures of the two subsystems should be equal at equilibrium for a diathermal wall.

Mechanical equilibrium.   Now we consider an isolated system consisting of two simple subsystems separated by a moveable diathermic partition that does not change shape. Therefore, $dV' + dV'' = 0, dU' + dU'' = 0,$

and $dn_i' = dn_i'' = 0$. Invoking the maximum entropy principle (Eq. (1.23)) and writing Eqs. (1.24) and (1.25) with the above constraints,

$$dS = \left(\frac{1}{T'} - \frac{1}{T''}\right)dU' + \left(\frac{P'}{T'} - \frac{P''}{T''}\right)dV' = 0 \qquad (1.29)$$

Since U' and V' are independent variables, each of the two coefficients, $(1/T' - 1/T'')$ and $(P'/T' - P''/T'')$ must be zero. Therefore,

$$T' = T''$$

$$P' = P''. \qquad (1.30)$$

Note that the equality of pressures can also be derived from the force balance. The equilibrium with a moveable adiabatic wall is an indeterminate problem.

Chemical equilibrium.   In this case, the partition of the isolated composite system is permeable to some or all of the species and the partition can be either rigid or moveable. The constraint equations are $d(U' + U'') = 0$, $d(V' + V'') = 0$, and $d(n_i' + n_i'') = 0; i = 1, \ldots, c$ when all species can freely pass through the moveable partition. Similarly to the previous two cases, we invoke the maximum entropy principle, and use Eqs. (1.24) and (1.25),

$$dS = \left(\frac{1}{T'} - \frac{1}{T''}\right)dU' + \left(\frac{P'}{T'} - \frac{P''}{T''}\right)dV' - \sum_{i=1}^{c}\left(\frac{\mu_i'}{T'} - \frac{\mu_i''}{T''}\right)dn_i' = 0. \quad (1.31)$$

Since U', V', and n_i' are independent of each other, then the coefficients of dU', dV', and dn_i' must be zero for dS to be zero, which provides the conditions for chemical equilibrium,

$$T' = T''$$

$$P' = P''$$

$$\mu_i' = \mu_i'' \qquad i = 1, \ldots, c. \qquad (1.32a)$$

The flow of species between the subsystems results in the equality of chemical potentials at equilibrium, and therefore the equilibrium is classified as chemical equilibrium.

For a rigid partition, if some of the components can pass freely between the two subsystems and other components cannot go through the partition, one can readily show that

$$T' = T''$$

$$\mu_i'(T', P', n_1', n_2', \ldots, n_c') = \mu_i''(T', P'', n_1'', n_2'', \ldots, n_c'')$$

$$i = 1, \ldots, l, \ l \leqslant c. \tag{1.32b}$$

The difference $(P'' - P')$ is the osmotic pressure. In Eq. (1.32b), l is the number of components that can pass freely between the two subsystems.

Mathematical properties of *U* and *S*

The additivity property of entropy and internal energy of subsystems demands that both S and U of the subsystems be first-order homogeneous functions of extensive properties which define the subsystem. From the definition of the first-order homogeneous property for U, one can write

$$U(\lambda S, \lambda V, \lambda n_1, \lambda n_2, \ldots, \lambda n_c) = \lambda U(S, V, n_1, n_2, \ldots, n_c) \tag{1.33}$$

where λ is a positive parameter.

Differentiating Eq. (1.33) with respect to λ,

$$\left(\frac{\partial U}{\partial \lambda S}\right)_{\lambda V, \lambda \underline{n}} \left(\frac{d\lambda S}{d\lambda}\right) + \left(\frac{\partial U}{\partial \lambda V}\right)_{\lambda S, \lambda \underline{n}} \left(\frac{d\lambda V}{d\lambda}\right) + \sum_{i=1}^{c} \left(\frac{\partial U}{\partial \lambda n_i}\right)_{\lambda S, \lambda V, \lambda n_i} \left(\frac{d\lambda n_i}{d\lambda}\right)$$

$$= U(S, V, n_1, \ldots, n_c) \tag{1.34}$$

or

$$\left(\frac{\partial U}{\partial \lambda S}\right) S + \left(\frac{\partial U}{\partial \lambda V}\right) V + \sum_{i=1}^{c} \left(\frac{\partial U}{\partial \lambda n_i}\right) n_i = U(S, V, n_1, \ldots, n_c), \tag{1.35}$$

where for brevity we have dropped the parameters that are held constant.

For $\lambda = 1$, Eq. (1.35) becomes

$$\left(\frac{\partial U}{\partial S}\right) S + \left(\frac{\partial U}{\partial V}\right) V + \sum_{i=1}^{c} \left(\frac{\partial U}{\partial n_i}\right) n_i = U(S, V, n_1, \ldots, n_c). \tag{1.36}$$

The partial derivatives of the first, second, and third terms on the left side are T, $-P$, and μ_i, respectively. Therefore, the expression for U is

$$U = TS - PV + \sum_{i=1}^{c} \mu_i n_i \ . \tag{1.37}$$

Of all the parameters on the right side of Eq. (1.37), only μ_i does not have an absolute value, which as expected makes U also not have an absolute value. The chemical potential's relative value will be discussed later.

From Eq. (1.37), one can derive the expression for S,

$$S = \frac{U}{T} + \left(\frac{P}{T}\right) V - \sum_{i=1}^{c} \frac{\mu_i}{T} n_i \ . \tag{1.38}$$

Gibbs-Duhem equation

The intensive properties of a phase—temperature, T, pressure, P, and the chemical potentials, μ_i—are not independent of each other. The Gibbs-Duhem equation provides the relationship between these variables. Let us write Eq. (1.37) in differential form:

$$dU = TdS + SdT - PdV - VdP + \sum_{i=1}^{c} \mu_i dn_i + \sum_{i=1}^{c} n_i d\mu_i. \tag{1.39}$$

Comparison of the above equation with Eq. (1.21) provides the Gibbs-Duhem equation,

$$SdT - VdP + \sum_{i=1}^{c} n_i d\mu_i = 0 \ . \tag{1.40}$$

Equation (1.40) demonstrates that T, P, and μ_i of a phase are not independent and provides the relation between them. We will use this equation extensively in Chapters 2 and 4.

For a single-component system, the Gibbs-Duhem equation is written as

$$SdT - VdP + nd\mu = 0 \tag{1.41}$$

or

$$d\mu = -sdT + vdP, \tag{1.42}$$

where s and v are the molar entropy and molar volume, respectively. At constant temperature, Eq. (1.42) becomes

$$(d\mu = vdP)_T. \tag{1.43}$$

Integrating Eq. (1.43) from pressure P' to a pressure P'',

$$\mu(P'', T) = \mu(P', T) + \left(\int_{P'}^{P''} v\,dP \right)_T . \tag{1.44}$$

Now, for the first time we observe the relation between the chemical potential and pressure–volume relationship. This implies that the change in the chemical potential can be determined from an equation of state, as we will discuss in Chapter 3.

Other fundamental equations

The enthalpy, H, Helmholtz free energy, A, and the Gibbs free energy, G, are respectively defined as

$$H = U + PV \tag{1.45}$$

$$A = U - TS \tag{1.46}$$

$$G = A + PV = U - TS + PV = H - TS. \tag{1.47}$$

Later we will see the usefulness of the above fundamental equations.

Combining Eqs. (1.45), (1.46), (1.47), and (1.37), one obtains

$$\boxed{H = TS + \sum_{i=1}^{c} n_i \mu_i} \tag{1.48}$$

$$\boxed{A = -PV + \sum_{i=1}^{c} n_i \mu_i} \tag{1.49}$$

$$\boxed{G = \sum_{i=1}^{c} n_i \mu_i} \tag{1.50}$$

where μ_i is given by Eq. (1.20). Note that Eqs. (1.45) to (1.50) apply to both closed and open systems. Let us write the differential form of Eq. (1.45):

$$dH = dU + PdV + VdP. \tag{1.51}$$

Combining Eqs. (1.51) and (1.21) provides

$$\boxed{dH = TdS + VdP + \sum_{i=1}^{c} \mu_i dn_i} , \tag{1.52}$$

from which one may infer that

$$H = H(S, P, n_1, n_2, \ldots, n_c). \tag{1.53}$$

Equation (1.53) in differential form is written as

$$dH = \left(\frac{\partial H}{\partial S}\right)_{P,\underline{n}} dS + \left(\frac{\partial H}{\partial P}\right)_{S,\underline{n}} dP + \sum_{i=1}^{c}\left(\frac{\partial H}{\partial n_i}\right)_{S,P,n_i} dn_i. \qquad (1.54)$$

Comparison of Eqs. (1.52) and (1.54) leads to

$$\mu_i = \left(\frac{\partial H}{\partial n_i}\right)_{S,P,n_i}. \qquad (1.55)$$

Similarly, we can write the differential forms of Eqs. (1.46) and (1.47) and combine the results with Eq. (1.21) to obtain

$$\boxed{dA = -SdT - PdV + \sum_{i=1}^{c} \mu_i dn_i} \qquad (1.56)$$

$$\boxed{dG = -SdT + VdP + \sum_{i=1}^{c} \mu_i dn_i}, \qquad (1.57)$$

which imply that

$$A = A(T, V, n_1, n_2, \ldots, n_c) \qquad (1.58)$$

$$G = G(T, P, n_1, n_2, \ldots, n_c). \qquad (1.59)$$

Note that A is a function of $T, V, n_1, n_2, \ldots, n_c$ and G is a function of $T, P, n_1, \ldots, n_c$. Such dependency in terms of measurable variables makes A and G the natural choices for the processes at constant T and V, and at constant T and P, respectively. In particular, G is best suited for describing the state of a fluid because it has the most convenient independent variables—all easily measurable.

Writing differential forms of Eqs. (1.58) and (1.59) and comparing the results with Eqs. (1.56) and (1.57) leads to

$$\mu_i = \left(\frac{\partial U}{\partial n_i}\right)_{S,V,n_i} = \left(\frac{\partial A}{\partial n_i}\right)_{T,V,n_i} = \left(\frac{\partial G}{\partial n_i}\right)_{T,P,n_i}. \qquad (1.60)$$

Internal energy minimum principle

The entropy maximum principle can be stated equivalently in terms of the internal energy minimum principle. In the following, we present the mathematical proof.

The entropy maximum principle can be mathematically stated as

$$\left(\frac{\partial S}{\partial \underline{X}}\right)_{U,V,\underline{n}} = 0 \tag{1.61}$$

and

$$\left(\frac{\partial^2 S}{\partial \underline{X}^2}\right)_{U,V,\underline{n}} < 0 \tag{1.62}$$

where the negativity of the second derivative shows that the extremum is a maximum. In the above equations, $S, U, V, \underline{n}$ are extensive variables of the composite system, and $\underline{X}$ correspond to the extensive variables of the subsystems that maximize the entropy, S, at equilibrium. Note that for a composite system composed of two subsystems primed and double-primed,

$$\underline{X} = (U', V', n_1', n_2', \ldots, n_c', U'', V'', n_1'', n_2'', \ldots, n_c'') \tag{1.63}$$

and

$$\underline{n} = (n_1, n_2, \ldots, n_c),\, n_i = n_i' + n_i'',\, i = 1, \ldots, c. \tag{1.64}$$

Since $U = U(S, \underline{X})$

then

$$dU = \left(\frac{\partial U}{\partial S}\right)_{\underline{X}} dS + \left(\frac{\partial U}{\partial \underline{X}}\right)_{S} d\underline{X}. \tag{1.65}$$

According to the partial derivatives rule (to be discussed shortly),

$$\left(\frac{\partial U}{\partial \underline{X}}\right)_{S} = -\left(\frac{\partial S}{\partial \underline{X}}\right)_{U} \bigg/ \left(\frac{\partial S}{\partial U}\right)_{\underline{X}}. \tag{1.66}$$

Since $(\partial S/\partial U)_{\underline{X}} = 1/T$ and at equilibrium $(\partial S/\partial \underline{X})_U = 0$ (see Eq. (1.61)),

$$\left(\frac{\partial U}{\partial \underline{X}}\right)_{S} = 0, \tag{1.67}$$

which states that U has an extremum at equilibrium (in the above and subsequent equations, V and $\underline{n}$ are kept constant). We have to prove that the extremum is a minimum. Let $\eta = (\partial U/\partial \underline{X})_S$. Since $U = U(S, \underline{X})$, then

$$\eta = \eta(U, \underline{X}). \tag{1.68}$$

Writing Eq. (1.68) in differential form,

$$d\eta = \left(\frac{\partial \eta}{\partial U}\right)_{\underline{X}} dU + \left(\frac{\partial \eta}{\partial \underline{X}}\right)_{U} d\underline{X}, \tag{1.69}$$

and dividing it by $d\underline{X}$ and holding S constant,

$$\left(\frac{\partial \eta}{\partial \underline{X}}\right)_{S} = \left(\frac{\partial \eta}{\partial U}\right)_{\underline{X}}\left(\frac{\partial U}{\partial \underline{X}}\right)_{S} + \left(\frac{\partial \eta}{\partial \underline{X}}\right)_{U}. \tag{1.70}$$

Using $\eta = (\partial U/\partial \underline{X})_S$, $(\partial U/\partial \underline{X})_U$ can be written as

$$\left(\frac{\partial \eta}{\partial \underline{X}}\right)_{U} = \frac{\partial}{\partial \underline{X}}\left[\left(\frac{\partial U}{\partial \underline{X}}\right)_{S}\right]_{U} \tag{1.71}$$

and
$$\frac{\partial}{\partial \underline{X}}\left[\left(\frac{\partial U}{\partial \underline{X}}\right)_{S}\right]_{U} = -\frac{\partial}{\partial \underline{X}}\left[\frac{(\partial S/\partial \underline{X})_{U}}{(\partial S/\partial U)_{\underline{X}}}\right]_{U}$$

$$= -\frac{(\partial^2 S/\partial \underline{X}^2)_{U}}{(\partial S/\partial U)_{\underline{X}}} + \frac{(\partial S/\partial \underline{X})_{U}(\partial^2 S/\partial \underline{X}\partial U)}{(\partial S/\partial U)^2_{\underline{X}}} \tag{1.72}$$

Using $(\partial U/\partial S)_{\underline{X}} = T$ and Eq. (1.72), Eq. (1.70) can be written as

$$\left(\frac{\partial^2 U}{\partial \underline{X}^2}\right)_{S} = \eta(\partial \eta/\partial U)_{\underline{X}} - T\left(\frac{\partial^2 S}{\partial \underline{X}^2}\right)_{U} + T^2(\partial S/\partial \underline{X})(\partial^2 S/\partial \underline{X}\partial U). \tag{1.73}$$

At equilibrium, both η and $(\partial S/\partial \underline{X})_{U}$ are zero and $(\partial^2 S/\partial \underline{X}^2)$ is negative (see Eq. (1.62)). Therefore,

$$(\partial^2 U/\partial \underline{X}^2)_{S} = -T(\partial^2 S/\partial \underline{X}^2)_{U}, \tag{1.74}$$

$$(\partial^2 U/\partial \underline{X}^2)_{S} > 0. \tag{1.75}$$

Equation (1.75) together with Eq. (1.67) provide the internal energy minimum principle. In other words at equilibrium, the internal energy U must be a minimum at constant S, V, and $n_i(i = 1, \ldots, c)$. The equilibrium criteria in terms of the H, A, and G can be shown to be the following:

- The enthalpy, H, must be a minimum at constant S, P, and $n_i(i = 1, \ldots, c)$.

- The Helmholtz free energy, A, must be a minimum at constant T, V, and $n_i (i = 1, \ldots, c)$.

- The Gibbs free energy, G, must be a minimum at constant T, P, and $n_i (i = 1, \ldots, c)$.

The mathematical proof for the H, A, and G minima principle is straightforward. Note that the minimum Gibbs free energy principle is very convenient to apply because pressure, temperature, and the total mole numbers of species i are held constant.

Next, we derive the relation between partial derivatives that was used in the mathematical derivation of the internal energy minimum principle.

Relation between partial derivatives of implicit functions. Let η be a continuous differentiable function of three variables x, y, and z:

$$\eta = \eta(x, y, z). \tag{1.76}$$

Then
$$d\eta = (\partial\eta/\partial x)_{y,z}dx + (\partial\eta/\partial y)_{x,z}dy + (\partial\eta/\partial z)_{x,y}dz. \tag{1.77}$$

Let us keep η and z constant and divide the remaining two terms by dx, then

$$(\partial\eta/\partial x)_{y,z} + (\partial\eta/\partial y)_{x,z}(\partial y/\partial x)_{\eta,z} = 0. \tag{1.78}$$

Equation (1.78) can be rearranged as

$$\boxed{(\partial y/\partial x)_{\eta,z} = -(\partial\eta/\partial x)_{y,z}/(\partial\eta/\partial y)_{x,z}} \tag{1.79}$$

The change of variables given by Eq. (1.79) is very useful in thermodynamics. We have already used it in Eqs. (1.66) and (1.72). Next we derive another relationship between the partial derivatives.

Reciprocity relation. The reciprocity relation or the cross-differentiation identity is very useful in some thermodynamic derivations. Let us represent the partial derivatives in Eq. (1.77) by

$$\eta_x = (\partial\eta/\partial x)_{y,z}, \quad \eta_y = (\partial\eta/\partial y)_{x,z}, \quad \eta_z = (\partial\eta/\partial z)_{x,y} \tag{1.80}$$

Then
$$\boxed{\frac{\partial}{\partial y}(\eta_x)_{x,z} = \frac{\partial}{\partial x}(\eta_y)_{y,z}} \tag{1.81}$$

Chemical potential of a component in a mixture

We can use the above reciprocity relation to derive the expression for the evaluation of μ_i. Consider the expression for dG given by Eq. (1.57) and apply the reciprocity relation to the coefficients of dV and dn_i terms to obtain

$$(\partial V/\partial n_i)_{T,P,n_i} = (\partial \mu_i/\partial P)_{T,\underline{n}} \qquad (1.82)$$

and then define the partial molar volume as

$$\bar{V}_i = (\partial V/\partial n_i)_{T,P,n_i}. \qquad (1.83)$$

Combining Eqs. (1.82) and (1.83) provides

$$\boxed{(d\mu_i = \bar{V}_i dP)_{T,\underline{n}}}. \qquad (1.84)$$

Equation (1.84) is an important expression in the thermodynamics of phase equilibrium. It relates the chemical potential of component i in the mixture to the measurable properties of pressure, temperature, composition, and volume. For a one-component system Eq. (1.84) simplifies to $(d\mu = vdP)_T$, which was derived earlier (see Eq. (1.43)). The significance and usefulness of partial molar volume and other partial quantities are discussed next.

Partial molar quantities

Consider a single phase comprised of various components at constant temperature and pressure. Let E represent any extensive property of the phase such as V, U, or G. The partial molar quantity of any component i within the phase is defined as

$$\bar{E}_i = (\partial E/\partial n_i)_{T,P,n_i}, \qquad (1.85)$$

which gives the change in the extensive property of component i in the phase due to a small change in the amount of that component while temperature, T, pressure, P, and the amount of all other components are held constant. As an example, let us add a small amount of component i, Δn_i, at constant T and P to the container holding the phase. Then the partial molar volume, $\bar{V}_i$, of component i is given by

$$\bar{V}_i \approx (\Delta V/\Delta n_i)_{T,P,n_i}. \qquad (1.86)$$

Figure 1.3 shows the process for the determination of $\bar{V}_i$. In Fig. 1.3a, Δn_i moles of component i are added to the system. The result may be the

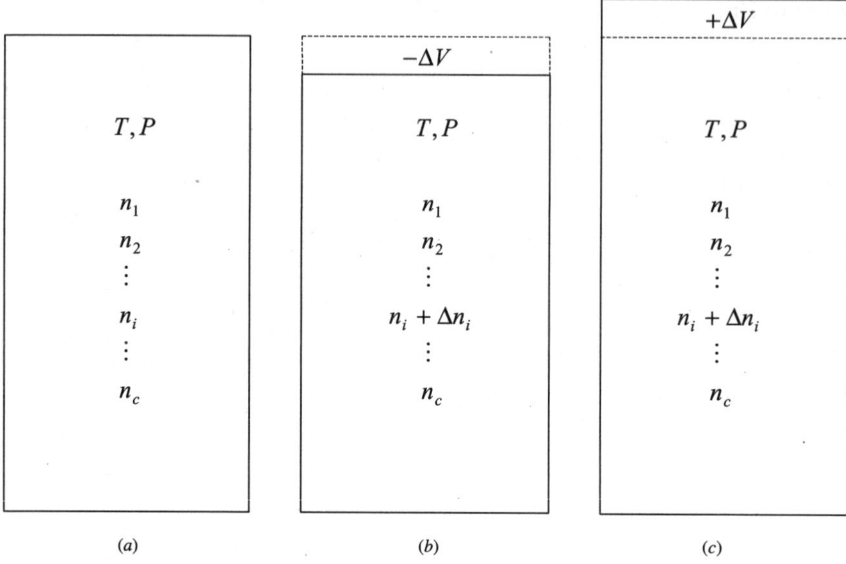

Figure 1.3 The addition of Δn_i to a system at constant T and P.

sketches in Fig. 1.3b or Fig. 1.3c. In Fig. 1.3b, after adding Δn_i moles, the volume decreases by ΔV. In Fig. 1.3c, after addition of Δn_i moles, the volume increases by ΔV. Either could happen in hydrocarbon mixtures. For the system in Fig. 1.3b, $\bar{V}_i = -\Delta V/\Delta n_i$ and for the system shown in Fig. 1.3c, $\bar{V}_i = +\Delta V/\Delta n_i$.

Similarly the partial molar enthalpy is defined as

$$\bar{H}_i \approx (\Delta H/\Delta n_i)_{T,P,\boldsymbol{n}_i} \tag{1.87}$$

Since V, U, S, H, A, and G are all extensive properties, then the extensive property, E, when expressed as a function of T, P, and n_i is a first-degree homogeneous function in $\underline{\boldsymbol{n}}$:

$$E(T, P, \lambda n_1, \lambda n_2, \ldots, \lambda n_c) = \lambda E(T, P, n_1, n_2, \ldots, n_c). \tag{1.88}$$

Differentiating the above equation with respect to λ at constant T and P,

$$\sum_{i=1}^{c} \left(\frac{\partial E}{\partial \lambda n_i}\right)_{T,P,\boldsymbol{n}_i} \left(\frac{d\lambda n_i}{d\lambda}\right) = E(T, P, n_1, n_2, \ldots, n_c). \tag{1.89}$$

For $\lambda = 1$, $\qquad E(T, P, n_1, \ldots, n_c) = \sum_{i=1}^{c} n_i \left(\frac{\partial E}{\partial n_i}\right)_{T,P,\boldsymbol{n}_i} \tag{1.90}$

Using the definition given by Eq. (1.85), then

$$E = \sum_{i=1}^{c} n_i \bar{E}_i .$$

(1.91)

Therefore, any extensive property can be calculated from its partial molar quantities; $V = \sum_{i=1}^{c} n_i \bar{V}_i$, $G = \sum_{i=1}^{c} n_i \bar{G}_i$, etc. The partial molar Gibbs free energy, $\bar{G}_i = (\partial G / \partial n_i)_{T,P,n_i}$ has a unique feature. From Eq. (1.60), and the definition of partial molar Gibbs free energy,

$$\mu_i = \bar{G}_i ,$$

(1.92)

which can also be obtained from Eq. (1.50).

The partial molar quantities could be either positive or negative which is particularly important in regards to partial molar volume. The physical meaning of negative partial molar volume of component i is that its addition at constant T and P results in a decrease in volume. Figure 1.4 shows the plot of measured partial molar volumes of C_2 and nC_7 in the C_2/nC_7 mixture from Wu and Ehrlich (1973). Note that the partial molar volume of the heavier hydrocarbon nC_7 is negative. Nega-

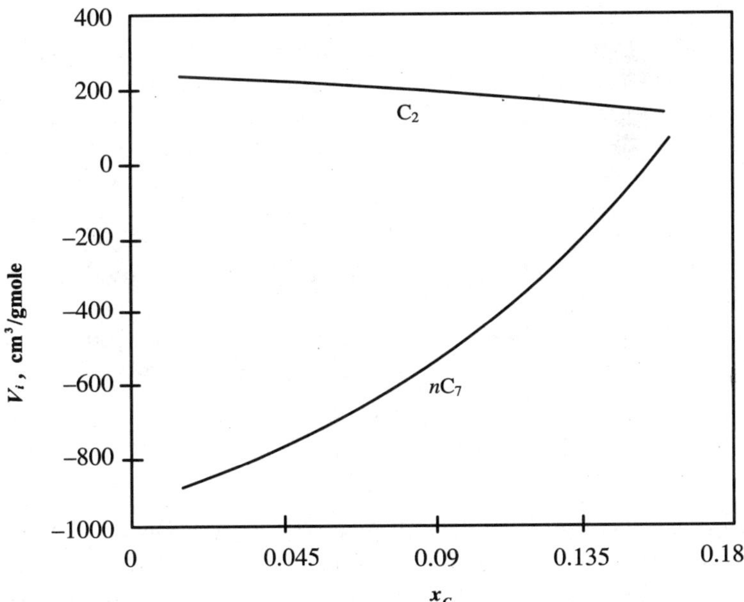

Figure 1.4   Measured partial molar volumes of C_2 and nC_7 at 80°C and 57 atm (data from Wu and Ehrlich, 1973).

tive partial molar volume implies that if a small amount of nC_7 is added to the system at constant T and P, the volume decreases. Another example of negative partial molar volume is the C_1/C_3 system at 100°F and 1,000 psia with a C_3 concentration of 60 weight percent in the overall system. The $\bar{V}_{C_3}$ in the liquid phase at the dewpoint at 100°F and 1000 psia is -3.52 ft^3/lbmole (Sage and Lacey, 1949). Partial molar quantities can be expressed both in terms of the quantity per mole or per unit mass. Therefore, in terms of mass, $\bar{V}_{C_3} = -0.08$ ft^3/lb.

Use of the partial molar volumes takes into account the volume change due to mixing, and there is no need to assume that the partial molar volume is equal to the molar volume of the pure component at the same temperature and pressure. In fact, the partial molar volume of some light components of reservoir fluids can be two times that of the pure component molar volume, and for some heavy components it can be negative. A significant change in volume due to mixing is a distinct feature of reservoir fluid systems. The change in volume due to mixing makes the use of equations of state very appropriate for reservoir fluids because such use readily takes into account the volume change. It is important to note that there is no relation between partial molar volume and the volume fraction of the component in the mixture. The error of assuming, partial molar volume = volume fraction, is sometimes seen in the literature.

Figure 1.5 shows the partial molar volumes of C_1 and C_3 vs. pressure in a mixture of C_1/C_3 with $x_{C_3} = 0.34$ (mole fraction) and at $T = 346$ K. The mixture in the entire range of pressure is in the gas state (see Fig. 1.6). The molar volumes of pure C_1 and C_3 are also graphed in Fig. 1.5 at $T = 346$ K and different pressures. The partial molar volumes are calculated from the Peng-Robinson equation of state (1976), which will be discussed in Chapter 3. The pure component molar volumes are from Starling (1973). The large difference between the partial molar volume and pure component molar volume provides strong evidence of the effect of mixing on the density.

The relationship between partial molar quantities is analogous to the relationships between the extensive variables. As an example,

$$\bar{G}_i = \mu_i = \bar{H}_i - T\bar{S}_i \qquad (1.93)$$

This equation can be obtained by differentiating Eq. (1.47) with respect to n_i at constant temperature, pressure and n_j. Next, we will define the fugacity, which has all the features of chemical potential, but unlike the chemical potential has an absolute value.

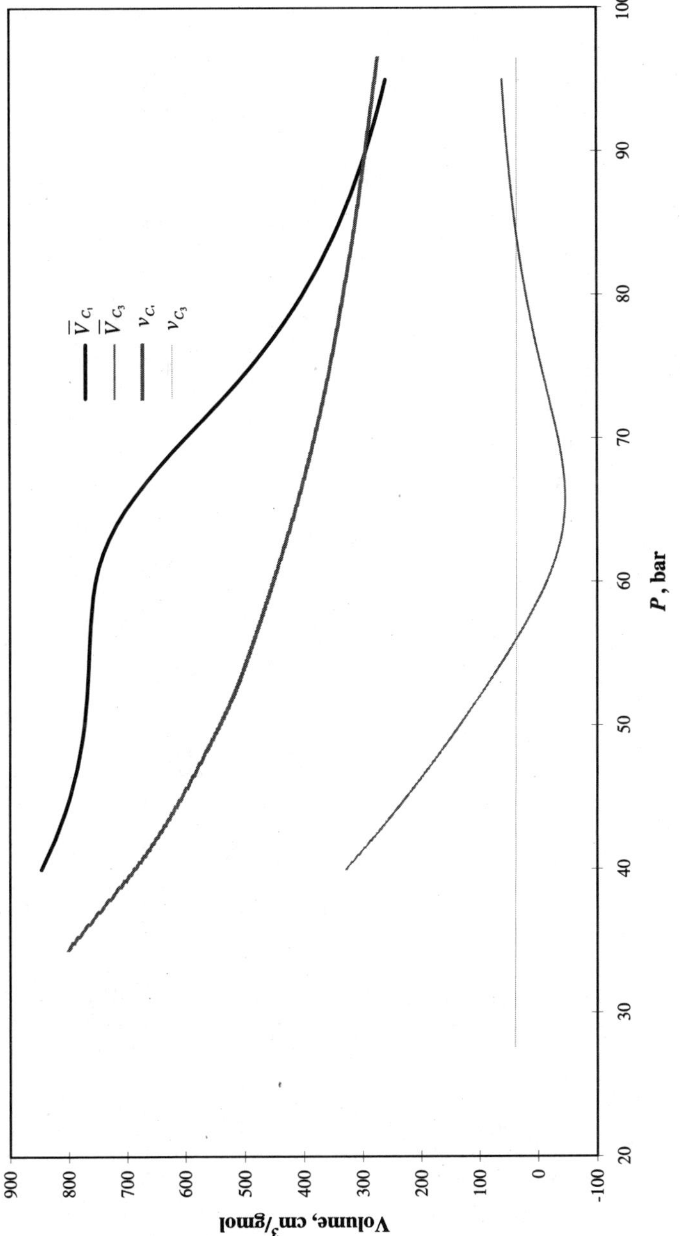

Figure 1.5 Partial molar and pure component molar volumes of the C_1/C_3 system at 346 K: $x_{C_1} = 0.34$.

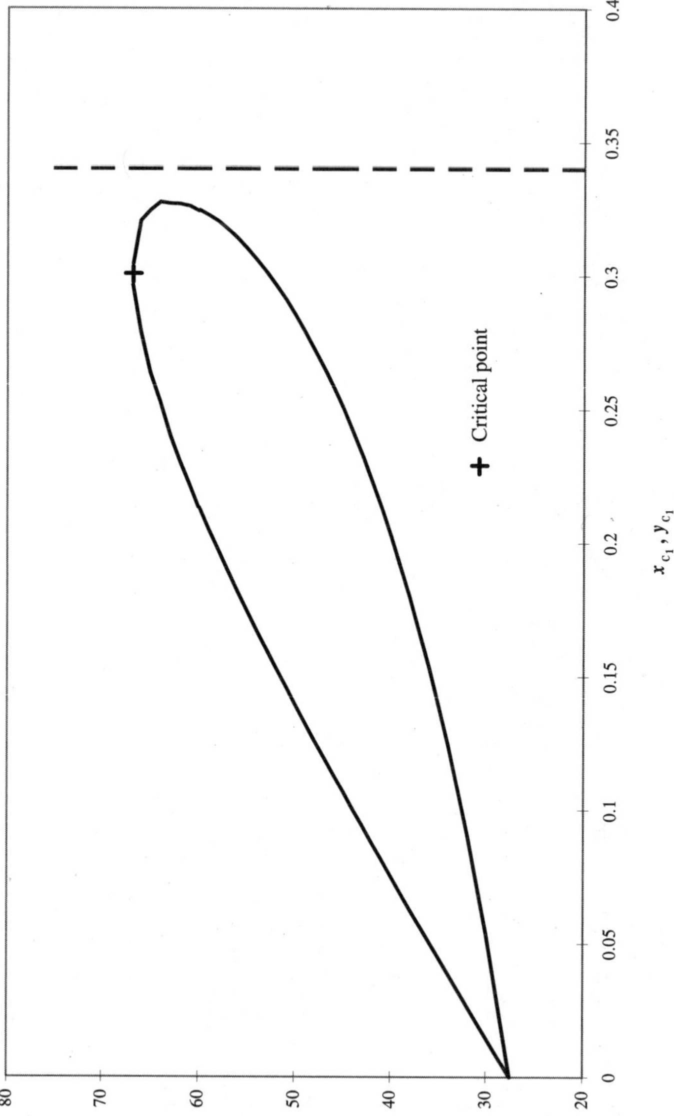

Figure 1.6 Phase diagram of the C_1/C_3 mixture at 346 K.

Fugacity

The fugacity, often represented by the symbol f, has the units of pressure. It is defined from the following relationship:

$$(d\mu_i = RTd\ln f_i)_{T,\underline{n}} \qquad i = 1, \ldots, c. \tag{1.94}$$

Another relationship is needed to complete the definition of fugacity:

$$\operatorname*{Lim}_{P\to 0} (f_i/x_i P) = 1, \tag{1.95}$$

where x_i is the mole fraction and f_i is the fugacity of component i in the mixture. The ratio $(f_i/x_i P)$ is called the fugacity coefficient, φ_i:

$$\varphi_i = f_i/x_i P \tag{1.96}$$

At low pressures, as $P \to 0$, $\varphi_i = 1$, and it is said that the fluid has an ideal behavior.

Let us derive the expression for calculating φ_i or f_i. Subtract $RTd\ln x_i P$ from both sides of Eq. (1.94):

$$(RTd\ln f_i - RTd\ln x_i P = d\mu_i - RTd\ln x_i P)_{T,\underline{n}} \qquad i = 1, \ldots, c \tag{1.97}$$

or $\qquad (RTd\ln \varphi_i = d\mu_i - RTd\ln P)_{T,\underline{n}} \qquad i = 1, \ldots, c. \tag{1.98}$

Note that on the right side of Eq. (1.98), the term $RTd\ln x_i$ is dropped since composition is held constant. Combining Eqs. (1.84) and (1.98), one writes

$$(RTd\ln \varphi_i = \bar{V}_i dP - RTd\ln P)_{T,\underline{n}} \qquad i = 1, \ldots, c. \tag{1.99}$$

Drop the subscripts T and $\underline{n}$ for brevity and integrate Eq. (1.99) from 0 to pressure P and combine the results with Eq. (1.95):

$$RT \ln \varphi_i = \int_0^P (\bar{V}_i - RT/P)dP. \tag{1.100}$$

This equation provides φ_i or f_i from the volumetric data. However, since we will later use a pressure-explicit EOS, it is preferrable to have the integral in terms of volume. For this purpose, the following derivations may be necessary. The EOS in the form

$$Z = PV/nRT \tag{1.101}$$

can be differentiated at constant temperature and moles:

$$dZ = (VdP + PdV)/nRT. \tag{1.102}$$

Multiplying both sides by $(nRT)/(PV)$ results in

$$dZ/Z = dP/P + dV/V. \tag{1.103}$$

Now consider the expression for dA given by Eq. (1.56). From the reciprocity relation,

$$-(\partial P/\partial n_i)_{T,V,\underline{n}} = (\partial \mu_i/\partial V)_{T,\underline{n}}. \tag{1.104}$$

Combining Eqs. (1.98) and (1.103),

$$RTd\ln\varphi_i = d\mu_i - RT[dZ/Z - dV/V]. \tag{1.105}$$

Dividing Eq. (1.105) by dV and once more showing the parameters that are held constant for clarity, one writes

$$RT\left(\frac{\partial \ln\varphi_i}{\partial V}\right)_{T,\underline{n}} = \left(\frac{\partial \mu_i}{\partial V}\right)_{T,\underline{n}} - \frac{RT}{Z}\left(\frac{\partial Z}{\partial V}\right)_{T,\underline{n}} + \frac{RT}{V}. \tag{1.106}$$

Combining Eqs. (1.106) and (1.104) results in

$$d\ln\varphi_i = \left[-\frac{1}{RT}\left(\frac{\partial P}{\partial n_i}\right)_{T,V,n_i} + \frac{1}{V}\right]dV - d\ln Z. \tag{1.107}$$

Integration of Eq. (1.107) provides

$$\ln\varphi_i = \int_0^P \left[\frac{1}{V} - \frac{1}{RT}\left(\frac{\partial P}{\partial n_i}\right)_{T,V,n_i}\right]dV - \ln Z. \tag{1.108}$$

The integration limits in terms of volume are

$$\boxed{\ln\varphi_i = \int_V^\infty \left[\frac{1}{RT}\left(\frac{\partial P}{\partial n_i}\right)_{T,V,n_i} - \frac{1}{V}\right]dV - \ln Z} \ . \tag{1.109}$$

The above equation provides φ_i or f_i in terms of volumetric properties to be discussed in Chapter 3.

Now let us examine the usefulness of Eq. (1.109) for vapor–liquid equilibria calculations. First we will establish that at equilibrium, instead of the equality of chemical potentials, one may equally write

$$f_i'(T, P, \boldsymbol{x}') = f_i''(T, P, \boldsymbol{x}'') \tag{1.110}$$

where $\boldsymbol{x}' = (x_1', x_2', \ldots, x_{c-1}')$ and $\boldsymbol{x}'' = (x_1'', x_2'', \ldots, x_{c-1}'')$. From Eq. (1.94) (see Problem 1.15),

$$\mu_i'(T, P, \boldsymbol{x}') = \mu_i^0(T, P) + RT \ln \frac{f_i'(T, P, \boldsymbol{x}')}{f_i^0(T, P)} \tag{1.111}$$

and

$$\mu_i''(T, P, \boldsymbol{x}'') = \mu_i^0(T, P) + RT \ln \frac{f_i''(T, P, \boldsymbol{x}'')}{f_i^0(T, P)}. \tag{1.112}$$

At equilibrium $\mu_i'(T, P, \boldsymbol{x}') = \mu_i''(T, P, \boldsymbol{x}'')$; then this equation together with Eqs. (1.111) and (1.112), provide Eq. (1.110). From the definition of the fugacity coefficient according to Eq. (1.96), one can derive

$$\varphi_i' x_i' = \varphi_i'' x_i'' \qquad i = 1, \ldots, c. \tag{1.113}$$

Now if we represent the gas phase by prime and the liquid phase by double-primes, and the composition of the gas and liquid phases by y_i and x_i, respectively,

$$K_i = y_i/x_i = \varphi_i^L/\varphi_i^V \tag{1.114}$$

Equation (1.114) relates the vapor–liquid equilibrium ratio, K_i, to the ratio of fugacity coefficients. The fugacity coefficients can be obtained from the volumetric properties given by an EOS. However, as Eq. (1.109) demands, the volumetric data are required from zero pressure to pressure P of the system at constant temperature and composition. Therefore, the EOS should represent the volumetric behavior over the whole range.

Next we define ideal and nonideal fluids and the representation of the corresponding chemical potentials.

Ideal and nonideal fluids

Ideal gas. An ideal gas is defined as the fluid that obeys the equation

$$PV = nRT. \tag{1.115}$$

For a multicomponent system, n is given by

$$n = n_1 + n_2 + \cdots + n_c. \tag{1.116}$$

Note that the partial molar volume, $\bar{V}_i$, of component i in an ideal gas mixture is simply

$$\bar{V}_i = (\partial V/\partial n_i)_{T,P,n_i} = RT/P \qquad i = 1, \ldots, c. \tag{1.117}$$

From Eq. (1.100), we notice that the fugacity coefficient of all the components of an ideal gas mixture is $\varphi_i = 1$, and, therefore, the fugacity of component i in an ideal gas mixture is equal to its partial pressure, P_i,

$$f_i = y_i P = P_i. \qquad i = 1, \ldots, c. \tag{1.118}$$

Using Eqs. (1.94) and (1.118) one obtains

$$\mu_i(T, P, y) = \mu_i^0(T, P^0) + RT \ln P_i/P^0, \tag{1.119}$$

where $\mu_i^0(T, P^0)$ is the chemical potential of pure component i at the reference state of pressure P^0 and temperature T. Note that for pure-component ideal gas at temperature T and pressure $P, f(T, P) = P$ from Eq. (1.118).

Ideal solution. An ideal solution whether gas, liquid, or solid obeys the following equation:

$$V(T, P, \underline{n}) = \sum_{i=1}^{c} n_i v_i(T, P). \tag{1.120}$$

Equation (1.120) states that when n_i moles of component i at temperature T and pressure P are mixed there will be no volume change on mixing. This is a very serious restriction for certain fluid mixtures and as we will show over and over; as an example, reservoir fluids do not obey Eq. (1.120). The partial molar volume of component i from Eq. (1.120) is

$$\bar{V}_i = v_i(T, P), \tag{1.121}$$

and substitution of the above value into Eq. (1.100) results in

$$f_i(T, P, \boldsymbol{x}) = x_i f_i(T, P). \tag{1.122}$$

Combining Eqs. (1.111) and (1.122) (note that $f_i^0(T, P)$ and $f_i(T, P)$ are the same), one obtains the expression for chemical potential of component i in an ideal solution,

$$\boxed{\mu_i(T, P, \boldsymbol{x}) = \mu_i^0(T, P) + RT \ln x_i} \qquad i = 1, \ldots, c. \tag{1.123}$$

Now let us examine the heat of mixing of an ideal solution. In other words, is there any heat of mixing when several different species at constant temperature and pressure are mixed? In order to answer this question, divide Eq. (1.123) by T and take the derivative with respect to T while holding P and $\underline{n}$ constant:

$$\frac{\partial}{\partial T}(\mu_i/T)_{P,\underline{n}} = \frac{\partial}{\partial T}(\mu_i^0/T)_{P,\underline{n}}. \tag{1.124}$$

Using the reciprocity relation on the coefficients of the first and third terms on the right side of Eq. (1.57) provides

$$(\partial S/\partial n_i)_{T,P,n_i} = (\partial \mu_i/\partial T)_{P,\underline{n}} = -\bar{S}_i. \tag{1.125}$$

Combining the above equation with Eq. (1.93) gives

$$\mu_i = \bar{H}_i + T(\partial \mu_i/\partial T)_{P,\underline{n}}. \tag{1.126}$$

Equation (1.126) can be written as

$$\frac{\partial}{\partial T}(\mu_i/T) = -\bar{H}_i/T^2. \tag{1.127}$$

For a pure substance at temperature T and pressure P,

$$\frac{\partial}{\partial T}(\mu_i^0/T) = -h_i/T^2. \tag{1.128}$$

Equations (1.124), (1.127), and (1.128) give

$$H(T, P, \underline{n}) = \sum_{i=1}^{c} n_i \bar{H}_i = \sum_{i=1}^{c} n_i h_i, \tag{1.129}$$

which implies that there is no enthalpy of mixing at temperature T, and pressure P. Since $\Delta V_{mix} = 0$, and since the mixing process is carried out at constant pressure, then

$$\Delta U_{mix} = \Delta H_{mix} = Q = 0. \tag{1.130}$$

Therefore, for an ideal fluid, there is no heat of mixing. Due to nonideality, mixing of hydrocarbons at constant temperature and pressure may, however, result in heating or cooling.

Nonideal solution. In the understanding of a nonideal solution, excess functions are often defined. Excess functions are thermodynamic properties of solutions that are in excess of those of an ideal solution at the same pressure, temperature, and composition. For an ideal solution, therefore, all excess functions are zero. Let us show all thermodynamic properties by E and all excess functions by E^E; then

$$E^E = E \text{ (real solution at } T, P, \underline{n}) - E \text{ (ideal solution at } T, P, \underline{n}). \tag{1.131}$$

The excess Gibbs free energy, excess enthalpy, and excess volume are defined as

$$G^E = G \text{ (real solution at } T, P, \underline{n}) - G \text{ (ideal solution at } T, P, \underline{n}),$$
(1.132)

$$H^E = H \text{ (real solution at } T, P, \underline{n}) - H \text{ (ideal solution at } T, P, \underline{n}),$$
(1.132)

$$V^E = V \text{ (real solution at } T, P, \underline{n}) - V \text{ (ideal solution at } T, P, \underline{n}).$$
(1.133)

The relations between excess functions are exactly the same as those of total thermodynamic functions,

$$G^E = H^E - TS^E$$
(1.134)

$$H^E = U^E + PV^E$$
(1.135)

Similarly, partial molar excess functions are defined in the same manner as the partial molar quantities,

$$\bar{E}_i^E = (\partial E^E / \partial n_i)_{T,P,n_i}.$$
(1.136)

Since E^E is a first-degree homogeneous function in mole numbers, one can readily show that,

$$E^E = \sum_{i=1}^{c} n_i \bar{E}_i^E.$$
(1.137)

Activity coefficient

In the expression for the chemical potential of an ideal solution given by Eq. (1.123), a modification may be necessary to describe nonideal solutions, or real solutions. The modified expression is

$$\mu_i(T, P, \boldsymbol{x}) = \mu_i^0(T, P) + RT \ln \gamma_i(T, P, \boldsymbol{x}) x_i,$$
(1.138)

where γ_i is called the activity coefficient which is generally a function of temperature, T, composition, $\boldsymbol{x}$, and pressure, P. The magnitude of γ_i depends on the magnitude of μ_i^0 (T, P), which is also unknown. To complete the definition of γ_i, we define the condition under which γ_i becomes equal to unity. There are two conventions:

(I)
$$\mu_i(T, P, \boldsymbol{x}) = \mu_i^0(T, P) + RT \ln \gamma_i(T, P, \boldsymbol{x}) x_i,$$
(1.139)
$$\gamma_i \to 1 \quad \text{as} \quad x_i \to 1 \qquad i = 1, \ldots, c$$

which means all of the components of a real solution approach ideal behavior as $x_i \to 1$. This implies that all the components in the pure state are in the same state as the mixture (i.e., if the mixture is liquid, components in the pure state at temperature, T, and pressure, P, are also in the liquid state). The activity coefficients obeying the above relationship are known as symmetrically organized.

$$(II) \qquad \mu_i(T, P, \boldsymbol{x}) = \mu_i^*(T, P) + RT \ln \gamma_i^*(T, P, \boldsymbol{x}) x_i \qquad (1.140)$$

$$\gamma_i^* \to 1 \quad \text{as} \quad x_i \to 0 \qquad i = 1, \ldots, a \quad a < c.$$

For $i = a + 1, \ldots, c$ Eq. (1.139) applies, and the components are identified as solvents. The components belonging to the indexing of $i = 1, \ldots, a$ are known as solutes. When the solutes are not in the same state as the solution, the second convention becomes useful. This is often the case for liquid solutions of noncondensable gases. Activity coefficients defined by convention II are unsymmetrical. The relationship between γ_i and γ_i^* will be discussed later in Chapter 5. Note that as $x_i \to 0$ the corresponding γ_i which is shown by γ_i^∞ has a defined value; as $x_i \to 0$, the corresponding γ_i^* which is shown by $\gamma_i^{*,\infty} \to 1$, as defined above.

Relation between γ_i and G^E.   Let us differentiate Eq. (1.132) with respect to n_i while holding T, P, and $\boldsymbol{n}_i$ constant. The result is,

$$\bar{G}_i^E = \mu_i \text{ (real solution at } T, P, \boldsymbol{x}) - \mu_i \text{ (ideal solution at } T, P, \boldsymbol{x}).$$
$$(1.141)$$

From $(d\mu_i = RTd \ln f_i)_{T,\underline{\boldsymbol{n}}}$, and Eqs. (1.123) and (1.141), one may derive

$$\bar{G}_i^E = RT \ln \frac{f_i(T, P, \boldsymbol{x})}{x_i f_i^0(T, P)}. \qquad (1.142)$$

By comparing Eq. (1.111) and Eq. (1.138) (without the prime superscript),

$$\boxed{f_i(T, P, \boldsymbol{x}) = \gamma_i(T, P, \boldsymbol{x}) x_i f_i^0(T, P)} . \qquad (1.143)$$

Combining Eqs. (1.142) and (1.143) results in

$$\bar{G}_i^E = RT \ln \gamma_i, \qquad (1.144)$$

and from Eq. (1.137)

$$\boxed{G^E = RT \sum_{i=1}^{c} n_i \ln \gamma_i} . \qquad (1.145)$$

Equation (1.145) shows that the excess Gibbs free energy and the activity coefficients are related. An alternative form of Eq. (1.145) is given by

$$g^E = RT \sum_{i=1}^{c} x_i \ln \gamma_i \,,$$

(1.146)

where g^E is the molar excess Gibbs free energy. Next, we will examine the effect of pressure and temperature on γ_i.

Pressure and temperature derivative of γ_i.   Let us divide Eq. (1.138) by T and take the derivative of the resulting expression with respect to temperature while holding P and $\mathbf{x}$ constant.

$$\frac{\partial}{\partial T}\left(\frac{\mu_i(T, P, \mathbf{x})}{T}\right)_{P,\mathbf{x}} = \frac{\partial}{\partial T}\left(\frac{\mu_i^0(T, P)}{T}\right)_{P} + R\frac{\partial}{\partial T}(\ln \gamma_i)_{P,\mathbf{x}}$$

(1.147)

Combining the above equation, with Eqs. (1.126), and (1.127),

$$\left(\frac{\partial \ln \gamma_i}{\partial T}\right)_{P,\mathbf{x}} = -\frac{\bar{H}_i - h_i}{RT^2} = -\frac{\bar{H}_i^E}{RT^2} \,.$$

(1.148)

The above equation provides the effect of temperature on the activity coefficients. The effect of pressure on γ_i can be obtained by taking the derivative of Eq. (1.138) with respect to pressure at constant T and $\mathbf{x}$:

$$\bar{V}_i = v_i + RT\left(\frac{\partial \ln \gamma_i}{\partial P}\right)_{T,\mathbf{x}} \,.$$

(1.149)

Rearrangement of Eq. (1.149) results in

$$\left(\frac{\partial \ln \gamma_i}{\partial P}\right)_{T,\mathbf{x}} = \frac{\bar{V}_i - v_i}{RT} = \frac{\bar{V}_i^E}{RT} \,.$$

(1.150)

Figure 1.5 shows $\bar{V}_i$ and v_i for C_1/C_3 vs. pressure at 346 K. This figure implies that the effect of the pressure on γ_i may be very pronounced.

Next, we will briefly discuss the activity coefficient models.

Activity coefficient models.   Prausnitz, Lichtenthaler, and de Azevedo (1986) provide an excellent presentation of activity coefficient models and theories of solutions extensively in Chapters 6 and 7 of their book. Here, we briefly review the basic concepts beyond various activity coefficient models.

Margules activity coefficient equations.   The expressions for the activity coefficients for a binary system using this model are

$$\ln \gamma_1 = \frac{A}{RT} x_2^2 \qquad (1.151)$$

$$\ln \gamma_2 = \frac{A}{RT} x_1^2, \qquad (1.152)$$

where A is a function of temperature. Consequently, $\bar{V}_1^E = 0$ and $\bar{V}_2^E = 0$, and the volume change on mixing is zero, i.e., $V^E = 0$.

Van Laar activity coefficient equations.   The van Laar activity coefficient model is based on the assumption that $V^E = 0$ and $S^E = 0$. Therefore, $G^E = U^E$. In other words, the model allows for heat of mixing but does not allow for volume change on mixing. Van Laar used the van der Waals EOS to calculate G^E and then Eq. (1.144) can be used to obtain the activity coefficients. The predicted activity coefficients for a binary system are

$$\ln \gamma_1 = \frac{A^1}{\left[1 + \dfrac{A^1}{B^1}\dfrac{x_1}{x_2}\right]^2} \qquad (1.153)$$

and
$$\ln \gamma_2 = \frac{B^1}{\left[1 + \dfrac{B^1}{A^1}\dfrac{x_2}{x_1}\right]^2}, \qquad (1.154)$$

where parameters A^1 and B^1 are functions of temperature and properties of pure components. This model does not account for the effect of pressure on γ.

Scatchard-Hildebrand regular-solution activity coefficients.   Hildebrand (1929) defined a regular solution as the mixture in which components mix with no excess entropy provided there is no volume change on mixing. Scatchard in an independent work arrived at the same conclusion. The definition of regular solutions (Hildebrand and Scott, 1950) is in line with van Laar's assumption that the excess entropy and the excess volume of mixing are negligible. Scatchard and Hildebrand used an approach different from van Laar's to calculate G^E. They defined parameter C as

$$C \equiv \frac{\Delta u^V}{v^L}, \qquad (1.155)$$

where Δu^V is the molar internal energy change upon isothermal vaporization of the saturated liquid to the ideal-gas state, and v^L is the saturated liquid molar volume. Note that $\Delta u^V = u^G(T, P^0) - u^L(T, P)$ where P^0 may be zero pressure. Then the solubility parameter δ, is defined as

$$\delta = C^{1/2}. \tag{1.156}$$

The solubility parameter has a clear physical meaning; when the difference between the solubility parameters of two substances is small, one can dissolve the other appreciably. The difference in solubility parameters is a measure of the solubility power. As an example, the solubility parameter of asphaltenes, heptane, and toluene are, 9.5, 7.5, 8.9 $(\text{cal/cm}^3)^{0.5}$, respectively (see Problem 5.10, Chapter 5). The solubility of asphaltenes in heptane is very low, and toluene dissolves the asphaltenes. In Chapter 3, the expression for the solubility parameter from the equation of state, will be presented (see Problem 3.10, Chapter 3).

Scatchard and Hildebrand obtained U^E from Eq. (1.155) for binary mixtures. The final results are

$$\ln \gamma_1 = \frac{v_1 \Phi_2^2}{RT} [\delta_1 - \delta_2]^2 \tag{1.157}$$

and
$$\ln \gamma_2 = \frac{v_2 \Phi_1^2}{RT} [\delta_1 - \delta_2]^2, \tag{1.158}$$

where the subscript L has been dropped from the v_1 and v_2 quantities. In Eqs. (1.157) and (1.158), Φ_1 and Φ_2 are the volume fraction of components 1 and 2, respectively. The volume fractions are defined by

$$\Phi_1 = \frac{x_1 v_1}{x_1 v_1 + x_2 v_2} \tag{1.159}$$

$$\Phi_2 = \frac{x_2 v_2}{x_1 v_1 + x_2 v_2}. \tag{1.160}$$

Unlike mass and mole fractions, the meaning of which is very clear, volume fractions do not have a clear physical meaning.

For three- and higher-component mixtures, the equation for the activity coefficients from the regular-solution theory is

$$\ln \gamma_i = \frac{v_i}{RT} (\delta_i - \bar{\delta})^2, \tag{1.161}$$

where

$$\bar{\delta} = \sum_{i=1}^{c} \Phi_i \delta_i \quad \text{and} \quad \Phi_i = \frac{x_i v_i}{\sum_{j=1}^{c} x_j v_j} \tag{1.162}$$

Flory-Huggins polymer solution activity coefficients. When the molecules of one component are much larger than the molecules of the other components in the mixture (i.e., polymers in solvents), the assumption of $S^E = 0$ may not be appropriate. For such systems it is found that H^E may be assumed to be zero since V^E is assumed to be zero (i.e., $U^E = 0$). From $G^E = H^E - TS^E$, then $G^E = -TS^E$ assuming $H^E = 0$. On this basis, activity coefficient models are proposed (Flory, 1953). The activity-coefficient model from the Flory-Huggins polymer solution theory has been used in the petroleum industry for asphaltene precipitation (see Chapter 5). Prausnitz *et al.* (1986) provide details of that model.

Legendre transformation

Let η be a continuously differentiable function of variables $X_1, X_2, \ldots, X_c, X_{c+1}$ and $X_{c+2}, \eta = \eta(X_1, X_2, \ldots, X_c, X_{c+1}, X_{c+2})$. Then $d\eta$ can be written as

$$d\eta = \sum_{i=1}^{c+2} \left(\frac{\partial \eta}{\partial X_i}\right)_{X_i} dX_i = \sum_{i=1}^{c+2} C_i dX_i, \tag{1.163}$$

where the coefficients $C_i = C_i(\underline{X})$. Let us define ζ as

$$\zeta = -\sum_{i=1}^{c+2} C_i X_i + \eta. \tag{1.164}$$

The differential of the above equation results in

$$d\zeta = -\sum_{i=1}^{c+2} X_i dC_i. \tag{1.165}$$

Therefore,

$$\zeta = \zeta(C_1, C_2, \ldots, C_{c+2}) = \zeta[(\partial\eta/\partial X_1)_{X_1}, \ldots, (\partial\eta/\partial X_{c+2})_{X_{c+2}}]. \tag{1.166}$$

Note that in the above transformation, the variables $X_1, X_2, \ldots, X_{c+2}$ are transformed into $(\partial\eta/\partial X_1)_{X_1}, (\partial\eta/\partial X_2)_{X_2}, \ldots$ and $(\partial\eta/\partial X_{c+2})_{X_{c+2}}$, respectively. This kind of transformation, which is very useful in transforming some nonlinear differential equations to linear differential equations (Courant and Hilbert, 1962), is called the Legendre transformation. Geometrically, the surface in the $X_1, X_2, \ldots, X_{c+2}$, η-space, which is a point set, is transformed into tangent-plane coordinates. For example, in the case of a curve in a 2-D space, instead of $\eta(X)$ representation, one could have $\eta(d\eta/dX)$, where instead of X, $(d\eta/dX)$, the tangent, is the independent variable. Figure 1.7 shows $\eta(X)$ and $\eta(dY/dX)$ in the point and tangent spaces. In the above transformations, there is no need to transform all the independent variables into their derivative variables. One could change only one or two variables in the group of $c+2$ variables. In this case, the transformation is called the first and the second Legendre transformation with respect to the particular variables.

Let us go back to the internal energy and perform Legendre transformation on its independent variables, and write

$$U = U(X_1, X_2, \ldots, X_{c+2}).\tag{1.167}$$

The first Legendre transformation on the first variable is written as

$$U^{(1)} = U^{(1)}(C_1, X_2, \ldots, X_{c+2}) = U - C_1 X_1.\tag{1.168}$$

The second Legendre transformation with respect to the first and second variables is

$$U^{(2)} = U^{(2)}(C_1, C_2, X_3, \ldots, X_{c+2}) = U - C_1 X_1 - C_2 X_2.\tag{1.169}$$

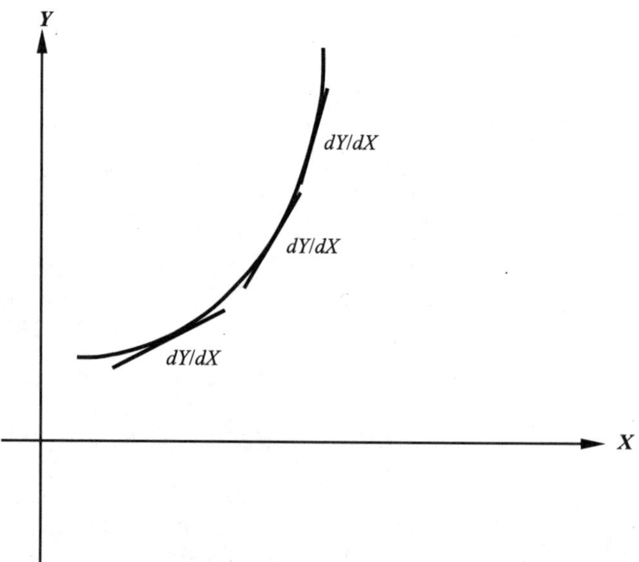

Figure 1.7   Depiction of Y vs. X, and dY/dX.

Finally, the total Legendre transformation of U is

$$U^{(c+2)} = U^{(c+2)}(C_1, C_2, \ldots, C_{c+2}) = U - \sum_{i=1}^{c+2} C_i X_i. \qquad (1.170)$$

The kth Legendre transformations of U and its differential are

$$\boxed{U^{(k)} = U - \sum_{i=1}^{k} C_i X_i}, \qquad (1.171)$$

$$\boxed{dU^{(k)} = -\sum_{i=1}^{k} X_i dC_i + \sum_{i=k+1}^{c+2} C_i dX_i}. \qquad (1.172)$$

Let us write the first and second Legendre transformations for $U = U(S, V, n_1, \ldots, n_c)$.

$$U^{(1)} = U - TS = A \qquad (1.173)$$

$$U^{(2)} = U - TS - (-P)V = U - TS + PV = G \qquad (1.174)$$

$$dU^{(1)} = -SdT - PdV + \sum_{i=1}^{c} \mu_1 dn_i = dA \qquad (1.175)$$

$$dU^{(2)} = -SdT + VdP + \sum_{i=1}^{c} \mu_i dn_i = dG \qquad (1.176)$$

Therefore, A and G are the first and second partial Legendre transformations of U. The total Legendre transformation of U is

$$U^{(c+2)} = U - TS - (-P)V - \sum_{i=1}^{c} \mu_i n_i = U - TS + PV - \sum_{i=1}^{c} \mu_i n_i \qquad (1.177)$$

$$dU^{(c+2)} = -SdT + VdP - \sum_{i=1}^{c} n_i d\mu_i = 0. \qquad (1.178)$$

Equation (1.178) is the Gibbs-Duhem equation.

Let us denote the kth Legendre transformation for a thermodynamic function $y^{(0)}$ by $y^{(k)}$; $y^{(k)}$ could represent the kth Legendre transformation of any of the four thermodynamic functions U, H, A, and G that we have defined. We can then write

$$y^{(0)} = y^{(0)}(X_1, X_2, \ldots, X_{c+2}) \qquad (1.179)$$

$$y^{(k)} = y^{(k)}(C_1, C_2, \ldots, C_k, X_{k+1}, \ldots, X_{c+2}) \qquad (1.180)$$

$$dy^{(k)} = -\sum_{i=1}^{k} X_i dC_i + \sum_{i=k+1}^{c+2} C_i dX_i. \qquad (1.181)$$

Note that Eq. (1.181) is the same as Eq. (1.172).

From Eq. (1.181), for $i \geq k + 1$,

$$C_i = (\partial y^{(k)}/\partial X_i)_{C_1,\ldots,C_k,X_{k+1},\ldots,X_{i-1},X_{i+1},\ldots,X_{c+2}}. \tag{1.182}$$

Writing the expressions for $dy^{(k-1)}$, $dy^{(k-2)}$, $dy^{(k-3)}$, $dy^{(1)}$, and $dy^{(0)}$, one readily establishes

$$
\begin{aligned}
C_k &= \left(\frac{\partial y^{(k-1)}}{\partial X_k}\right)_{C_1,\ldots,C_{k-1},X_{k+1},\ldots,X_{c+2}} = \left(\frac{\partial y^{(k-2)}}{\partial X_k}\right)_{C_1,\ldots,C_{k-2},X_{k-1},X_{k+1},\ldots,X_{c+2}} \\
&= \cdots = \left(\frac{\partial y^{(1)}}{\partial X_k}\right)_{C_1,X_2,\ldots,X_{k-1},X_{k+1},\ldots,X_{c+2}} = \left(\frac{\partial y^{(0)}}{\partial X_k}\right)_{X_1,\ldots,X_{k-1},X_{k+1},\ldots,X_{c+2}}.
\end{aligned}
$$

$$\tag{1.183}$$

The above set of equations will be used frequently in the manipulations of Legendre transformations in Chapter 4.

In Eq. (1.182), when $i = c + 2$ and $k = c + 1$,

$$C_{c+2} = (\partial y^{(c+1)}/\partial X_{c+2})_{C_1,\ldots,C_{c+1}}. \tag{1.184}$$

Let us assume that $y^{(0)} = U(S, V, n_1, \ldots, n_c)$. Then from the dU expression, $C_1 = T$, $C_2 = -P$, $C_3 = \mu_1, \ldots$, and $C_{c+1} = \mu_{c-1}$. For a single phase according to the Gibbs-Duhem expression (see Eq. (1.40)), $c + 1$ intensive variables $T, P, \mu_1, \ldots, \mu_{c-1}$ define the system. Therefore C_{c+2} is fixed, and there could be no variation of C_{c+2} with respect to X_{c+2}; i.e.,

$$C_{c+2,c+2} = (\partial^2 y^{(c+1)}/\partial^2 X_{c+2})_{C_1,\ldots,C_{c+1}} = 0. \tag{1.185}$$

The above relationship will be used in establishing the criteria of phase stability in Chapter 4.

Next we will discuss Jacobian transformations.

Jacobian transformation

Jacobians, which are simply functional determinants, have long been used to simplify derivations of functions in thermodynamics (Crawford, 1950; Carroll, 1965). Our interest in Jacobian transformation is for its great simplicity and usefulness in transformation of the various derivations of the Legendre transformation of thermodynamic functions. Without the Jacobian transformation, the interrelation between the derivations of Legendre transformation become very complicated.

Let us define the general $(c+2)$ dimensional Jacobian J as

$$J = \frac{\partial(Z_1, Z_2, \ldots, Z_{c+2})}{\partial(X_1, X_2, \ldots, X_{c+2})}$$

$$= \begin{vmatrix} (\partial Z_1/\partial X_1) & (\partial Z_1/\partial X_2) & \cdots & (\partial Z_1/\partial X_{c+2}) \\ \vdots & \vdots & \vdots & \vdots \\ (\partial Z_{c+2}/\partial X_1) & (\partial Z_{c+2}/\partial X_2) & \cdots & (\partial Z_{c+2}/\partial X_{c+2}) \end{vmatrix} \qquad (1.186)$$

where dependent variables $Z_1, \ldots, Z_{c+2}$ are functions of independent variables $X_1, \ldots, X_{c+2}$. In the above determinant, we have dropped the variables that are held constant in the derivatives. As an example, $(\partial Z_1/\partial X_1)$ should be written as $(\partial Z_1/\partial X_1)_{X_2,\ldots,X_{c+2}}$, where variables X_2 to X_{c+2} are kept constant. There are well-known rules in the manipulation of determinants. These include (1) sign rule, (2) reduction properties, and (3) transformation properties. As an example, the sign of J will change when a neighboring pair of Zs or Xs are interchanged. The reduction property is useful for the type of derivatives we have in mind. Whenever a common variable occurs between Zs and Xs, a reduction takes place. As an example if $Z_1 = X_1$ and $Z_2 = X_2$, then from Eq. (1.186),

$$\frac{\partial(Z_1, Z_2, Z_3, \ldots, Z_{c+2})}{\partial(X_1, X_2, X_3, \ldots, X_{c+2})} = \frac{\partial(X_1, X_2, Z_3, \ldots, Z_{c+2})}{\partial(X_1, X_2, X_3, \ldots, X_{c+2})} = \frac{\partial(Z_3, \ldots, Z_{c+2})}{\partial(X_3, \ldots, X_{c+2})}\Bigg|_{X_1,X_2}$$

$$(1.187)$$

Note that in Eq. (1.187), X_1 and X_2 must remain constant. In. Eq. (1.187), the order of the Jacobian has been reduced from $(c+2)$ to c. The above process can be reversed, going from order c to $c+2$, which is also true from Eq. (1.187). The transformation properties of Jacobians is useful in the manipulation of Legendre transformation derivatives and can be utilized by the introduction of a new set of independent variables $\underline{Y} = (Y_1, Y_2, \ldots, Y_{c+2})$,

$$\frac{\partial(Z_1, Z_2, \ldots, Z_{c+2})}{\partial(X_1, X_2, \ldots, X_{c+2})} = \frac{\dfrac{\partial(Z_1, Z_2, \ldots, Z_{c+2})}{\partial(Y_1, Y_2, \ldots, Y_{c+2})}}{\dfrac{\partial(X_1, X_2, \ldots, X_{c+2})}{\partial(Y_1, Y_2, \ldots, Y_{c+2})}} \qquad (1.188)$$

which is simply the ratio of the two Jacobians.

Now we will use the Jacobian transformation to provide a relationship between the derivatives of higher-order Legendre transformations to the derivatives of a lower-order one. Let us denote the derivative of

$y^{(k-1)}$ with respect to X_k while holding $C_1, C_2, \ldots, C_{k-1}, X_{k+1}, \ldots, X_{c+2}$ constant by $y_k^{(k-1)}$; i.e., $y_k^{(k-1)} = (\partial y^{(k-1)}/\partial X_k)_{C_1,\ldots,C_{k-1},X_{k+1},\ldots,X_{c+2}}$. The second derivative with respect to X_k is shown by $y_{kk}^{(k-1)}$, which is $(\partial^2 y^{(k-1)}/\partial X_k^2)_{C_1,\ldots,C_{k-1},X_{k+1},\ldots,X_{c+2}}$. From Eq. (1.183),

$$y_{kk}^{(k-1)} = (\partial C_k/\partial X_k)_{C_1,\ldots,C_{k-1},X_{k+1},\ldots,X_{c+2}}. \tag{1.189}$$

From Eq. (1.187), and the sign rule of determinants,

$$y_{kk}^{(k-1)} = \frac{\partial(C_1, C_2, \ldots, C_{k-1}, C_k, X_{k+1}, \ldots, X_{c+2})}{\partial(C_1, C_2, \ldots, C_{k-1}, X_k, X_{k+1}, \ldots, X_{c+2})}, \tag{1.190}$$

which can be also written as

$$y_{kk}^{(k-1)} = \frac{\partial(C_{k-2}, C_{k-1}, C_k)}{\partial(C_{k-2}, C_{k-1}, X_k)}\bigg|_{C_1,C_2,\ldots,C_{k-3},X_{k+1},\ldots,X_{c+2}} \tag{1.191}$$

According to Eq. (1.188),

$$y_{kk}^{(k-1)} = \frac{\partial(C_{k-2}, C_{k-1}, C_k)/\partial(C_{k-2}, X_{k-1}, X_k)}{\partial(C_{k-2}, C_{k-1}, X_k)/\partial(C_{k-2}, X_{k-1}, X_k)}\bigg|_{C_1,C_2,\ldots,C_{k-3},X_{k+1},\ldots,X_{c+2}} \tag{1.192}$$

Equation (1.192) can now be written in a determinant form,

$$y_{kk}^{(k-1)} = \frac{\begin{vmatrix} \left(\dfrac{\partial C_{k-2}}{\partial C_{k-2}}\right)_{X_{k-1},X_k} & \left(\dfrac{\partial C_{k-2}}{\partial X_{k-1}}\right)_{C_{k-2},X_k} & \left(\dfrac{\partial C_{k-2}}{\partial X_k}\right)_{C_{k-2},X_{k-1}} \\ \left(\dfrac{\partial C_{k-1}}{\partial C_{k-2}}\right)_{X_{k-1},X_k} & \left(\dfrac{\partial C_{k-1}}{\partial X_{k-1}}\right)_{C_{k-2},X_k} & \left(\dfrac{\partial C_{k-1}}{\partial X_k}\right)_{C_{k-2},X_{k-1}} \\ \left(\dfrac{\partial C_k}{\partial C_{k-2}}\right)_{X_{k-1},X_k} & \left(\dfrac{\partial C_k}{\partial X_{k-1}}\right)_{C_{k-2},X_k} & \left(\dfrac{\partial C_k}{\partial X_k}\right)_{C_{k-2},X_{k-1}} \end{vmatrix}}{\begin{vmatrix} \left(\dfrac{\partial C_{k-2}}{\partial C_{k-2}}\right)_{X_{k-1},X_k} & \left(\dfrac{\partial C_{k-2}}{\partial X_{k-1}}\right)_{C_{k-2},X_k} & \left(\dfrac{\partial C_{k-2}}{\partial X_k}\right)_{C_{k-2},X_{k-1}} \\ \left(\dfrac{\partial C_{k-1}}{\partial C_{k-2}}\right)_{X_{k-1},X_k} & \left(\dfrac{\partial C_{k-1}}{\partial X_{k-1}}\right)_{C_{k-2},X_k} & \left(\dfrac{\partial C_{k-1}}{\partial X_k}\right)_{C_{k-2},X_{k-1}} \\ \left(\dfrac{\partial X_k}{\partial C_{k-2}}\right)_{X_{k-1},X_k} & \left(\dfrac{\partial X_k}{\partial X_{k-1}}\right)_{C_{k-2},X_k} & \left(\dfrac{\partial X_k}{\partial X_k}\right)_{C_{k-2},X_{k-1}} \end{vmatrix}}. \tag{1.193}$$

On the right side we have dropped the variables $C_1, C_2, \ldots, C_{k-3}, X_{k+1}, \ldots, X_{c+2}$ which are held constant. One could readily establish the values of various elements of the two determinants. We will first write those that are either 0 or 1.

$$(\partial C_{k-2}/\partial C_{k-2}) = 1, (\partial C_{k-2}/\partial X_{k-1})_{C_{k-2},X_k} = (\partial C_{k-2}/\partial X_k)_{C_{k-2},X_k} = 0$$

$$(\partial X_k/\partial X_k) = 1, (\partial X_k/\partial C_{k-2})_{X_{k-1},X_k}$$

$$= (\partial C_{k-2}/\partial X_{k-1})_{C_{k-2},X_k}$$

$$= (\partial C_{k-2}/\partial X_k)_{C_{k-2},X_{k-1}} = (\partial X_k/\partial X_{k-1})_{C_{k-2},X_k} = 0$$

The derivative of a function with respect to itself is 1 and the derivative of a function while the function itself is held constant is zero. With the above entries, $y_{kk}^{(k-1)}$ simplifies to,

$$y_{kk}^{(k-1)} = \frac{\begin{vmatrix} \left(\dfrac{\partial C_{k-1}}{\partial X_{k-1}}\right)_{C_{k-2},X_k} & \left(\dfrac{\partial C_{k-1}}{\partial X_k}\right)_{C_{k-2},X_{k-1}} \\[12pt] \left(\dfrac{\partial C_k}{\partial X_{k-1}}\right)_{X_{k-2},X_k} & \left(\dfrac{\partial C_k}{\partial X_k}\right)_{C_{k-2},X_{k-1}} \end{vmatrix}}{\left(\dfrac{\partial C_{k-1}}{\partial X_{k-1}}\right)_{C_{k-2},X_k}} \Bigg|_{C_1,C_2,\ldots,C_{k-3},\ldots,X_{k+1},\ldots,X_{c+2}}$$

$$(1.194)$$

Note that we have shown the parameters that are held constant on the right side of Eq. (1.194). Now we can use the expression for $dy^{(k-2)}$ to obtain the elements of the above equation. The reason for using $dy^{(k-2)}$ is that $C_1,\ldots,C_{k-3}$ are held constant for all four elements (see Eq. (1.183))

$$(\partial C_{k-1}/\partial X_{k-1})_{C_1,C_2,\ldots,C_{k-3},C_{k-2},X_k,\ldots,X_{c+2}} = y_{k-1,k-1}^{(k-2)}$$

$$(\partial C_{k-1}/\partial X_k)_{C_1,C_2,\ldots,C_{k-3},C_{k-2},X_{k-1},X_{k+1},\ldots,X_{c+2}} = y_{k-1,k}^{(k-2)}$$

$$(\partial C_k/\partial X_{k-1})_{C_1,C_2,\ldots,C_{k-3},C_{k-2},X_k,\ldots,X_{c+2}} = y_{k,k-1}^{(k-2)}$$

$$(\partial C_k/\partial X_k)_{C_1,C_2,\ldots,C_{k-3},C_{k-2},X_{k-1},X_{k+1},\ldots,X_{c+2}} = y_{k,k}^{(k-2)}$$

$$(1.195)$$

From the reciprocity relationship (see Eq. (1.81)), $y_{k,k-1}^{(k-2)} = y_{k-1,k}^{(k-2)}$. Substitution of the above results in Eq. (1.194) leads to

$$\boxed{y_{kk}^{(k-1)} = y_{kk}^{(k-2)} - (y_{k,k-1}^{(k-2)})^2/y_{k-1,k-1}^{(k-2)}}. \qquad (1.196)$$

In Chapter 4 we will provide an alternative derivation of the above equation, and discuss the fact that $y_{kk}^{(k-1)}$ goes to zero before $y_{kk}^{(k-2)}$ goes to zero. Equation (1.196) is a useful relationship between the derivatives of $y^{(k-1)}$ and the derivatives of $y^{(k-2)}$. For $k = 3$, then

$$y^{(2)}_{33} = y^{(1)}_{33} - (y^{(1)}_{32})^2 / y^{(1)}_{22}.\tag{1.197}$$

If $y^{(0)} = U$, then $y^{(1)} = A$ and $y^{(2)} = G$. Eq. (1.197) provides the relationship between the derivatives of Gibbs free energy and the Helmholtz free energy.

Maxwell's relations

Maxwell's relations can be readily obtained from the Jacobian transforms. However, we will obtain them by applying the reciprocity relations to the differential expressions of dU, dH, dA, and dG, given by Eqs. (1.21), (1.52), (1.56), and (1.57). The full set of Maxwell's relations is:

$$(\partial T/\partial V)_{S,\underline{n}} = -(\partial P/\partial S)_{V,\underline{n}}\tag{1.198}$$

$$-(\partial S/\partial P)_{T,\underline{n}} = (\partial V/\partial T)_{P,\underline{n}}\tag{1.199}$$

$$(\partial T/\partial P)_{S,\underline{n}} = (\partial V/\partial S)_{P,\underline{n}}\tag{1.200}$$

$$(\partial S/\partial V)_{T,\underline{n}} = (\partial P/\partial T)_{V,\underline{n}}.\tag{1.201}$$

Note that in Eqs. (1.199) and (1.201), on the right side the variables are only P, T, V, and $\underline{n}$, all directly measurable. We will use these relationships in Chapter 3 in the derivation of the expressions for specific heat quantities c_P and c_V.

Examples and theory extension

Example 1.1 Prove that any substance in a mixture tends to pass from the region of higher to the region of lower chemical potential. Assume the effect of gravity to be negligible.

Solution Consider a composite system (shown below) consisting of two subsystems primed and double-primed. The partition could be either rigid or moveable and permeable to component i. Suppose the temperature and the pressure are the same in both subsystems.

μ'_i	μ''_i		$d(U' + U'') = 0$
T	T		$d(n'_i + n''_i) = 0 \quad i = 1, \ldots, c$
P	P		$d(V' + V'') = 0$

The change of the entropy of the composite system, dS, is given by

$$dS = dS' + dS''.$$

From Eq. (1.22),

$$dS = -\sum_{i=1}^{c}(1/T)(\mu_i' - \mu_i'')dn_i'.$$

If the system has not reached equilibrium, then $dS > 0$. Assume $dn_i' < 0$ then $dn_i'' > 0$ which implies that component i is flowing from the primed subsystem to the double-primed subsystem. Since $dS > 0$, and $dn_i' < 0$, then $(\mu_i' - \mu_i'') > 0$; $\mu_i' > \mu_i''$. Therefore, diffusion of component i will be in the direction of $\mu_i' > \mu_i''$.

Example 1.2 It is often useful to write the expressions for thermodynamic functions in terms of the variable set $(n_1, n_2, \ldots, n_{c-1}, n)$ instead of $(n_1, n_2, \ldots, n_{c-1}, n_c)$ where $n = \sum_{i=1}^{c} n_i$. In the new variable set,

$$dU = TdS - PdV + \sum_{i=1}^{c-1}(\mu_i - \mu_c)dn_i + \mu_c dn$$

and

$$dG = -SdT + VdP + \sum_{i=1}^{c-1}(\mu_i - \mu_c)dn_i + \mu_c dn.$$

From the above two equations, derive the expression for du and dg.

Solution For molar quantities, we simply fix $n = 1$; therefore

$$\boxed{du = Tds - Pdv + \sum_{i=1}^{c-1}(\mu_i - \mu_c)dx_i}$$

and

$$\boxed{dg = -sdT + vdP + \sum_{i=1}^{c-1}(\mu_i - \mu_c)dx_i}\,,$$

where $x_i = n_i/n$ is the mole fraction of component i, and therefore $\sum_{i=1}^{c} x_i = 1$, i.e., $x_c = 1 - \sum_{i=1}^{c-1} x_i$.

Example 1.3 Reservoir fluids are unique in comparison to the fluids that chemical engineers are used to. The unique features of reservoir fluids are due to components such as C_1 and C_2. Let us consider a mixture of 95 gmoles C_2 and 5 gmoles of nC_7 at $80°C$ and 74.5 atm. Let us add 2 gmoles of nC_7 to the mixture at the same pressure and temperature. Calculate the volume change (decrease!) due to the addition of 2 gmoles of nC_7. Partial molar volume data are available from Wu and Ehrlich (1973). Later in Chapter 3, we will show how to predict these values from an equation of state:

Data (from Wu, and Ehrlich, 1973):

$$\bar{V}_{C_2}(80°C,\ 74.5\ \text{atm},\ x_{C_7} = 0.05) = 2.3 \times 10^2\ \text{cm}^3/\text{gmole}$$

$$\bar{V}_{C_7}(80°C,\ 74.5\ \text{atm},\ x_{C_7} = 0.05) = -8.1 \times 10^2\ \text{cm}^3/\text{gmole}$$

$$\bar{V}_{C_2}(80°C,\ 74.5\ \text{atm},\ x_{C_7} = 0.068) = 2.21 \times 10^2\ \text{cm}^3/\text{gmole}$$

$$\bar{V}_{C_7} = (80°C,\ 74.5\ \text{atm},\ x_{C_7} = 0.068) = -7.1 \times 10^2\ \text{cm}^3/\text{gmole}$$

Solution   From $V = \sum_{i=1}^2 n_i \bar{V}_i$, one can calculate the volume before and after addition of 2 gmoles of nC_7 to the mixture. Note that after adding 2 gmoles of nC_7 the composition of the mixture changes; before addition $x_{C_7} = 0.05$, after addition $x_{C_7} = 7/102 \approx 0.068$.

$$V_{before} = (95)(2.3 \times 10^2) + (5)(-8.1 \times 10^2) = 17{,}800\ \text{cm}^3$$

$$V_{after} = (95)(2.21 \times 10^2) + (7)(-7.1 \times 10^2) = 16{,}025\ \text{cm}^3$$

Therefore from the addition of 2 gmoles of nC_7 at constant temperature and pressure, the volume decreases by 1,775 cm^3.

Let us now calculate the volume of pure C_2 and pure nC_7 and compare it with the original mixture—all at the same temperature and pressure. The pure component data are

$$v_{C_2}(80°C,\ 74.5\ \text{atm}) \approx 2.3 \times 10^2\ \text{cm}^3/\text{gmole (from Starling, 1973)}$$

$$v_{C_7}(80°C,\ 74.5\ \text{atm}) \approx 1.6 \times 10^2\ \text{cm}^3/\text{gmole (from Katz, } et\ al.,\ 1959)$$

$$V_{pure\ C_2} = 95 \times 2.3 \times 10^2 = 21{,}850\ \text{cm}^3$$

$$V_{pure\ C_7} = 5 \times 1.6 \times 10^2 = 800\ \text{cm}^3$$

$$\text{The total volume before mixing} = 22{,}650\ \text{cm}^3$$

$$\text{The total volume after mixing} = 17{,}800\ \text{cm}^3$$

Therefore, the volume change due to mixing is 4,845 cm^3. There is a volume decrease of about 22 percent due to mixing.

Example 1.4   Derive the following expressions for the entropy and Gibbs free energy of mixing at constant temperature and pressure for an ideal gas mixture:

$$\Delta G_{mix} = +RT \sum_{i=1}^c n_i \ln y_i$$

$$\Delta S_{mix} = -R \sum_{i=1}^c n_i \ln y_i.$$

Solution The expression for the ΔG_{mix} is

$$\Delta G_{mix} = G(T, P, \underline{n}) - \sum_{i=1}^{c} n_i g_i(T, P) = \sum_{i=1}^{c} n_i \mu_i(T, P, \underline{y}) - \sum_{i=1}^{c} n_i g_i(T, P),$$

where $g_i(T, P)$ is the molar Gibbs free energy of pure component i at temperature T and pressure P.

The chemical potential of component i in an ideal gas mixture and in an ideal pure gas can be written as

$$\mu_i(T, P, y) = \mu_i^0(T, P^0) + RT \ln y_i P - RT \ln P^0,$$

$$\mu_i(T, P) = \mu_i^0(T, P^0) + RT \ln P - RT \ln P^0.$$

Combining the above equations,

$$\boxed{\Delta G_{mix} = +RT \sum_{i=1}^{c} n_i \ln y_i}.$$

Note that since $\ln y_i < 0$, then ΔG_{mix} is negative. We can relate ΔG_{mix} to ΔS_{mix} through

$$G(T, P, \underline{n}) = H(T, P, \underline{n}) - TS(T, P, \underline{n})$$

$$G(T, P) = H(T, P) - TS(T, P) = \sum_{i=1}^{c} n_i h_i(T, P) - T \sum_{i=1}^{c} n_i s_i(T, P),$$

where $G(T, P)$ is the Gibbs free energy of various components before mixing and $G(T, P, \underline{n})$ is the Gibbs free energy after mixing.

Substracting the above two equations,

$$\Delta G_{mix} = \Delta H_{mix} - T\Delta S_{mix}$$

or $$\Delta S_{mix} = (\Delta H_{mix} - \Delta G_{mix})/T.$$

From Eq. (1.130), $\Delta H_{mix} = 0$; therefore,

$$\boxed{\Delta S_{mix} = -R \sum_{i=1}^{c} n_i \ln y_i}.$$

Note that the entropy of mixing in an ideal gas mixture is positive since $\ln y_i$ is negative.

Example 1.5 The Helmholtz free energy of a mixture at temperature T and pressure P can be expressed as

$$\boxed{A = \int_{V}^{\infty} (P - nRT/V)dV - RT \sum_{i=1}^{c} n_i \ln(V/n_i RT) + \sum_{i=1}^{c} n_i(u_i^0 - Ts_i^0)},$$

where u_i^0 is the ideal gas molar internal energy at temperature T and s_i^0 is the molar entropy of pure component i at temperature T and pressure $P^0 = 1$ atm. Derive the above expression.

Solution The expression for dA at constant temperature and composition (see Eq. (1.56)) is given by $(dA = -PdV)_{T,\underline{n}}$. Let A represent the Helmholtz free energy of the mixture at T, P, and $\underline{n}$. Then

$$A - A^0 = -\int_{P^0}^{P} PdV = -\int_{V^0}^{V} PdV.$$

The above equation can be written as

$$A - A^0 = -\int_{V^0}^{\infty} PdV - \int_{\infty}^{V} PdV.$$

Note that from V^0 to ∞ (i.e., from P^0 to pressure zero), the fluid is an ideal gas. Therefore, the first term on the right side can be computed for an ideal gas. Let us add and subtract $\int_{\infty}^{V}(nRT/V)dV$ to and from the above equation:

$$A - A^0 = -\int_{V^0}^{\infty} PdV - \int_{V}^{\infty}(nRT/V)dV + \int_{V}^{\infty}(nRT/V)dV + \int_{V}^{\infty} PdV.$$

The term $\int_{V}^{\infty}(nRT/V)dV = \int_{V}^{V^0}(nRT/V)dV + \int_{V^0}^{\infty}(nRT/V)dV$ and, therefore,

$$A - A^0 = nRT \ln V^0/V - \int_{\infty}^{V}(P - nRT/V)dV$$

$A^0 = U^0 - TS$; $U^0 = \sum_{i=1}^{c} n_i u_i^0$, and $S^0 = \Delta S_{mix}^0 + \sum_{i=1}^{c} n_i s_i^0$ (see Example 1.4), where u_i^0 and s_i^0 are the ideal gas molar internal energy and entropy of pure component i at T and P^0, respectively. From Example 1.4,

$$S^0 = -R \sum_{i=1}^{c} n_i \ln y_i + \sum_{i=1}^{c} n_i s_i^0$$

The next step is the derivation of an expression for $nRT \ln (V^0/V)$. At total pressure of P^0, the partial pressure of component i is P_i^0; $P_i^0 V^0 = n_i RT$.

$$RT \sum_{i=1}^{c} n_i \ln V/(n_i RT) = RT \sum_{i=1}^{c} n_i \ln V/(P_i^0 V^0)$$

$$= nRT \ln(V/V^0) - RT \sum_{i=1}^{c} n_i \ln P_i^0$$

From the above equations,

$$A = \int_{V}^{\infty}(P - nRT/V)dV - RT \sum_{i=1}^{c} n_i \ln V/(n_i RT)$$

$$+ \sum_{i=1}^{c} n_i(u_i^0 - Ts_i^0) + RT \sum_{i=1}^{c} n_i \ln y_i - RT \sum_{i=1}^{c} n_i \ln P_i^0.$$

The last two terms on the right side are $RT \sum_{i=1}^{c} n_i \ln P^0$ since $P_i^0/y_i = P^0$. For $P^0 = 1$ atm,

$$A = \int_{V}^{\infty}(P - nRT/V)dV - RT \sum_{i=1}^{c} n_i \ln V/(n_i RT) + \sum_{i=1}^{c} n_i(u_i^0 - Ts_i^0).$$

Note that the reference pressure for the ideal gas entropy is, therefore, 1 atm.

Example 1.6 Derive the expression for the mixture entropy, S, given by

$$S = \int_V^\infty \left[(nR/V) - (\partial P/\partial T)_{V,\underline{n}}\right]dV + R\sum_{i=1}^c n_i \ln V/(n_i RT) + \sum_{i=1}^c n_i s_i^0.$$

Solution From Eq. (1.201), $[dS = (\partial P/\partial T)_{V,\underline{n}}dV]_{T,\underline{n}}$ and, therefore, $S(T, P, \underline{n}) - S^0(T, P^0, \underline{n}) = \int_{P^0}^P (\partial P/\partial T)_{V,\underline{n}}dV = \int_{V^0}^V (\partial P/\partial T)_{V,\underline{n}}dV$. The integral on the right side can be written as $S - S^0 = \int_{V^0}^\infty (\partial P/\partial T)_{V,\underline{n}}dV + \int_\infty^V (\partial P/\partial T)_{v,\underline{n}}dV$. Adding and subtracting $\int_V^\infty (nR/V)dV$ to and from the right side, $S - S^0 = \int_V^\infty [(nR/V) - (\partial P/\partial T)_{V,\underline{n}}]dV - \int_V^\infty (nR/V)dV + \int_{V^0}^\infty (\partial P/\partial T)_{V,\underline{n}}dV$.

But $\int_V^\infty (n\bar{R}/V)dV = \int_V^{V^0}(nT/V)dV + \int_{V^0}^\infty (nR/V)dV$; therefore, $S - S^0 = \int_V^\infty [(nR/V) - (\partial P/\partial T)_{V,\underline{n}}]dV + \int_{V^0}^\infty [(\partial P/\partial T)_{V,\underline{n}} - (nR/V)]dV - \int_V^{V^0}(nR/V)dV$.

In the above equation, the second term of the right side is zero, and the third term is $-nR \ln V^0/V$. The second term is zero because from volume V^0 to ∞, the ideal gas describes the fluid; $PV = nRT$, and $(\partial P/\partial T)_{V,\underline{n}} = nR/V$.

$$S - S^0 = \int_V^\infty \left[(nR/V) - (\partial P/\partial T)_{V,\underline{n}}\right]dV - nR \ln V^0/V.$$

From Example 1.5, $nR \ln V^0/V = -R\sum_{i=1}^c n_i \ln P_i^0 - R\sum_{i=1}^c n_i \ln V/(n_i RT)$ and from Example 1.4, $S^0 = \sum_{i=1}^c n_i s_i^0 - R\sum_{i=1}^c n_i \ln y_i$. Combining the above three equations and assuming $P^0 = 1$ atm, the sought expression is obtained. Note that here, similarly to the previous example, $P^0 = 1$ atm corresponds only to the ideal gas state for entropy.

Example 1.7 *Derivation of Raoult's law* Suppose a multicomponent mixture is in the gas–liquid equilibrium state. Under what assumptions, can one write,

$$x_i P_i^{sat}(T) = y_i P\,?$$

In the above equation, x_i and y_i are the liquid and vapor phase mole fractions, respectively, and P is the pressure. $P_i^{sat}(T)$ is the vapor pressure at temperature T for pure component i.

Solution At equilibrium, one can write

$$f_i^L(T, P, \boldsymbol{x}) = f_i^V(T, P, \boldsymbol{y})$$

For an ideal solution $f_i^L(T, P, \boldsymbol{x}) = x_i f_{pure\,i}^L(T, P)$ and $f_i^L(T, P, \boldsymbol{y}) = y_i f_{pure\,i}^V(T, P)$. For an ideal gas $f_{pure\,i}^V(T, P) = P$, since $\varphi_{pure\,i}(T, P) = 1$. Now assume that the liquid phase of pure component i at T has a vapor pressure $P_i^{sat}(T)$; at $P_i^{sat}(T), f_{pure\,i}^L(T, P_i^{sat}) = f_{pure\,i}^V(T, P_i^{sat})$ and if the gas phase is ideal, then $f_{pure\,i}^V(T, P_i^{sat}) = P_i^{sat}(T)$. If we assume that the effect of pressure on $f_{pure\,i}^L$ is negligible, then $f_{pure\,i}^L(T, P) = P_i^{sat}(T)$. All these assumptions lead to the derivation of Raoult's Law.

Example 1.8 Liquefied petroleum gases (LPG) are mixtures of propane (C_3) and n-butane (nC_4). Use Raoult's law to determine the bubblepoint and dewpoint pressures of a mixture of 50 percent C_3 and 50 percent nC_4 (equimolar mixture) at $150°F$. The vapor pressure data are $P_{C_3}^{sat}(150°F) = 350$ psia, $P_{C_4}^{sat}(150°F) \approx 105$ psia.

Solution At the dewpoint,

$$\sum_{i=1}^{c} x_i = 1$$

From Raoult's law derived in Example 1.7, $x_i = y_i P / P_i^{sat}(T)$ and, therefore $P_{dew} = 1/\sum_{i=1}^{c} y_i / P_i^{sat}(T)$.
   At the bubblepoint,

$$\sum_{i=1}^{c} y_i = 1$$

$$y_i = x_i P_i^{sat}(T)/P, \qquad P_{bubble} = \sum_{i=1}^{c} x_i P_i^{sat}(T).$$

From these two equations, $P_{bubble} = 227.5$ psia and $P_{dew} = 162$ psia. These calculated results are very reasonable. When the hydrocarbon species are of the same size, Raoult's law performs well. However, when they are of different size and properties due to nonidealities, Raoult's law performs poorly.

Example 1.9 *Gibbs phase rule (flat interface)* Derive the phase rule $F = c + 2 - p$, where F is the number of degrees of freedom, c is the number of components, and p is the number of phases. Assume the interface between the phases is flat.

Solution The criteria of chemical equilibrium of multicomponent systems with c components and p phases are (see Eqs. (1.32a))

$$T^{(1)} = T^{(2)} = \cdots = T^{(p)}$$

$$P^{(1)} = P^{(2)} = \cdots = P^{(p)}$$

$$\mu_1^{(1)} = \mu_1^{(2)} = \cdots = \mu_1^{(p)}$$

$$\vdots$$

$$\mu_c^{(1)} = \mu_c^{(2)} = \cdots = \mu_c^{(p)}.$$

There are $(p-1)$ temperature and $(p-1)$ pressure equations. The number of equations for the chemical potentials is $c(p-1)$. Therefore, the total number of equations is $(c+2)(p-1)$. For each phase the intensive variables are related by the Gibbs-Duhem equations (see Eq. (1.40)). Therefore, the number of independent intensive variables of each phase is $(c+1)$; T, P, and $\mu_1, \mu_2, \ldots, \mu_{c-1}$. The degrees of freedom, F, is defined as the difference between the number of independent intensive variables of the system [which is $p(c+1)$] and the number of equations between these variables [which is $(c+2)(p-1)$].

Therefore,

$$\boxed{F = p(c + 1) - (p - 1)(c + 2) = c + 2 - p}.$$

The above relationship can be used to describe the state of a particular phase of a composite system; it provides the number of independent intensive variables for specifying the system. Since chemical potential of a phase is a function of temperature, pressure, and mole fraction of components 1 to $(c - 1)$, then one may use $(c - 1)$ mole fractions instead of c chemical potential for every phase.

Let us now give two examples of the phase rule. In the first example, consider a two-component system consisting of three fluid phases; therefore, $F = 2 + 2 - 3 = 1$. If we specify the temperature, the system is then defined; we are not free to specify both temperature and pressure of a three-phase two-component composite system. In the second example, we consider a two-phase two-component system; $F = 2$. If we specify the temperature and pressure, each of the phases are specified. In other words, if we change the amount of each of the two components, as long as the system is in two phase, the composition of each phase remains unchanged.

The phase rule presented above is valid if all the components are present in all phases. It should be modified when some of the components are absent in one or more phases. We will see in Chapter 2 that the above phase relationship should be modified when the interface between the phases is curved. A modification is also needed when there is influence of gravity on equilibrium.

Example 1.10 *Derivatives with respect to mole numbers and mole fractions* Derivatives with respect to mole numbers and mole fractions and the relation between them often result in confusion. This example is designed to avoid such a confusion. (a) Consider the simple derivatives $(\partial w_1/\partial x_1)$ and $(\partial w_1/\partial n_1)$ for binary and ternary systems (w is the weight fraction, x is the mole fraction, and n is the number of moles). Derive the expressions for $(\partial w_1/\partial x_1)$ and $(\partial w_1/\partial n_1)$. (b) What is the relationship between $(\partial f_i/\partial n_j)_{T,P,n_1,\ldots,n_{j-1},n_{j+1},\ldots,n_c}$, $(\partial f_i/\partial n_j)_{T,P,n_1,\ldots,n_{j-1},n_{j+1},\ldots,n_{c-1},n}$, and $(\partial f_i/\partial x_j)_{T,P,x_1,\ldots,x_{j-1},x_{j+1},\ldots,x_{c-1}}$? The subscript $n = \sum_{i=1}^{c} n_i$.

Solution   (a) Let us consider first the binary system.

$$w_1 = \frac{x_1 M_1}{x_1 M_1 + x_2 M_2} = \frac{x_1 M_1}{\bar{M}},$$

where M_1 is the molecular weight of component 1 and $\bar{M}$ is the average molecular weight.

$$\frac{\partial w_1}{\partial x_1} = \frac{M_1 \bar{M} - x_1 M_1 (\partial \bar{M}/\partial x_1)}{\bar{M}^2} \qquad \text{(E1.10.1a)}$$

The key term is the derivative $(\partial \bar{M}/\partial x_1)$; from

$$\bar{M} = x_1 M_1 + x_2 M_2 = x_1 M_1 + (1 - x_1)M_2$$

$$\boxed{\frac{\partial \bar{M}}{\partial x_1} = M_1 - M_2}\,.$$

Note that in the above derivative x_2 cannot be held constant since $x_1 + x_2 = 1$. Combining the expression for $(\partial w_1/\partial x_1)$ and $(\partial \bar{M}/\partial x_1)$,

$$\boxed{\frac{\partial w_1}{\partial x_1} = \frac{M_1 \bar{M} - x_1 M_1(M_1 - M_2)}{\bar{M}^2} = \frac{M_1 M_2}{\bar{M}^2}}\,.$$

Now let us calculate $(\partial w_1/\partial n_1)$. Here $\partial w_1/\partial n_1|_{n_2}$ and $\partial w_1/\partial n_1|_n$ (where $n = n_1 + n_2$) are not the same. The latter derivative is related to $(\partial w_1/\partial x_1)$, but the former is not.

From

$$n\bar{M} = n_1 M_1 + n_2 M_2,$$

the derivative with respect to n_1 at constant n_2 is

$$n\frac{\partial \bar{M}}{\partial n_1}\bigg|_{n_2} + \bar{M} = M_1,$$

and, therefore,

$$\frac{\partial \bar{M}}{\partial n_1}\bigg|_{n_2} = \frac{M_1 - \bar{M}}{n}$$

Note that since n is not constant we cannot write $\partial \bar{M}/\partial x_1|_{n_2} = M_1 - \bar{M}$. The derivative of $n\bar{M}$ at constant n is given by

$$n\frac{\partial \bar{M}}{\partial n_1}\bigg|_n = M_1 + M_2\frac{\partial n_2}{\partial n_1}\bigg|_n\,.$$

Since $n = n_1 + n_2$ then $(\partial n_2/\partial n_1)_n = -1$, and, therefore,

$$n\frac{\partial \bar{M}}{\partial n_1}\bigg|_n = M_1 - M_2.$$

Since n is constant then $n(\partial \bar{M}/\partial n_1)|_n = \partial \bar{M}/\partial x_1$.

Now let us consider a ternary system. One first can write

$$w_1 = \frac{x_1 M_1}{\bar{M}},$$

where $\bar{M} = x_1 M_1 + x_2 M_2 + x_3 M_3$. The expression for $\partial w_1 / \partial x_1$ is given by Eq. (E.1.10.1a). The main task is finding the derivative $\partial \bar{M} / \partial x_1$. In a ternary system $x_1 + x_2 + x_3 = 1$; therefore, $x_3 (= 1 - x_2 - x_1)$ is not an independent variable. Hence,

$$\left. \frac{\partial \bar{M}}{\partial x_1} \right|_{x_2} = M_1 - M_3 \qquad\qquad \text{(E1.10.2a)}$$

and

$$\left. \frac{\partial w_1}{\partial x_1} \right|_{x_2} = \frac{M_1 \bar{M} - x_1 M_1 (M_1 - M_3)}{\bar{M}^2} = \frac{(1 - x_2) M_1 M_3 + x_2 M_1 M_2}{\bar{M}^2}.$$

The calculation of $(\partial w_1 / \partial n_1)_{n, n_2}$ is based on $n_3 (= n - n_1 - n_2)$ not being an independent variable. From

$$n\bar{M} = n_1 M_1 + n_2 M_2 + n_3 M_3 = n_1 M_1 + n_2 M_2 + (n - n_1 - n_2) M_3,$$

$$n \left. \frac{\partial \bar{M}}{\partial n_1} \right|_{n, n_2} = M_1 - M_3$$

or

$$\left. \frac{\partial \bar{M}}{\partial x_1} \right|_{x_2} = M_1 - M_3$$

which is the same as given by Eq. (E1.10.2a). The expressions for $(\partial w_1 / \partial n_1)_{n_2, n_3}$ and $(\partial w_1 / \partial n_1)_{n, n_2}$ are readily calculated.

(b) The fugacity of component i in a mixture can be expressed by,

$$f_i = f_i(T, P, x_1, x_2, \ldots, x_{c-1}).$$

The differential of f_i at constant T and P is

$$df_i = \sum_{k=1}^{c-1} \left. \frac{\partial f_i}{\partial x_k} \right|_{x_k} dx_k \qquad\qquad \text{(E1.10.1b)}$$

where $x_k = (x_1, x_2, \ldots, x_{k-1}, x_{k+1}, \ldots, x_{c-1})$. The mole fraction of component k is a function of mole numbers of all the components

$$x_k = x_k(n_1, n_2, \ldots, n_c).$$

The differential of x_k is expressed by

$$dx_k = \sum_{l=1}^{c} \left. \frac{\partial x_k}{\partial n_l} \right|_{n_l} dn_l$$

where $\boldsymbol{n}_l = (n_1, n_2, \ldots, n_{l-1}, n_{l+1}, \ldots, n_c)$. Substituting dx_k from the above expression into the expression for df_i,

$$df_i = \sum_{k=1}^{c-1} \left.\frac{\partial f_i}{\partial x_k}\right|_{x_k} \sum_{l=1}^{c} \left.\frac{\partial x_k}{\partial n_l}\right|_{n_l} dn_l.$$

Now divide the above equations by ∂n_j while holding $(n_1, n_2, \ldots, n_{j-1}, n_{j+1}, \ldots, n_c)$ constant

$$\left(\frac{\partial f_i}{\partial n_j}\right)_{n_j} = \sum_{k=1}^{c-1} \left.\frac{\partial f_i}{\partial x_k}\right|_{x_k} \left(\frac{\partial x_k}{\partial n_j}\right)_{n_j} \tag{E1.10.2b}$$

The derivative $(\partial x_k / \partial n_j)_{n_j}$ can be readily calculated from

$$x_k = \frac{n_k}{n_1 + n_2 + \cdots + n_c} \qquad k = 1, \ldots, c$$

$$\left.\frac{\partial x_k}{\partial n_j}\right|_{n_j} = -\frac{n_k}{n^2} \qquad \text{for } j \neq k \tag{E1.10.3b}$$

$$\left.\frac{\partial x_k}{\partial n_j}\right|_{n_j} = \frac{n - n_k}{n^2} \qquad \text{for } j = k \tag{E1.10.4b}$$

Combining Eqs. (E1.10.2b) to (E1.10.4b) provides the relation between mole-number and mole-fraction derivatives of fugacity of component i

$$\boxed{\left.\frac{\partial f_i}{\partial n_j}\right|_{n_j} = -\sum_{k=1}^{c-1} \frac{x_k}{n} \left.\frac{\partial f_i}{\partial x_k}\right|_{x_k} + \frac{1}{n} \left.\frac{\partial f_i}{\partial x_j}\right|_{x_j}}. \tag{E1.10.5b}$$

In a similar manner by writing $x_k = x_k(n_1, n_2, \ldots, n_{c-1}, n)$ in differential form, and further manipulations, we obtain,

$$\boxed{\left(\frac{\partial f_i}{\partial n_j}\right)_{n_1, n_2, \ldots, n_{j-1}, n_{j+1}, \ldots, n} = \frac{1}{n} \left(\frac{\partial f_i}{\partial x_j}\right)_{x_j}}.$$

Note that according to Eq. (E1.10.5b), we need an array of fugacity derivatives with respect to mole fractions to calculate a single mole-number derivative.

Example 1.11 *Net heat of transport of component i*   In thermodynamics of irreversible processes, the net heat of transport of component i is defined by,

$$Q_i^* = Q_i - \bar{H}_i$$

where Q_i is the energy transported across a given reference plane per mole of diffusing component i, in an isothermal process. Show that Q_i^* corresponds to the absorption of heat.

Solution We will follow Denbigh (1951), and consider a certain bounded region at constant temperature and pressure. Suppose dn_i moles of component i diffuse across the boundary. The energy flow across the boundary will be $Q_i dn_i$. In order to keep the pressure and temperature of region constant, d_-Q heat may be absorbed by the system from the surrounding and d_-W work may be done by the region. The change in internal energy is

$$dU = -Q_i dn_i + d_-Q - d_-W.$$

The change in internal energy is $-\bar{U}_i dn_i$, and the change in volume is $-\bar{V}_i dn_i$, where $\bar{U}_i$ and $\bar{V}_i$ are the partial molar internal energy and partial molar volume, respectively, and $d_-W = -P\bar{V}_i dn_i$. Therefore, $d_-Q = (Q_i - \bar{U}_i - P\bar{V}_i)dn_i$. From $\bar{H}_i = \bar{U}_i + P\bar{V}_i$,

$$d_-Q = (Q_i - \bar{H}_i)dn_i = Q_i^* dn_i.$$

Or, $Q_i^* = d_-Q/dn_i$, which is the net heat that must be absorbed per mole of diffusing component i to keep the pressure and temperature of the region constant.

Problems

1.1 Derive the criteria of: (1) thermal equilibrium; (2) mechanical equilibrium; and (3) chemical equilibrium from the internal energy minimum principle. These criteria were derived from the entropy maximum principle at the beginning of this chapter.

1.2 Show that the heat added to a closed system at constant volume results in an increase of the internal energy, U, while the heat added at constant pressure results in an increase in the enthalpy.

1.3 Consider the following two liquids separated by a membrane that is permeable to component i. Assume that these liquids are an ideal solution. Derive the following expression for the osmotic pressure

$$\pi = P' - P = -(RT/\bar{v}_i)\ln x_i,$$

where $\bar{v}_i$ is the average molar volume of pure component i and x_i is the mole fraction of component i in the liquid mixture.

Pure comp. i at T, P liquid	mixture containing comp. i at T, P' liquid

1.4 Consider a mixture of C_1/C_3 with a composition of $x_{C_1} = 0.34$ and $x_{C_3} = 0.66$

(mole fractions). Calculate the volume of 1 kg of this mixture at 346 K and 60 bar using (1) the partial molar volumes of Fig. 1.5 and (2) pure component data of Fig. 1.5 and $v_{mix} = \sum_{i=1}^{c} x_i v_{pure\,i}$. What is the percentage difference between the two and which is more accurate?

1.5   At constant T and P, the Gibbs-Duhem equation simplifies to $\sum_{i=1}^{c} n_i d\mu_i = 0$. Analogous to chemical potential μ_i are partial molar properties such as $\bar{V}_i$, $\bar{H}_i$, and $\bar{S}_i$. Show that for all partial molar quantities such as $\bar{V}_i$, $\bar{H}_i$, and $\bar{S}_i$, at constant T and P one can also write

$$\sum_{i=1}^{c} n_i d\bar{V}_i = 0, \ \sum_{i=1}^{c} n_i d\bar{H}_i = 0, \ \text{and} \ \sum_{i=1}^{c} n_i d\bar{S}_i = 0.$$

1.6   The partial molar volumes can be calculated from the experimental data of $V(x_1)$ of binary mixtures from

$$\bar{V}_1 = v - x_2(\partial v/\partial x_2)_{T,P}$$

$$\bar{V}_2 = v - x_1(\partial v/\partial x_1)_{T,P}$$

where v is the molar volume of the mixture. Derive the above relationships.

1.7   Show that the partial molar enthalpy in a binary mixture can be calculated from

$$\bar{H}_1 = h - x_2(\partial h/\partial x_2)_{T,P}$$

$$\bar{H}_2 = h - x_1(\partial h/\partial x_1)_{T,P},$$

where h is the molar enthalpy of the mixture.

1.8   A student was preparing a cylinder of natural gas for his use in the laboratory. In the course of preparing the gas cylinder, he observed some unusual behavior. The cylinder free from liquid at 75°F, and 2005 psia, was cooled to 32°F. As a result of lower temperature, some condensation took place and he removed about 500 cm^3 of liquid from the cylinder. He then brought back the temperature of the cylinder to 75°F, the initial temperature. But the pressure increased to 2060 psia, 55 psi higher than before liquid removal. How would you explain this observation?

1.9   Derive the following expressions:

$$U = \int_{V}^{\infty} [P - T(\partial P/\partial T)_{V,\underline{n}}]dV + \sum_{i=1}^{c} n_i u_i^0$$

$$H = \int_{V}^{\infty} [P - T(\partial P/\partial T)_{V,\underline{n}}]dV + PV + \sum_{i=1}^{c} n_i u_i^0$$

$$\mu_i = \int_{V}^{\infty} [(\partial P/\partial n_i)_{T,V,n_i} - RT/V]dV - RT \ln V/(n_i RT) + RT + \mu_i^0 - Ts_i^0.$$

Note: In the derivation of μ_i one can use $\mu_i - \mu_i^0 = [\partial(A - A^0)/\partial n_i]_{T,V,n_i}$.

1.10 From $\mu_1 = (\partial G/\partial n_1)_{T,P,n_2\ldots,n_c}$, use the Jacobian transformation, to show that

$$\mu_1 - \mu_c = (\partial G/\partial n_1)_{T,P,n_2,\ldots,n_{c-1},n} = (\partial g/\partial x_1)_{T,P,x_2,\ldots,x_{c-1}},$$

where $n = \sum_{i=1}^{c} n_i$.

1.11 Show that along an isotherm:

$$\int_0^P \bar{V}_i dP = \int_V^{\infty} (\partial P/\partial n_i)_{T,V,n_i} dV.$$

Hint: Use $\bar{V}_i = (\partial V/\partial n_i)_{T,P,n_i} = -[(\partial P/\partial n_i)_{T,V,n_i}]/[(\partial P/\partial V)_{T,P,\underline{n}}]$ to derive the above relationship.

1.12 Derive the Gibbs-Duhem equation in the following forms:

$$-SdT + VdP - \sum_{i=1}^{c-1} n_i d(\mu_i - \mu_c) - nd\mu_c = 0$$

$$- sdT + vdP - \sum_{i=1}^{c-1} x_i d(\mu_i - \mu_c) - d\mu_c = 0,$$

where $n = \sum_{i=1}^{c} n_i$.

1.13 (a) Show that for excess properties, the Gibbs-Duhem equation takes the following form,

$$-S^E dT + V^E dP + \sum_{i=1}^{c} n_i dG_i^E = 0.$$

(b) Show that at constant T and P, the activity coefficients are not independent from each other,

$$\left(\sum_{i=1}^{c} n_i d \ln \gamma_i = 0 \right)_{T,P}.$$

1.14 Derive the following expression for the activity coefficient of component i at infinite dilution γ_i^{∞} (that is, γ_i as $x_i \to 0$),

$$\gamma_i^{\infty} = \varphi_i^{\infty}/\varphi_{pure\ i},$$

where both φ_i^{∞}, and $\varphi_{pure\ i}$ are at temperature T, and pressure P. *Hint*: Use $RT \ln f_i/f_i^0(T,P) = RT \ln \gamma_i x_i$, where $f_i^0(T,P) \equiv f_{pure\ i}$, from which $RT \ln(\varphi_i x_i)/\varphi_i^0 = RT \ln \gamma_i x_i$ can be readily established.

1.15 Derive the following expression

$$\mu_i(T,P,\underline{n}) = \mu_i(T,P) + RT \ln \frac{f_i(T,P,\underline{n})}{f_i(T,P)}.$$

Hint: You may combine the following equations:

$$\mu_i(T, P, \underline{n}) = \mu_i(T, P^o, \underline{n}) + RT \ln \frac{f_i(T, P, \underline{n})}{f(T, P^o, \underline{n})}$$

$$f_i(T, P^o, \underline{n}) = x_i f_i(T, P^o)$$

$$\mu_i(T, P^o, \underline{n}) = \mu_i(T, P^o) + RT \ln x_i$$

$$\mu_i(T, P) = \mu_i(T, P^o) + RT \ln \frac{f_i(T, P)}{f_i(T, P^o)}$$

in your derivation. Note that P^o is chosen to be a low enough pressure at which the fluid has an ideal behavior.

1.16 Derive the Maxwell relations from the Jacobian transforms.

References

1. Callen, H.B.: *Thermodynamics and an Introduction to Thermostatics*, 2d ed., John Wiley & Sons, New York, 1985.
2. Carroll, B.: "On the Use of Jacobians in Thermodynamics," *J. of Chemical Education*, p. 218, April 1965.
3. Courant, R., and D. Hilbert : *Methods of Mathematical Physics, Vol II: Partial Differential Equations*, Interscience Publishers, John Wiley & Sons, New York, 1962.
4. Crawford, F.H.: "Thermodynamic Relations in n-Variable Systems in Jacobian Form: Part I, General Theory and Application to Understand Systems," *Proceedings of the American Academy of Arts and Sciences*, vol. 78, p. 165, 1950.
5. Denbigh, K.: *The Principles of Chemical Equilibrium*, 3d ed., Cambridge University Press, London, 1971.
6. Denbigh, K.G.: *The Thermodynamics of Steady State*, John Wiley and Sons, New York, 1951.
7. Flory, P.J.: *Principles of Polymer Chemistry*, Cornell University Press, Ithaca, NY, 1953.
8. Gibbs, J.W.: *Collected Works— Vol I— Thermodynamics*, Yale University Press, New Haven, Connecticut, 1957.
9. Hildebrand, J.H.: "Solubility: XII. Regular Solutions" *J. Am. Chem. Soc.*, vol. 51, p. 66, 1929.
10. Hildebrand, J.H., and R.L. Scott: *The Solubility of Nonelectrolytes*, 3d ed., Reinhold Publishing Corporation, New York, 1950.
11. Katz, D.L., D. Cornell, R. Kobayashi, F. Poettmann, J. Vary, J. Elenbaas and C.F. Weinaug: *Handbook of Natural Gas Engineering*, McGraw-Hill Book Co., New York, 1959.
12. Peng, D.Y., and D.B. Robinson: "A New Two-Constant Equation of State," *Ind. Eng. Chem. Fund.* vol. 15, No. 1, p. 59, 1976.
13. Prausnitz, J.M., R.N. Lichtenthaler, and E.G. de Azevedo: *Molecular Thermodynamics of Fluid-Phase Equilibria*, 2d ed., Prentice-Hall Inc., Englewood Cliffs, NJ, 1986.
14. Sage, B.H., and W.N. Lacey: *Volumetric and Phase Behavior of Hydrocarbons*, Chapter VIII, Stanford University Press, Stanford, CA, 1949.
15. Starling, K.E.: *Fluid Thermodynamic Properties for Light Petroleum Systems*, Gulf Publishing Company, 1973.
16. Tisza, L.: *Generalized Thermodynamics*, M.I.T. Press, Cambridge, MA, 1966.
17. Wu, P.C., and P. Ehrlich: "Volumetric Properties of Supercritical Ethane-n-Heptane Mixtures: Molar Volumes and Partial Molar Volumes," *AIChE J.* p. 553, May 1973.

2

General theory of phase equilibria and irreversible phenomena in hydrocarbon reservoirs

Generally, in phase-equilibria calculations of hydrocarbon reservoirs, the effects of gravity and the curvature of the interface between the phases are neglected. Under certain conditions and for certain classes of problems, however, compositional variation in hydrocarbon reservoirs may become important because of the gravitational field and the effects of interface curvature on phase properties and composition. The main significance of both gravity and interface curvature in hydrocarbon reservoirs is due to the multicomponent nature and nonideal behavior of reservoir fluids, which makes them distinct from fluid mixtures commonly encountered in many disciplines. In some hydrocarbon-reservoir applications, we cannot invoke the Gibbs criterion of equilibrium because of nonzero entropy production, even at the stationary state. As an example, when there is a temperature gradient in a subsystem or a composite system, dG cannot become zero to establish equilibrium. On the other hand, as we will see shortly, we can have a pressure gradient within a subsystem and still have equilibrium.

In this chapter, we will first derive the criteria of equilibrium both in a gravitational field and for a composite system with a curved interface between subsystems. Then, we will derive explicit expressions for the effect of interface curvature on the saturation pressure in single- and multicomponent mixtures. Next, irreversible thermodynamics will be introduced in connection with the stationary state in hydrocarbon reservoirs. Then, the effect of a temperature gradient on composition variation both with and without natural convection will be formulated.

Equilibrium condition under the influence of gravity

The work term in Eq. (1.4) of Chapter 1 includes only the expansion or compression contribution; $d_-W = -PdV$ represents the work of expansion (or compression) done by the system on the surroundings. Now consider a mass m undergoing both expansion (or compression), and change in position in the vertical direction (see Fig. 2.1). In order to raise mass m to some height dz, a certain amount of work must be done; $d_-W = mgdz$, provided there is no change in the volume of mass m. When there are both change in the volume of mass m and displacement in the vertical direction by the distance, dz, the work term is simply

$$d_-W = -PdV + mgdz. \tag{2.1}$$

The expression for dU for a closed system is given by

$$dU = TdS - PdV + mgdz. \tag{2.2}$$

One may derive expressions for dH, dA, and dG in a gravity field as

$$dH = TdS + VdP + mgdz \tag{2.3}$$

$$dA = -SdT - PdV + mgdz \tag{2.4}$$

$$dG = -SdT + VdP + mgdz. \tag{2.5}$$

Now let us apply the Gibbs criterion of equilibrium: dG must vanish at equilibrium. As a result, the independent terms on the right side of Eq. (2.5) must vanish. Pressure, P, and vertical position, z, are not independent. Therefore at equilibrium,

$$dT = 0 \tag{2.6}$$

and

$$VdP + mgdz = 0. \tag{2.7}$$

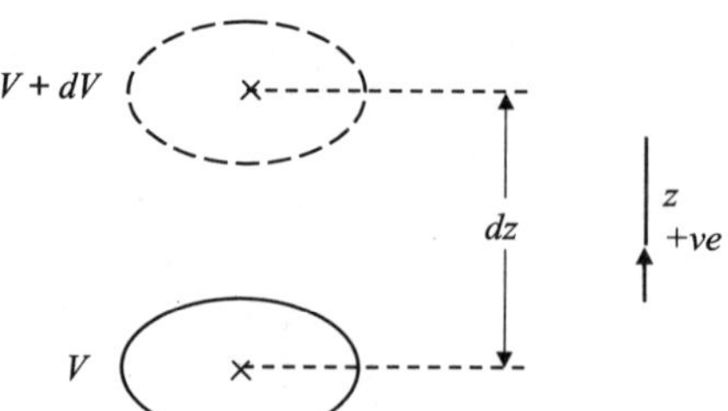

Figure 2.1 Schematic of work performed due to gravity for a single-component fluid.

Equation (2.6) states that the temperature T must be the same everywhere in the system. Since $\rho = m/V$, then Eq. (2.7) becomes

$$dP = -\rho g dz, \tag{2.8}$$

which is the expression for the hydrostatic head.

In the above derivations, we defined our system to be closed, a definition that is applicable to single-component systems. For a multicomponent fluid, when the mass m is displaced from z to $z + dz$, at the new position we expect transfer of components across the wall of volume of fixed mass m, and we have to take various other work terms into account. We will first move mass m from z to $z + dz$; because of the change in pressure, there will be a change in the volume of mass m. The corresponding work terms are similar to those of a single-component system: $-PdV$ and $mgdz$. At $z + dz$, because of transfer of various components across the system, there is the chemical work (see Chapter 1), which is given by $\mu_i dn_i$ for each component i (see Figure 2.2). The work associated with the corresponding mass displacement, dm_i for component i is $zgdm_i$. Since $m_i = n_i M_i$, where M_i is the molecular weight of component i, $zdm_i = M_i zdn_i$. For all the components, the work when the system is brought to $z + dz$ is

$$d_-W' = \sum_{i=1}^{c} (\mu_i + M_i zg)dn_i. \tag{2.9}$$

Adding Eq. (2.9) to Eq. (2.2) provides the expression for dU when mass m is displaced from z to $z + dz$ for a multicomponent fluid:

$$dU = TdS - PdV + mgdz + \sum_{i=1}^{c} (\mu_i + M_i zg)dn_i. \tag{2.10}$$

Note that for a single-component system, the last term on the right side of Eq. (2.10) is absent (see Eq. (2.2)).

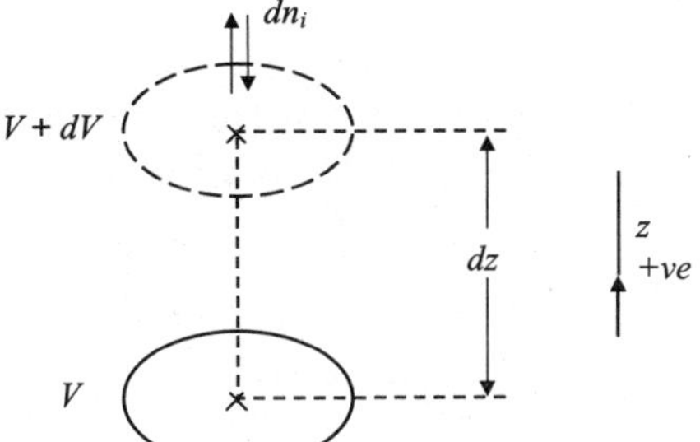

Figure 2.2 Schematic of work performed due to gravity for a multicomponent fluid.

The corresponding expression for dG is

$$dG = -SdT + VdP + mgdz + \sum_{i=1}^{c}(\mu_i + M_i zg)dn_i. \qquad (2.11)$$

At equilibrium, dG must vanish. Since z and P are dependent, then

$$dT = 0$$

$$\mu_i + M_i gz = 0 \qquad i = 1, \ldots, c$$

$$VdP + mgdz = 0. \qquad (2.12)$$

The first and third expressions in Eq. (2.12) provide the same results as previously for the single-component system. The second expression provides the Gibbs sedimentation expression,

$$\boxed{(d\mu_i = -M_i gdz)_T} \qquad i = 1, \ldots, c. \qquad (2.13)$$

The equilibrium concept of gravity segregation leads to the expression that, in a multicomponent fluid column under isothermal conditions, the chemical potential of the ith component, μ_i, is a function of position, z, according to the above differential equation. Equation (2.13) provides both composition and pressure as a function of depth, as we will see next. From $(d\mu_i = RTd\ln f_i)_T$ and Eq. (2.13),

$$(RTd\ln f_i = -M_i gdz)_T \qquad i = 1, \ldots, c. \qquad (2.14)$$

Equation (2.14) can be derived using a different approach by minimizing the total Helmholtz free energy of the system when it has a continuous variation (Aavatsmark, 1995; Wheaton, 1991).

Integrating Eq. (2.14) from a reference depth of zero to z,

$$\boxed{f_i = f_i^o \exp\left[-\frac{M_i}{RT}gz\right]} \qquad i = 1, \ldots, c. \qquad (2.15)$$

Equation (2.15) provides the fugacity of component i in a given phase as a function of position; given composition and pressure at the reference point, one can calculate both the composition and pressure at any point in the vertical direction.

$$F_i = f_i(T, P, y_1, y_2, \ldots, y_{c-1}) - f_i^o(T, P^o, y_1^o, y_2^o, \ldots, y_{c-1}^o)$$

$$\times \exp\left(-\frac{M_i}{RT}gz\right) = 0 \qquad i = 1, \ldots, c, \qquad (2.16)$$

subject to the constraint equation,

$$\sum_{i=1}^{c} y_i = 1. \tag{2.17}$$

The unknowns are pressure and composition at the desired depth. Newton's method can be used to solve the above system of nonlinear equations. Note that in Eqs. (2.16) and (2.17), there is no need to write the expression for hydrostatic head. One may relate the pressure and composition at any point to those at the reference point from Eqs. (2.16) and (2.17).

Conditions for pronounced compositional variation

Let us write $d\mu_i$ from Eq. (2.13) in terms of the independent variables, $P, y(y_1, \ldots, y_{c-1})$:

$$\left[d\mu_i = \left(\frac{\partial \mu_i}{\partial P}\right)_y dP + \sum_{j=1}^{c-1} \left(\frac{\partial \mu_i}{\partial y_j}\right)_{P,y_j} dy_j \right]_T. \tag{2.18}$$

Combining $(d\mu_i = \bar{V}_i dP)_{T,y}$, Eqs. (2.13) and (2.18) and the hydrostatic expression given by Eq. (2.8), one obtains

$$\sum_{j=1}^{c-1} \left(\frac{\partial \mu_i}{\partial y_j}\right)_{P,T,y_j} \left(\frac{dy_j}{dz}\right) = (\rho \bar{V}_i - M_i)g \qquad i = 1, \ldots, c-1. \tag{2.19a}$$

Note that in Eq. (2.19a), $i = 1, \ldots, c-1$ and not $i = 1, \ldots, c$ as in Eq. (2.13). Eq. (2.13) gives y_j and P; Eq. (2.19) gives only dy_j/dz but not dP/dz. The above equation can also be written as

$$\begin{pmatrix} \partial \mu_1/\partial y_1 & \partial \mu_1/\partial y_2 & \cdots & \partial \mu_1/\partial y_{c-1} \\ \partial \mu_2/\partial y_1 & \partial \mu_2/\partial y_2 & \cdots & \partial \mu_2/\partial y_{c-1} \\ \vdots & \vdots & \cdots & \vdots \\ \partial \mu_{c-1}/\partial y_1 & \partial \mu_{c-1}/\partial y_2 & \cdots & \partial \mu_{c-1}/\partial y_{c-1} \end{pmatrix} \begin{pmatrix} dy_1/dz \\ dy_2/dz \\ \vdots \\ dy_{c-1}/dz \end{pmatrix}$$

$$= g \begin{pmatrix} \rho \bar{V}_1 - M_1 \\ \rho \bar{V}_2 - M_2 \\ \vdots \\ \rho \bar{V}_{c-1} - M_{c-1} \end{pmatrix} \tag{2.19b}$$

For the sake of brevity we have dropped the subscripts of T, P, and y_i from the elements of the matrix above. According to Eq. (2.19b), compositional variation in a multicomponent column could be pronounced when (1) the term $(\rho \bar{V}_i - M_i)$ is large or (2) the determinant of the matrix above is small. The term $(\rho \bar{V}_i - M_i)$ for asphaltene micelles in a crude oil could be very large; the micelle molecular weight could be of the order of several thousand and even over 20,000, as we will discuss in Chapter 5. The molecular weights of other species, especially lighter components, range from 16 to perhaps 800. The determinant of the matrix in Eq. (2.19b) becomes very small near the critical point, and becomes zero at the critical point as we will show in Chapter 4. (For the special case of a two-component system, the results will be shown later in this chapter.) Therefore, compositional segregation and saturation-pressure variation in an oil column can be very pronounced when asphaltene materials are present. Compositional variation in a gas column is enhanced when a gas condensate fluid is in the critical region.

Figure 2.3 shows schematically the variation of pressure and saturation pressure with depth in both the oil leg and the gas cap. Saturation pressure variation of Fig. 2.3 is due to compositional grading. Kingston and Niko (1975) report considerable bubblepoint-pressure variation for the Brent and Stafford reservoirs of the North Sea. The bubblepoint pressure gradient in the oil zone is reported to be 3.6 and 4.0 psi/ft, respectively, for these reservoirs. Some field data show even a greater bubblepoint-pressure decrease with depth, greater than 5 psi/ft.

Figure 2.4 shows another schematic of the variation of pressure and saturation pressure with depth. Note that there is no gas–oil contact. At the top, there is the gas phase, and at the bottom, there is the liquid phase. In such a case, the critical temperature is less than the reservoir temperature at the top and more than the reservoir temperature at the bottom; the reservoir pressure is higher than dewpoint pressure at the top it is also higher than the bubblepoint pressure at the bottom. Oil-field examples of such a behavior are given by Neveux and Sathikumar (1988) and by Espach and Fry (1951).

Figure 2.5 shows the variation of the molecular weight and the mole percent of C_{7+} of the East Painter Reservoir (Creek and Schrader, 1985). Figure 2.6 depicts the variation of the amount of C_1 as a function of depth. Figure 2.7 shows the variation of total production gas–oil ratio (GOR) as a function of depth.

Equilibrium condition for curved interfaces

The interface between different phases (e.g., gas and oil) may not be flat. As an example, in a capillary tube, the interface between phases may

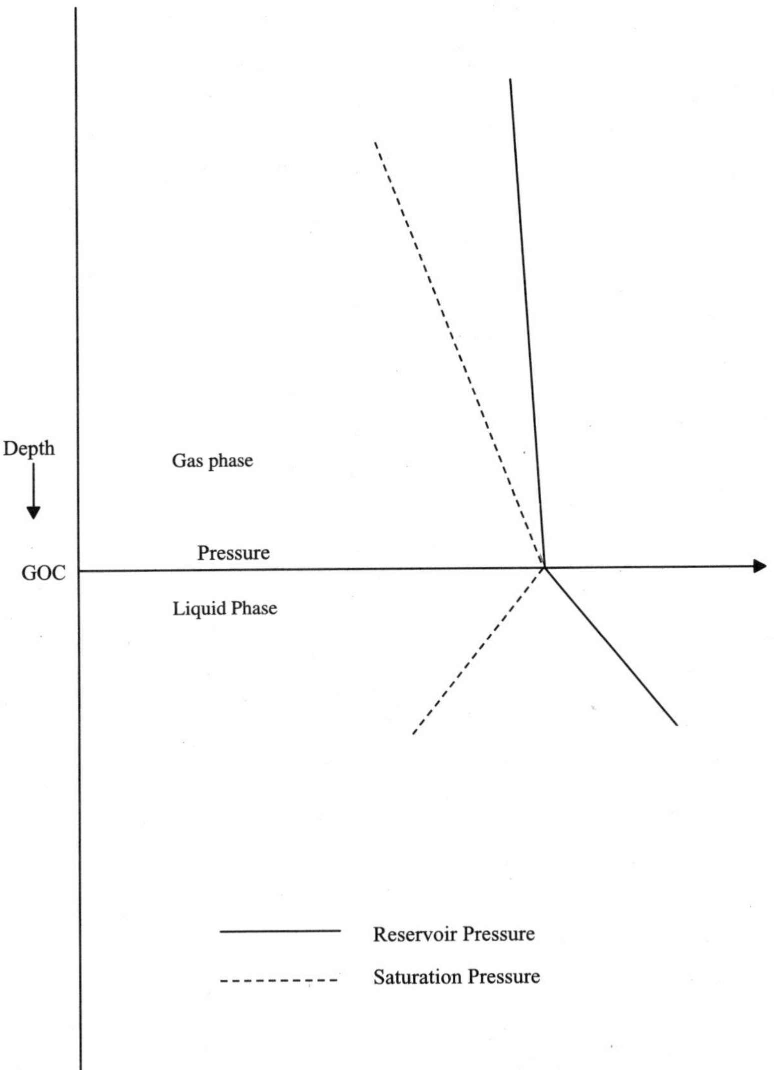

Figure 2.3   Schematic of depth vs. pressure and saturation pressure in the gas cap and oil column with a distinct gas–oil contact (GOC).

be curved. In the derivation of equilibrium condition, the effect of interface curvature should be taken into account. Our approach to account for the effect of interface curvature is to modify the expression for the work term.

Consider the two systems sketched in Figure 2.8. The work of expansion is $d_-W = -PdV$ for the system on the left. For the system on the

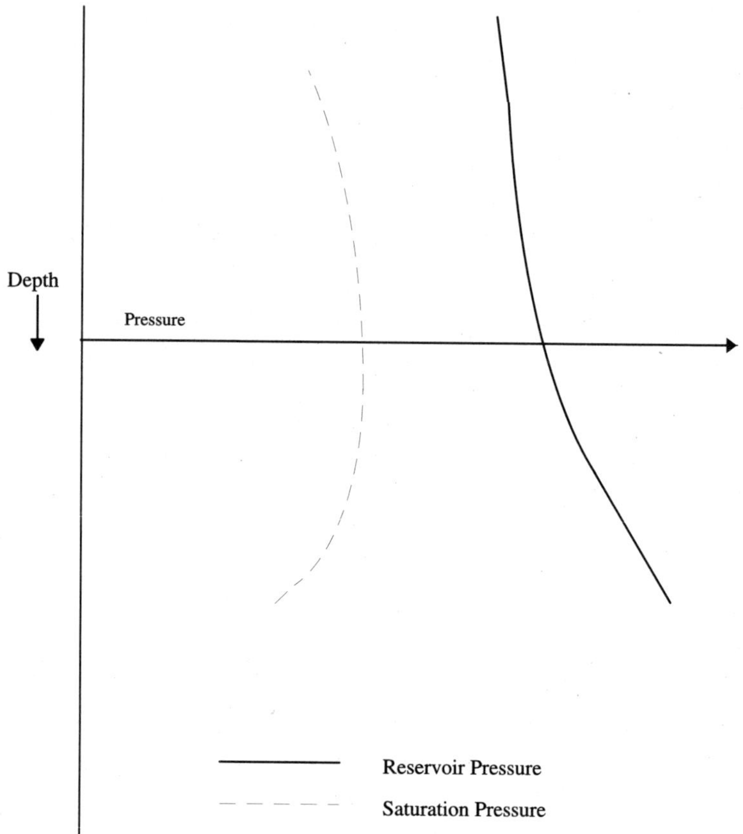

Figure 2.4 Schematic of depth vs. pressure and saturation pressure for a near-critical fluid.

right, as the bubble expands, the work expression contains two terms:

$$d_- W = -PdV + \sigma d\mathcal{A}, \tag{2.20}$$

where PdV is the contribution from expansion (or compression) and $\sigma d\mathcal{A}$ is the work required to increase the bubble surface area by $d\mathcal{A}$ (σ is the surface or interfacial tension). Explicit derivation of Eq. (2.20) will be presented shortly. Including $\sigma d\mathcal{A}$ in Eq. (1.21) of Chapter 1, the expression for dU is obtained:

$$dU = TdS - PdV + \sum_{i=1}^{c} \mu_i dn_i + \sigma d\mathcal{A}. \tag{2.21}$$

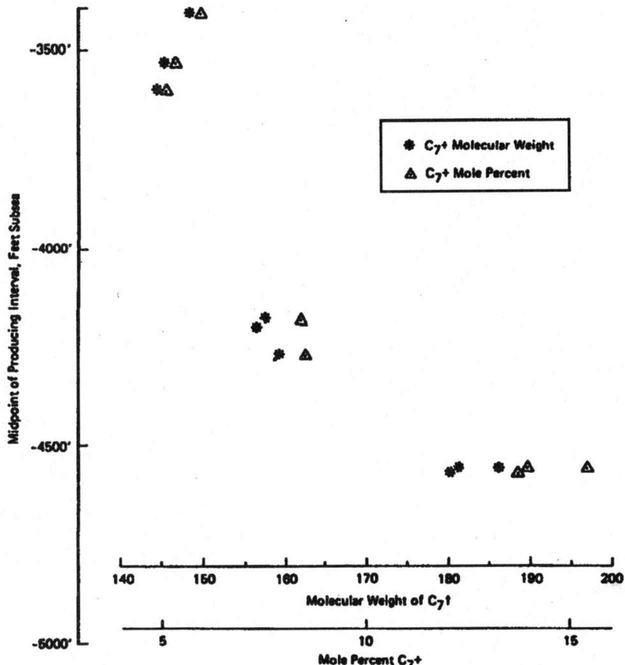

Figure 2.5 East Painter field variation of mole percent and molecular weight of C_{7+} with depth (adapted from Creek and Schrader, 1985).

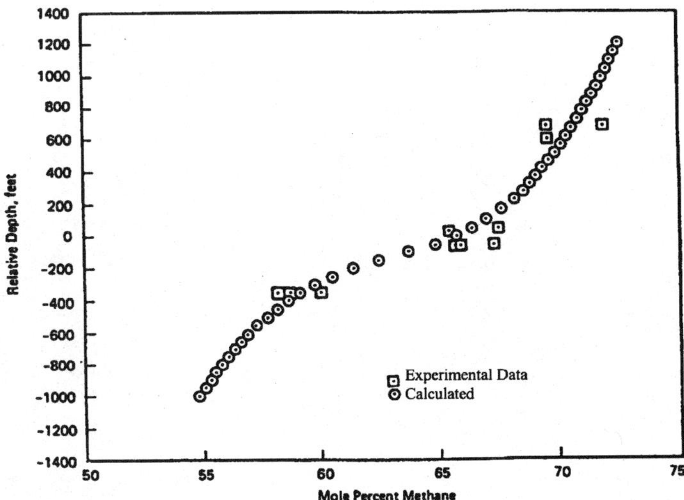

Figure 2.6 East Painter field comparison of calculated and experimental mole percent methane with depth (adapted from Creek and Schrader, 1985).

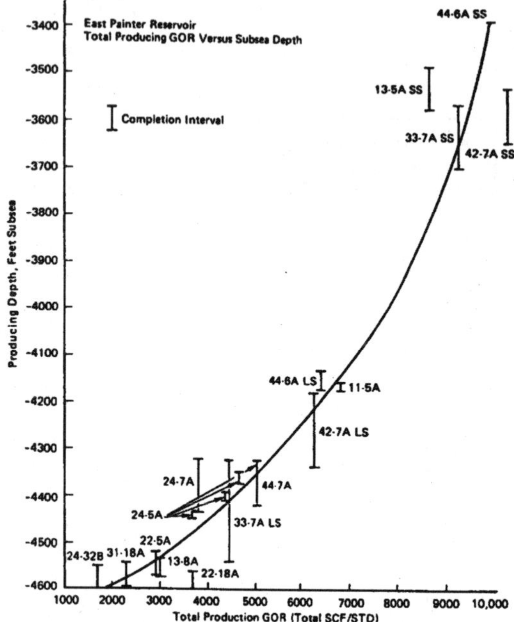

Figure 2.7 East Painter field variation of total producing GOR with depth (adapted from Creek and Schrader, 1985).

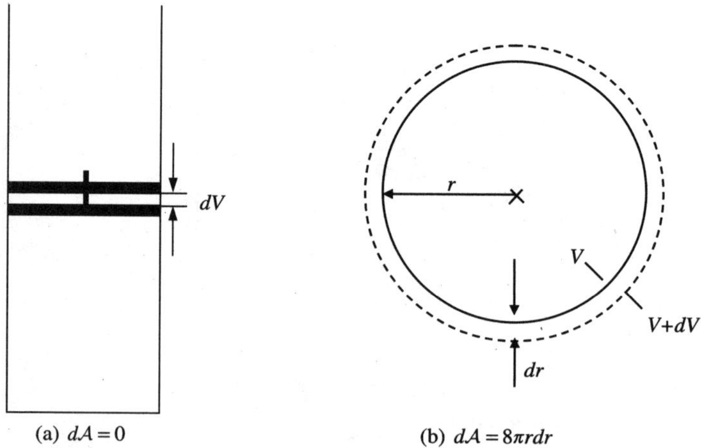

(a) $dA = 0$ (b) $dA = 8\pi r\,dr$

Figure 2.8 Change of interfacial surface due to expansion.

Similar expressions for dH, dA, and dG are given below.

$$dH = TdS + VdP + \sum_{i=1}^{c} \mu_i dn_i + \sigma d\mathcal{A} \tag{2.22}$$

$$dA = -SdT - PdV + \sum_{i=1}^{c} \mu_i dn_i + \sigma d\mathcal{A} \tag{2.23}$$

$$dG = -SdT + VdP + \sum_{i=1}^{c} \mu_i dn_i + \sigma d\mathcal{A} \tag{2.24}$$

Now consider a composite system consisting of a gas bubble (shown by superscript prime or by b superscript) and the surrounding liquid phase (shown by superscript double-primes or by L superscript) sketched in Fig. 2.9. The expression for dU of each phase depends on how one assigns the interface. If we assign the interface to the gas bubble, then

$$dU^b = T^b dS^b - P^b dV^b + \sum_{i=1}^{c} \mu_i^b dn_i^b + \sigma d\mathcal{A}^b \tag{2.25}$$

$$dU^L = T^L dS^L - P^L dV^L + \sum_{i=1}^{c} \mu_i^L dn_i^L \tag{2.26}$$

Alternatively, if the interface is assigned to the surrounding liquid phase and not as part of the gas bubble, the $\sigma d\mathcal{A}^b$ term should be part of dU^L and not part of dU^b. The interface could also be treated as a distinct entity; it can represent the surface phase or even a line phase when three phases with curved interfaces share a common line. However, for convenience and simplicity, we regard the interface as part of either the gas bubble or the surrounding liquid. Such a simplifi-

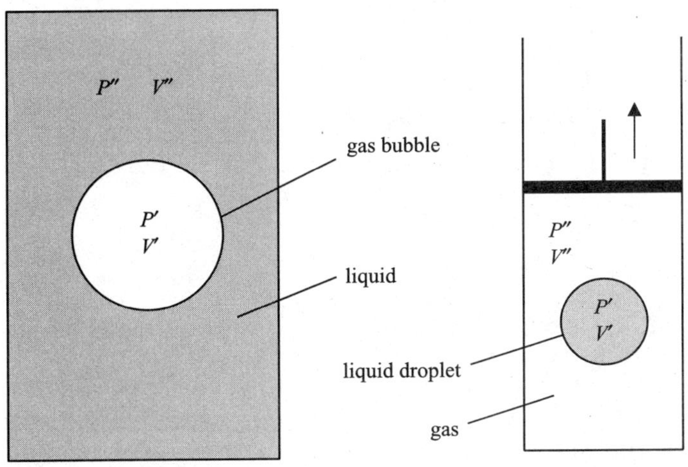

Figure 2.9   Gas bubble
Figure 2.10   Liquid droplet

cation is justified as long as the thickness of the interfacial region between the bulk phases is negligible in comparison with the radii of the interface. Physically, the boundary between the two bulk phases is a heterogeneous region with a thickness of a few molecular diameters. Other considerations should be taken into account for highly curved interfaces.

In order to establish the criteria of equilibrium when the interface between the phases is curved, let us consider the system of Fig. 2.9 at constant volume and constant temperature. The pressure inside the gas bubble is P' and the surrounding liquid pressure is P''. The volumes of the gas bubble and the surrounding liquid phase are V' and V'', respectively. The expressions for the differential Helmholtz free energy of the bubble and surrounding liquid for a single-component system are

$$dA' = -P'dV' + \sigma dA' + \mu'dn' \tag{2.27}$$

$$dA'' = -P''dV'' + \mu''dn'' \tag{2.28}$$

and the differential Helmholtz free energy of the total system, dA, then is

$$dA = dA' + dA'' = -P'dV' - P''dV'' + \sigma dA' + \mu'dn' + \mu''dn''. \tag{2.29}$$

Since $dV' = -dV''$, and $dn' = -dn''$,

$$dA = -(P' - P'')dV' + \sigma dA' + (\mu' - \mu'')dn'. \tag{2.30}$$

The necessary condition for the system to be at equilibrium is that dA must vanish. V' and A' are not independent of each other; therefore, from Eq. (2.30),

$$\boxed{\begin{aligned} \mu'(P') &= \mu''(P'') \\ P' - P'' &= \sigma dA'/dV' \end{aligned}} \; . \tag{2.31}$$

If we assume that the gas bubble has a spherical shape of radius r,

$$dA' = 8\pi r dr \tag{2.32}$$

$$dV' = 4\pi r^2 dr. \tag{2.33}$$

Combining Eqs. (2.31) through (2.33),

$$\boxed{P' - P'' = P_c = 2\sigma/r} \; . \tag{2.34}$$

Equation (2.34) is the well-known Young-Laplace equation of capillarity, which provides the condition for mechanical equilibrium of a curved interface.

For a multicomponent system at constant temperature,

$$dA' = -P'dV' + \sum_{i=1}^{c} \mu_i' dn_i' + \sigma d\mathcal{A}' \qquad (2.35)$$

$$dA'' = -P''dV'' + \sum_{i=1}^{c} \mu_i'' dn_i'' \qquad (2.36)$$

and with $V' + V'' = $ constant and $n_i' + n_i'' = $ constant $(i = 1, \ldots, c)$,

$$dA = dA' + dA'' = -(P' - P'')dV' + \sigma d\mathcal{A}' + \sum_{i=1}^{c} (\mu_i' - \mu_i'')dn_i'.$$

Again, the necessary condition for the system to be at equilibrium is that dA must vanish. From the first two terms, Eq. (2.31) can be obtained, and from the last term,

$$\boxed{\mu_i'(T, P', \underline{\boldsymbol{n}}') = \mu_i''(T, P'', \underline{\boldsymbol{n}}'')} \ . \qquad (2.37)$$

Equation (2.37) shows that for a curved interface, the chemical potentials of each component on both sides of the interface should be equal to achieve equilibrium. The chemical potentials, however, are evaluated at different pressures; the pressures on different sides of the interface are related by the second expression in Eq. (2.31). In Eq. (2.20), the work required to increase the surface area of the bubble was expressed by $\sigma d\mathcal{A}$. Defay and Prigogine (1966) derive this work term in a straightforward manner. Consider the system shown in Fig. 2.10, where a spherical liquid droplet of volume V' and surface area $\mathcal{A}'$ is surrounded by its vapor. The total volume of the system, V, is equal to $V' + V''$, where V'' is the volume of the vapor surrounding the liquid droplet. Suppose part of the droplet vaporizes and the piston moves upward to expand the system to volume $V + dV$. The work done by the system is, therefore,

$$d\_W = -P''dV = -P''d(V' + V'') = -P''dV' - P''dV''. \qquad (2.38)$$

Let us add and subtract $P'dV'$ from Eq. (2.38) and rearrange:

$$d\_W = -(P'' - P')dV' - P''dV'' - P'dV'. \qquad (2.39)$$

By using the Young-Laplace equation of capillarity, $-(P'' - P') = 2\sigma/r$, the expression for d_W becomes

$$d\_W = -P'dV' - P''dV'' + \sigma d\mathcal{A}, \qquad (2.40)$$

where $d\mathcal{A} = dV'/2r$. The first two terms on the right side of Eq. (2.40) are the work of expansion of the droplet and the surrounding liquid. The third term, σdA, is the work required to increase the surface area by dA.

Effect of curvature on saturation pressure: condensation and vaporization in porous media.

Equations (2.31) and (2.37) reveal that the interface curvature affects the equilibrium. We wish to derive an explicit expression for the effect of curvature on the saturation pressure in porous media. Consider the simple system sketched in Fig. 2.11a. In this figure, the diameter of capillary tube 1, d_1 is less than the diameter of capillary tube 2, d_2. The volume of the space above the capillary tubes where the piston is located is very large compared to the volume of capillary tubes 1 and 2. Now suppose a superheated gas at temperature T and pressure P is charged to the system. The saturation pressure of the fluid at temperature T with a flat interface is P_d^{∞}. If the pressure is raised by isothermal compression, the fluid can condense. The question is where would the liquid form first? To answer this question, we need to derive the expression for the effect of curvature on saturation pressure.

Consider a gas–liquid system where the liquid wets the solid surface. Based on mechanical and chemical equilibrium, one can write

$$P^G - P^L = 2\sigma/r \tag{2.41}$$

$$\mu_i^G = \mu_i^L \qquad i = 1, \ldots, c. \tag{2.42}$$

In the above equations, P^G is the pressure in the gas phase, P^L is the pressure in the liquid phase, σ is the interfacial tension at the gas–liquid interface, and r is the interface radius. Note that in Eq. (2.41), if we assume that the liquid completely wets the solid substrate, i.e., contact angle $\theta = 0$, then $r = d/2$. For $\theta > 0$, $r = d/2\cos\theta$ (d is the tube diameter).

For an equilibrium displacement by change of r, Eqs. (2.41) and (2.42) are written as

$$d(P^G - P^L) = d(2\sigma/r) \tag{2.43}$$

$$d\mu_i^G = d\mu_i^L \qquad i = 1, \ldots, c. \tag{2.44}$$

Under isothermal conditions, the Gibbs-Duhem equation for gas and liquid phases can be written as

$$-v^L dP^L + \sum_{i=1}^{c} x_i d\mu_i^L = 0 \tag{2.45}$$

$$-v^G dP^G + \sum_{i=1}^{c} y_i d\mu_i^G = 0. \tag{2.46}$$

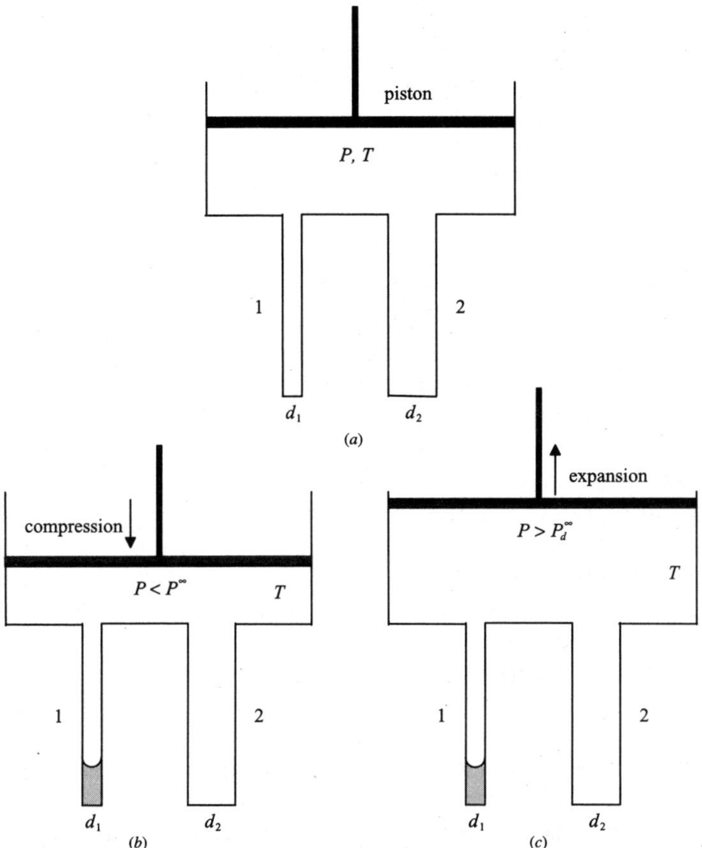

Figure 2.11 The influence of curvature on the vapor pressure of a pure substance and the dewpoint pressure of a hydrocarbon mixture: (a) system in the single-phase gaseous state. (b) For a pure substance, as pressure increases, gas may condense and the condensation will occur first in Tube 1. (c) For a hydrocarbon gas mixture with retrograde condensation behavior, as pressure decreases, liquid may form also in Tube 1 (liquid wets the substrate).

Subtracting Eq. (2.46) from Eq. (2.45), assuming that for the gas phase the composition y_i is constant, which is valid at the dewpoint (i.e., $d\mu_i^G = \bar{V}_i^G dP^G$), and using Eqs. (2.43) and (2.44) and the relationship $v^G = \sum_{i=1}^{c} y_i \bar{V}_i^G$,

$$dP^G = \frac{2d(\sigma/r)}{\left[1 - \sum\limits_{i=1}^{c} x_i \bar{V}_i^G / v^L\right]} = \frac{2d(\sigma/r)}{\left[1 - \sum\limits_{i=1}^{c} x_i \bar{V}_i^G \Big/ \sum\limits_{i=1}^{c} x_i \bar{V}_i^L\right]}. \tag{2.47}$$

The above equation provides the effect of interface curvature on the dewpoint pressure at constant temperature. Let us now turn our attention to a pure substance, for which Eq. (2.47) simplifies to

$$\boxed{dP^G = \frac{2d(\sigma/r)}{1 - (v^G/v^L)}} \tag{2.48}$$

The above simple equation, which appears in many texts, reveals in very clear terms that the vapor pressure decreases as the interface curvature increases, provided the liquid wets the solid substrate. The denominator is negative since for pure substances $v^G > v^L$, and $d(\sigma/r) = (\sigma/r)|_{r=R} - (\sigma/r)|_{r=\infty} = \sigma/R$ is positive (for a flat interface, $r = \infty$). Therefore, for the process of condensation sketched in Fig. 2.11a, since $dP^G < 0$, the condensation for a pure substance will first occur in tube 1 with $d_1 < d_2$, as shown in Fig. 2.11b.

Now let us examine Eq. (2.47). For hydrocarbon mixtures in the critical region, the denominator can be positive, $[1 - (\sum_{i=1}^{c} x_i \bar{V}_i^G)/v^L] > 0$, and therefore $dP^G > 0$. Consequently, the dewpoint pressure may increase as the interface curvature increases. For the system sketched in Fig. 2.11a, if a natural gas is introduced into the system at $T > T_c$, (where T_c is the critical temperature of the natural gas) in the course of expansion at constant temperature, the liquid might also first form in tube 1 (see Fig. 2.11c).

For a pure substance, the dewpoint and bubblepoint pressures are the same when the gas–liquid interface is flat, and therefore Eq. (2.48) can be used to study the effect of curvature on the saturation pressure, whether we approach the saturation pressure from the gas or liquid side. The substrate is assumed to wet the liquid, and the pressure in the liquid phase is less than that in the gas phase. For mixtures, dewpoint and bubblepoint (for a given composition) are different for a flat interface between the gas and liquid phases. Equation (2.47) applies to the dewpoint only. The expression for the change of bubblepoint pressure, i.e., vaporization, can be obtained from

$$\boxed{dP^G = \frac{2d(\sigma/r)}{\left[1 - v^G/\sum\limits_{i=1}^{c} y_i \bar{V}_i^L\right]} = \frac{2d(\sigma/r)}{\left[1 - \sum\limits_{i=1}^{c} y_i \bar{V}_i^G/\sum\limits_{i=1}^{c} y_i \bar{V}_i^L\right]}} \tag{2.49}$$

This equation is obtained by holding the composition of the liquid phase x_i constant (i.e., $d\mu_i^L = \bar{V}_i^L dP^L$). Since for a hydrocarbon mixture in the critical region, v^G might be less than $\sum_{i=1}^{c} y_i \bar{V}_i^L$, dP^G can be positive, which implies that the bubblepoint pressure can increase with an increase in curvature. Based on the above, when we lower the pressure of an undersaturated near-critical oil mixture in a porous medium, the tight pores may provide nucleation sites where liquid will form first. In Fig. 2.12, when the pressure of a compressed near-critical liquid is lowered, the gas bubble will first appear in tube 1 and then in tube 2 at a lower pressure, provided the liquid wets the solid; for $\theta = 0$, the gas should form spherical bubbles.

Next we will derive simple expressions for the vapor pressure of pure substances as a function of interface curvature.

Vapor pressure of pure substances.   For a pure substance at low and moderate pressures away from the critical point, often $v^G >> v^L$. Therefore Eq. (2.48) can be approximated by

$$-v^G dP^G = 2v^L d(\sigma/r).$$ (2.50)

If we further assume that the gas phase can be described by the ideal gas law, $P^G v^G = RT$, then

$$-RT d \ln P^G = 2v^L d(\sigma/r).$$ (2.51)

We can integrate Eq. (2.51) from $r = \infty$ to any r. The vapor pressure at r, P^G, is related to the vapor pressure of the flat interface, P^∞, through

$$RT \ln(P^\infty/P^G) = 2 \int_{r=\infty}^{r} v^L d(\sigma/r).$$ (2.52)

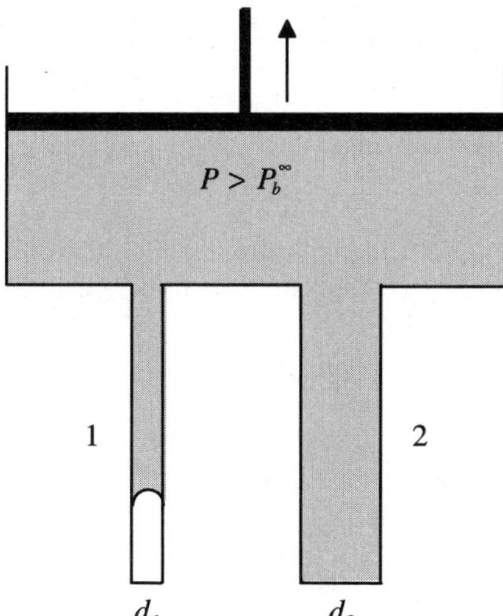

Figure 2.12 The influence of curvature on the bubblepoint of a hydrocarbon mixture.

If we further assume that the liquid molar volume does not change with pressure,

$$P^G = P^\infty \exp\left[-\frac{2\sigma}{r}\frac{v^L}{RT}\right].$$

(2.53)

Equation (2.53), known as the Kelvin equation, reveals that the vapor pressure, P^G, decreases with increasing interface curvature. So far, we have assumed that the substrate is liquid-wet or that the new phase forms as a bubble. For a droplet or when gas is the wetting phase, the effect of curvature on saturation pressure is formulated shortly. When the radius of a bubble or droplet becomes very small (say $r < 10^{-5}$ cm, for a pure substance), the interfacial tension may become a function of the radius (Defay and Prigogine, 1966). However, the derivation of Eq. (2.53) was not based on the assumption of the interfacial tension being independent of r.

Figure 2.13 provides a schematic of liquid- and gas-wet systems. Bubble and droplet formations are also sketched in parallel analogy with liquid-wet and gas-wet systems, respectively.

Effect of wettability. Our attention to the effect of curvature on the saturation pressure was directed to a bubble or, equivalently, when the liquid wets the solid surface. This is often true in hydrocarbon reservoirs in which gas is the nonwetting phase and oil is the wetting phase. However, in some other systems, gas may be the wetting phase and oil (or the liquid) may be the nonwetting phase. Even in rocks, liquid mercury is the nonwetting phase and air is the wetting phase. In such cases, the Young-Laplace equation of capillarity for a tube should be written as

$$P_c = P^G - P^L = \frac{2\sigma}{r} = \frac{\sigma\cos\theta}{d},$$

(2.54)

where d is the tube diameter and θ is the contact angle. Note that when gas wets the substrate or for a droplet, gas pressure is less than the liquid pressure because when $\theta > 90°$, $\cos\theta < 0$. For a droplet, or when $\theta = 180°$, one writes

$$dP^G = \frac{2d(\sigma/r)}{(v^G/v^L) - 1},$$

(2.55)

where r is the radius of the droplet. Equation (2.55) indicates that the vapor pressure of a pure substance increases as the interface curvature increases.

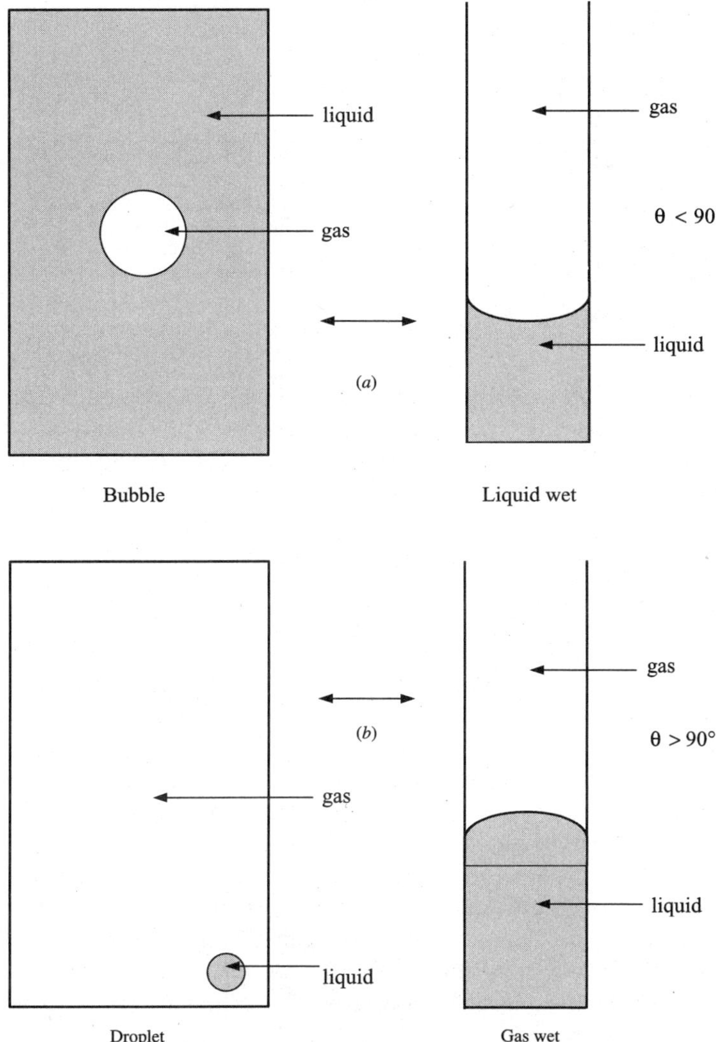

Figure 2.13   Parallel analogy between wettability, a bubble, and a droplet.

Effect of porous medium on phase behavior. Several authors have studied the effect of the porous medium on the phase behavior of reservoir fluid systems. Russian authors Trebin and Zadora (1968) report a strong influence of the porous medium on the dewpoint pressure and vapor–liquid equilibrium (VLE) of gas condensate systems. The porous medium used by these authors was a silica sand mixture (0.300 to 0.215 mm diameter) ground by a special cutter–pulverizer. Three different packings with permeabilities of 5.6, 0.612, and 0.111 darcies and

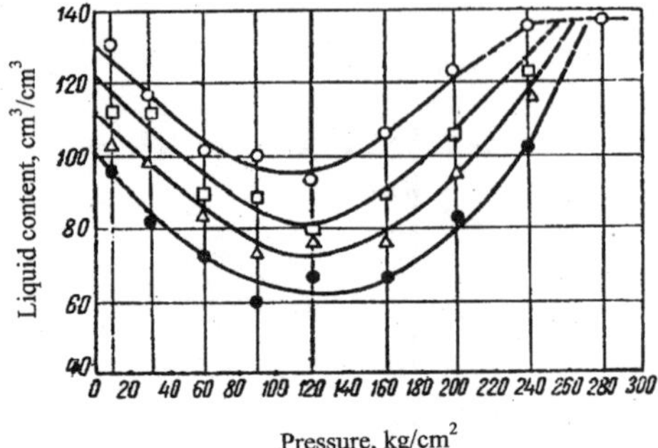

Figure 2.14 Effect of the surface area of the porous medium on condensation: ○ flat interface (PVT cell), □ 563, □ 1,307, and ■ 3,415 surface area cm^2/cm^3 of porous medium (adapted from Trebin and Zadora, 1968).

porosities of 34, 31.4, and 29.8 percent, respectively, were used. The calculated surface areas for these packings were 563, 1,307, and 3,415 cm^2/cm^3, respectively. Figure 2.14 shows the effect of the porous media on the liquid content of the produced fluids. This figure shows that as the surface area of the porous media increases, the dewpoint pressure increases. Trebin and Zadora report a 10 to 15 percent increase in the dewpoint pressure in the porous media of the type that was used in their work. When these authors increased the temperature, the effect of the porous media on VLE was decreased.

Tindy and Raynal (1966) measured the bubblepoint pressure of two reservoir crude oils in both an open space (PVT cell) and a porous medium with grain sizes in the range of 160 to 200 microns. The bubblepoint pressures of these two crude oils were higher in the porous medium than in a PVT cell by 7 and 4 kg/cm^2, respectively. Specifically, the bubblepoint pressure of one of the two crude oils measured at 80°C in a PVT cell was 121 kg/cm^2 and the bubblepoint pressure at the same temperature in a porous medium of 160 to 20 microns was 128 kg/cm^2. On the other hand, when these authors used a mixture of methane and n-heptane, they observed no differences in the saturation pressure. Sigmund *et al.* (1973) have also investigated the effect of the porous medium on phase behavior of model fluids. Their measurements on dewpoint and bubblepoint pressures showed no effect of the porous medium. The fluid systems used by these authors were C_1/nC_4 and C_1/nC_5. The smallest bead size used was 30 to 40 U.S. mesh. In Example 2.3 presented at the end of this chapter, the effect of interface curvature on dewpoint pressure and equilibrium phase composition will be examined.

Irreversible phenomena

In all of our derivations for equilibrium, we have made the fundamental assumption that the process is reversible. For an isolated system, the process is reversible when $dS = 0$: the entropy of the system does not change with time. A consequence of the reversibility assumption for all the equilibrium cases that we have studied (i.e., thermal, mechanical, and chemical) is that the temperature is the same throughout the composite system. In other words, the necessary condition to achieve equilibrium is the uniformity of temperature in the composite system. Under nonisothermal conditions, the process becomes irreversible and equilibrium cannot be achieved. Therefore, the Gibbs criterion of sedimentation (or segregation), $(d\mu_i = -M_i g dz)_T$, cannot be invoked when there is a temperature gradient in the system. In hydrocarbon reservoirs, there is often a temperature gradient of 1 to 4°F per 100 ft in the vertical direction. In some reservoirs, there is also a horizontal temperature gradient of 1 to 5°F/mile. These temperature gradients may have a strong influence on composition variation. The main purpose here is to briefly introduce the subject of thermodynamics of irreversible processes and to present the fundamental relationships for describing nonisothermal processes.

Let us examine the time variation of the entropy of a composite system (entropy change for the composite system from time t to $t + dt$), dS:

$$dS = (dS)_e + (dS)_i, \tag{2.56}$$

where $(dS)_e$ is the change in entropy supplied to the composite system from the exterior and $(dS)_i$ is the change in entropy produced inside the system (during time interval dt). An alternate statement of the Second Law (Postulates II, III, and IV of Chapter 1) is that $(dS)_i$ must be zero for reversible processes and positive for irreversible processes.

$$(dS)_i = 0 \quad \text{for reversible processes} \tag{2.57}$$

$$(dS)_i > 0 \quad \text{for irreversible processes} \tag{2.58}$$

According to the Second Law, the entropy production within the system, $(dS)_i$, cannot be negative. On the other hand, the entropy supplied to the system, $(dS)_e$, can be positive, zero, or negative. For an isolated system, $(dS)_e = 0$, and therefore

$$dS = (dS)_i. \tag{2.59}$$

Equation (2.59) implies that for a reversible process in an isolated system, $dS = 0$, and for an irreversible process, $dS > 0$. The mathematical expressions are

$$dS = 0 \quad \text{isolated system, reversible process} \tag{2.60}$$

$$dS > 0 \quad \text{isolated system, irreversible process.} \tag{2.61}$$

In Chapter 1, the entropy supplied to the system was defined by the relationship $(dS)_e = dQ/T$ for a closed system, where dQ is the heat supplied to the system at constant temperature T. Since the process was assumed reversible, then

$$dS = dQ/T, (dS)_i = 0 \text{ closed system.} \tag{2.62}$$

In summary, equilibrium thermodynamics addresses reversible processes for which there is no entropy production within the system. In thermodynamics of irreversible processes, the entropy production, $(dS)_i$, is formulated and then is related to the irreversible phenomena that may occur in the system.

Let us write the expression for the rate of entropy production in a system of volume V:

$$(dS/dt)_i = \int^V \sigma dV, \tag{2.63}$$

where σ is the entropy production per unit volume and time and it can be called the entropy production strength. There is no meaning to the destruction of entropy.

Entropy production strength. We will derive the expression for σ. Consider a small volume element $\Delta x \Delta y \Delta z$ within a given system. The entropy balance for this volume element can be stated as

$$\text{Rate of Entropy Change} = \text{Net Entropy Influx}$$
$$+ \text{Rate of Entropy Production.}$$

Let s denote the specific entropy (that is, entropy per unit mass), and $\vec{J^s}$ the entropy flux per unit time per unit surface area; then

$$\frac{\partial}{\partial t}[\rho s \Delta x \Delta y \Delta z] = J_x^s \Delta y \Delta z|_{x+\Delta x}^x + J_y^s \Delta x \Delta z|_{y+\Delta y}^y$$
$$+ J_z^s \Delta x \Delta y|_{z+\Delta z}^z + \sigma \Delta x \Delta y \Delta z \tag{2.64}$$

In the above equation J_x^s, J_y^s, and J_z^s are the total entropy fluxes in the x, y, and z directions, respectively, and ρ is the mass density. Dividing by $\Delta x \Delta y \Delta z$ and taking the limit as $\Delta x \to 0$, $\Delta y \to 0$, and $\Delta z \to 0$, and writing the result in vector notation,

$$\frac{\partial \rho s}{\partial t} = -\nabla \cdot (\vec{J^s}) + \sigma \tag{2.65}$$

Equation (2.65) is similar to any other balance expression, with the difference that there is a source term representing the production of entropy. A different form of the above equation can be presented in terms of total derivative of s:

$$\rho \frac{ds}{dt} = -\nabla \cdot (\vec{J}^s - \rho s \vec{v}) + \sigma. \qquad (2.66a)$$

In Eq. (2.66a), the total entropy flux $\vec{J}^s$ consists of three parts: (1) $\vec{q}^{cond}/T$, (2) $\sum_{i=1}^{c} \tilde{S}_i \vec{J}_i$, and (3) $\rho s \vec{v}$ which represent the contributions from conduction, diffusion, and convection, respectively ($\vec{q}^{cond}$ is the energy flux by conduction, $\vec{J}_i$ is the diffusion mass flux of component i to be described later in this chapter, and $\tilde{S}_i = \bar{S}_i/M_i$). Therefore, Eq. (2.66a) can be written as

$$\rho \frac{ds}{dt} = -\nabla \cdot \left(\frac{\vec{q}^{cond}}{T} + \sum_{i=1}^{c} \tilde{S}_i \vec{J}_i \right) + \sigma \qquad (2.66b)$$

The total derivative (ds/dt) represents the variation of s with respect to time and position; $s = s(t, x, y, z)$. The partial derivative $(\partial s/\partial t)$ represents the variation of s with respect to time at a fixed position. For any scalar quantity such as s,

$$\rho \frac{ds}{dt} = \frac{\partial \rho s}{\partial t} + \nabla \cdot \rho s \vec{v}. \qquad (2.67)$$

The above equation can be obtained by combining the mass balance expression $(\partial \rho/\partial t) = -\nabla \cdot (\rho \vec{v})$ and the expression $(ds/dt) = (\partial s/\partial t) + v \cdot \nabla s$. Combining Eqs. (2.65) and (2.67) results in the expression given by Eq. (2.66a). Now recall the expression for dS from Chapter 1,

$$T dS = dU + P dV - \sum_{i=1}^{c} (\mu_i/M_i) dm_i, \qquad (2.68)$$

where S, U, and V are the total entropy, internal energy, and volume, respectively, and m_i is the total mass of component i. Note that in the above equation, the chemical potential is defined as $\mu_i = (\partial U/\partial n_i)_{S,V,n_i}$; the molecular weight of component i, M_i, appears to make the chemical potential stay the same as defined in Chapter 1. The extensive quantities S, U, and V can be expressed in terms of specific quantities (quantity per unit mass): $S = ms$, $U = mu$, and $V = mv$, where $m = \sum_{i=1}^{c} m_i$. If we assume $m = 1$ in Eq. (2.68) and divide by dt,

$$T \frac{ds}{dt} = \frac{du}{dt} + P \frac{dv}{dt} - \sum_{i=1}^{c} \tilde{\mu}_i \frac{dw_i}{dt}, \qquad (2.69)$$

where $\tilde{\mu}_i = \mu_i/M_i$ and w_i is the mass fraction of component i. Equation (2.69) is based on the assumption of equilibrium; it is assumed that at a local level, the process is still an equilibrium process. When the volume element becomes large, the equilibrium assumption may not be valid. In irreversible thermodynamics, it is shown that for most transport processes, the equilibrium assumption is justified at the local level (de Groot and Mazur, 1984; Haase, 1969).

The continuity equation for species i and the energy balance are then used to provide the expression for σ. For the volume element, one can readily derive the mass-balance expression

$$\rho\frac{dw_i}{dt} = -\nabla \cdot \vec{J}_i \qquad i = 1, \dots, c. \tag{2.70}$$

In the above equation, $\vec{J}_i$ is the diffusion mass flux of component i.

The expression for energy balance is taken from Bird $et\ al.$ (see Table 18.3-1 of Bird $et\ al.$, 1960). These authors provide the details of the derivation of the energy-balance expression

$$\rho\frac{du}{dt} = -\nabla \cdot \left(\vec{q}^{\,cond} + \sum_{i=1}^{c} \tilde{H}_i\vec{J}_i\right) - (\vec{\vec{\tau}} + P\vec{\vec{\delta}}) : \nabla\vec{v} - \sum_{i=1}^{c} \vec{J}_i \cdot \vec{g} \tag{2.71}$$

In the above equation, in addition to the energy flux by conduction, $\vec{q}^{\,cond}$, the energy flux by interdiffusion, $(\sum_{i=1}^{c} \tilde{H}_i\vec{J}_i)$, is also taken into account ($\tilde{H}_i = H_i/M_i$), but the energy flux by concentration gradient (i.e., Dufour-effect) is neglected. Note that the sign of the last term is reversed from Bird $et\ al.$, (1960); this is done to be consistent with our section on the effects of gravity on equilibrium, where we have assumed the upward direction to be positive. In Eq. (2.71), the symbol "$\rightrightarrows$" represents a tensor; both $\vec{\vec{\tau}}$ and $\vec{\vec{\delta}}$ are tensors. The stress tensor $\vec{\vec{\tau}}$ has nine components $\tau_{xx}, \tau_{yy}, \tau_{zz}, \tau_{xy}, \tau_{xz}, \tau_{yx}, \tau_{yz}, \tau_{zx}$, and τ_{zy}, whereas the velocity vector $\vec{v}$ has three components v_x, v_y, and v_z. The unit tensor $\vec{\vec{\delta}}$ has also nine components; it can be represented by a 3×3 matrix with diagonal elements unity and nondiagonal elements zero (Appendix A of Bird $et\ al.$ (1960) provides a brief and useful description of vector and tensor operations). Note that $\nabla\vec{v}$ in Eq. (2.71) is the dyadic product of vector ∇ and vector $\vec{v}$; it is a tensor with nine components, $\partial v_x/\partial x$, $\partial v_x/\partial y$, $\partial v_x/\partial z$, $\partial v_y/\partial x$, $\partial v_y/\partial y$, $\partial v_y/\partial z$, $\partial v_z/\partial x$, $\partial v_z/\partial y$, and $\partial v_z/\partial z$. The double-dot operation ":" represents the scalar product of two tensors,

$$\vec{\vec{\tau}} : \nabla\vec{v} = \sum_{i=i}^{3}\sum_{j=1}^{3} \tau_{ij}\frac{\partial v_i}{\partial x_j}, \tag{2.72}$$

where i and j take on the values of $x, y,$ and z.

The term $P\vec{\vec{\delta}} : \nabla\vec{v}$ in Eq. (2.71) can be simplified considerably,

$$P\vec{\vec{\delta}} : \nabla\vec{v} = P(\nabla \cdot \vec{v}). \tag{2.73}$$

From the continuity equation, $(d\rho/dt) = -\rho(\nabla \cdot \vec{v})$. Therefore,

$$P\vec{\vec{\delta}} : \nabla\vec{v} = -(P/\rho)\frac{d\rho}{dt}. \tag{2.74}$$

Since $\rho = 1/v$, where v is the specific volume,

$$(1/\rho)(d\rho/dt) = -\rho\left(\frac{dv}{dt}\right), \tag{2.75}$$

and Eq. (2.74) simplifies to

$$P\vec{\vec{\delta}} : \nabla\vec{v} = (P\rho)\frac{dv}{dt}. \tag{2.76}$$

Combining Eqs. (2.71) and (2.76) gives

$$\rho\frac{du}{dt} = -\nabla \cdot \left(\vec{q}^{\,cond} + \sum_{i=1}^{c} \tilde{H}_i \vec{J}_i\right) - \vec{\vec{\tau}} : \nabla\vec{v} - P\rho\frac{dv}{dt} - \sum_{i=1}^{c} \vec{J}_i \cdot \vec{g}. \tag{2.77}$$

The above form of the energy equation is combined with Eqs. (2.69) and (2.70) to provide the expression for ds/dt,

$$\rho T\frac{ds}{dt} = -\nabla \cdot \vec{q}^{\,cond} - \nabla \cdot \left(\sum_{i=1}^{c} \tilde{H}_i \vec{J}_i\right) - \vec{\vec{\tau}} : \nabla v + \sum_{i=1}^{c} \tilde{\mu}_i(\nabla \cdot \vec{J}_i) - \sum_{i=1}^{c} \vec{J}_i \cdot \vec{g}. \tag{2.78}$$

The first term on the right side of Eq. (2.78) can be part of the following relationship,

$$\nabla \cdot (\vec{q}^{\,cond}/T) = \frac{1}{T}\nabla \cdot \vec{q}^{\,cond} - \left(\frac{1}{T^2}\right)(\vec{q}^{\,cond} \cdot \nabla T). \tag{2.79}$$

The second term on the right side of Eq. (2.78) can be written as:

$$-\nabla \cdot \left(\sum_{i=1}^{c} \tilde{H}_i \vec{J}_i - \sum_{i=1}^{c} \tilde{\mu}_i \vec{J}_i + \sum_{i=1}^{c} \tilde{\mu}_i \vec{J}_i\right) = -\nabla \cdot \left(T\sum_{i=1}^{c} \tilde{S}_i \vec{J}_i + \sum_{i=1}^{c} \tilde{\mu}_i \vec{J}_i\right). \tag{2.80}$$

Combining the above three expressions, and using,

$$\nabla \cdot \sum_{i=1}^{c}(\tilde{\mu}_i \vec{J}_i) = \sum_{i=1}^{c} \vec{J}_i \cdot \nabla\tilde{\mu}_i + \sum_{i=1}^{c} \tilde{\mu}_i(\nabla \cdot \vec{J}_i)$$

and a similar expression for $\nabla \cdot (T \sum_{i=1}^{c} \tilde{S}_i \vec{J}_i)$,

$$\rho \frac{ds}{st} = -\nabla \cdot \left[(\vec{q}^{cond}/T) + \sum_{i=1}^{c} (\tilde{S}_i \vec{J}_i) \right] - (1/T^2) \left[\vec{q}^{cond} + \sum_{i=1}^{c} \tilde{S}_i \vec{J}_i \right] \cdot \nabla T$$

$$- (1/T)\vec{\vec{\tau}} : \nabla \vec{v} - (1/T) \sum_{i=1}^{c} \vec{J}_i (\nabla \tilde{\mu}_i + \vec{g}). \tag{2.81}$$

From the comparison of Eqs. (2.66b) and (2.81),

$$\boxed{\begin{aligned} \sigma = -(1/T^2) &\left[\vec{q}^{cond} + \sum_{i=1}^{c} \tilde{S}_i \vec{J}_i \right] \cdot \nabla T - (1/T)\vec{\vec{\tau}} : \nabla v \\ &- (1/T) \sum_{i=1}^{c} \vec{J}_i \cdot (\nabla \tilde{\mu}_i + \vec{g}) \end{aligned}} \tag{2.82}$$

Note that the entropy production strength, according to the above equation consists of three terms. The first term arises from the temperature gradient, the second term is due to velocity gradients (see Eq. (2.85) below), and the third term is from the gradient of the chemical potential. Equation (2.82) implies that at stationary states (states in which the variables are independent of time), when temperature is time-independent but has a nonuniform value, σ is nonzero. We have shown that entropy production within the system is due to (1) temperature gradient, (2) velocity gradient, and (3) chemical potential gradient. One can also show that the terms on the right side of Eq. (2.82) are positive. Let us show that the second term is always ≥ 0.

In order to prove that $(-\vec{\vec{\tau}} : \nabla \vec{v})$ is positive, we need the expression for various stresses. For Newtonian fluids, the expression for stresses are (see Bird et $al.$, 1960)

$$\tau_{ii} = -\frac{2}{3} \cdot \frac{\partial v_i}{\partial x_i} + \frac{2}{3} \cdot \mu (\nabla \cdot \vec{v}) \qquad i = 1, 2, 3 \tag{2.83}$$

$$\tau_{ij} = \tau_{ji} = -\mu \left(\frac{\partial v_i}{\partial x_j} + \frac{\partial v_j}{\partial x_i} \right) \qquad i = 1, 2, 3, \ j = 1, 2, 3, i \neq j, \tag{2.84}$$

where i and j take on the values of x, y, and z.

Combining Eqs. (2.72), (2.83), and (2.84) results in the following sum of the squares:

$$(-\vec{\vec{\tau}} : \nabla \vec{v}) = \frac{1}{2} \mu \sum_{i=1}^{3} \sum_{j=1}^{3} \left[\left(\frac{\partial v_i}{\partial x_j} + \frac{\partial v_j}{\partial x_i} \right) - \frac{2}{3} (\nabla \cdot \vec{v}) \delta_{ij} \right]^2, \tag{2.85}$$

where $\delta_{ij} = 1$ for $i = j$, and $\delta_{ij} = 0$ for $i \neq j$. Therefore, it is established that the $(-\vec{\vec{\tau}} : \nabla \vec{v})$ term is always ≥ 0.

Let us also examine Eq. (2.82) for a convection-free system at isothermal conditions, where $\vec{v} = \vec{0}$, and $\nabla T = 0$. For the system to have no entropy production (that is, to be at equilibrium state), $\nabla \tilde{\mu}_i = -\vec{g}$. Since in a gravity field $g_x = g_y = 0$ and $g_z = g$, then $d\tilde{\mu}_i / dz = -g$ or $d\mu_i = -M_i g dz$, which is the same as Eq. (2.13). With the above background, we now switch to the expression for the total diffusion flux, which can be derived from the thermodynamics of irreversible processes.

Let us start by writing the expression for the mass flux, $\vec{m}_i$:

$$\vec{m}_i = \rho w_i \vec{v} + \vec{J}_i, \tag{2.86}$$

where ρ is the total mass density, w_i is the mass fraction of component i, $\vec{v}$ is the mass-average velocity, and $\vec{J}_i$ is the total diffusion mass flux (nonconvective) of component i. In the above equation, the first term on the right side is the convective mass flux, and the second term is the diffusive flux. Total diffusion flux, to our knowledge, has not yet been accounted for in any of the numerical simulation models for hydrocarbon reservoirs.

The expression for the total diffusion mass flux in a binary system for component 1 is given by Bird *et al.* (1960), and Ghorayeb and Firoozabadi (1999):

$$\vec{J}_1 = (-\eta^2 / \rho) M_1 M_2 D_{12}$$

$$\times \left[\left(\frac{\partial \ln f_1}{\partial \ln x_1} \right)_{T,P} \nabla x_1 + \frac{M_1 x_1}{RT} \left(\frac{\bar{V}_1}{M_1} - \frac{1}{\rho} \right) \nabla P + k_T \nabla \ln T \right], \tag{2.87}$$

where η is the total molar density, ρ is the mass density, D_{12} is the molecular diffusion coefficient of components 1, and 2, f is the fugacity, k_T is the thermal diffusion ratio of component 1 and M is the molecular weight. We assign the positive sign to k_T when component 1 moves to the hot region. For component 2, the thermal diffusion ratio is $-k_T$.

Three different diffusion processes are included in Eq. (2.87). The first term represents molecular diffusion. The second term is for pressure diffusion, which under the influence of gravity leads to gravity segregation. The last term represents thermal diffusion (Soret effect) which is the tendency of a convection-free mixture to separate under a temperature gradient. An interesting feature of Eq. (2.87) is that in the absence of a temperature gradient, it simplifies to the Gibbs segregation equation, which we will soon demonstrate. As one can see from Eq. (2.82), with thermal diffusion and even with negligible convection, the segregation equation $d\mu_i = -M_i g\, dz$ does not hold because equilibrium is not established.

In hydrocarbon reservoirs, the temperature distribution is often available from temperature measurements. Then for a 1D or a 2D space with two components, there are two unknowns at each point, x_1 and P (x_2 is not an independent variable since $x_2 = 1 - x_1$). For a 1D space, because of the absence of convection, the problem becomes very simple. Let us derive an explicit expression for dx_1/dz at steady state (i.e., stationary state) for a two-component system for a 1D problem.

Thermal diffusion and gravity segregation in 1D.   The combined effect of thermal diffusion and gravity segregation in the 1D vertical direction at steady state is governed by $J_{1z} = 0$. Using Eq. (2.87) and $dP/dz = -\rho g$, one obtains

$$\boxed{\frac{dx_1}{dz} = \left(\frac{\partial \ln x_1}{\partial \ln f_1}\right)_{T,P}\left[\frac{gx_1}{RT}(\rho \bar{V}_1 - M_1) + k_T \frac{d\ln T}{dz}\right].} \qquad (2.88)$$

An alternative form of Eq. (2.88) can be obtained by using $\rho = \sum_{i=1}^{2} x_i M_i / \sum_{i=1}^{2} x_i \bar{V}_i$:

$$\frac{dx_1}{dz} = \left(\frac{\partial \ln x_1}{\partial \ln f_1}\right)_{T,P}\left[\frac{(M_2/\bar{V}_2) - (M_1/\bar{V}_1)}{1/(x_1 \bar{V}_1) + 1/(x_2 \bar{V}_2)}\frac{g}{RT} - k_T \frac{d\ln T}{dz}\right]. \qquad (2.89)$$

The hydrostatic-head expression is given by

$$\frac{dP}{dz} = -\left[\frac{PM}{ZRT}\right]g, \qquad (2.90)$$

where the substitution $\rho = (PM/ZRT)$ has been made. Eqs. (2.89) and (2.90) are two first-order differential equations and their solution provides x_1 and P. Pressure and composition at a reference depth should be given. A numerical method, such as the Euler scheme, can be used to solve these two equations.

In Eq. (2.89), $(\partial \ln x_1/\partial \ln f_1)_{T,P}$, $\bar{V}_1$, $\bar{V}_2$, and in Eq. (2.90), Z, all can be obtained from an EOS, which will be discussed in Chapter 3.

When the temperature gradient is negligible, the last term in Eq. (2.88) drops out, and since

$$(\partial \ln x_1 / \partial \ln f_1)_{T,P} = \frac{1}{x_1}\left(\frac{\partial x_1}{\partial \ln f_1}\right),$$

then
$$\frac{dx_1}{dz} = \left(\frac{\partial x_1}{\partial \ln f_1}\right)_{T,P}\frac{g}{RT}(\rho \overline{V}_1 - M_1). \tag{2.91}$$

Equation (2.91) is the fugacity form of Eq. (2.19a) when the number of components is two. For a near-critical two-component fluid mixture $(\partial x_1 / \partial \ln f_1)_{T,P}$ is large, and therefore, (dx_1/dz) is also large (see the next section).

Before we move on to study the effect of the temperature gradient on natural convection and thermal diffusion in hydrocarbon reservoirs, we will briefly discuss the thermal diffusion ratio, k_T.

Thermal diffusion ratio, k_T. Thermal diffusion ratio, k_T, is a measure of thermal diffusion. The sign of k_T determines the direction of thermal diffusion. Another defined parameter for thermal diffusion is the thermal diffusion factor, α, which is defined as

$$\alpha = \frac{k_T}{x_1 x_2} \tag{2.92}$$

for a binary mixture. The parameter α is nearly independent of composition for low-pressure gases. There has been some confusion in the literature in representing the direction of α either in experiment or theory. For example, the experimental data of Rutherford and Roof (1959) for the mixture C_1/nC_4 show that C_1 (component 1) goes to the hot region when $\alpha > 0$ and nC_4 (component 2) with $\alpha < 0$ goes to the cold region. In contrast, in the theoretical models of Rutherford (1963) and Kempers (1989), the component with $\alpha > 0$ goes to the cold region. In order to avoid this confusion, as was stated earlier, component 1 with $\alpha > 0$ is assigned to segregate to the hot region.

In low-pressure gaseous mixtures and ideal liquid mixtures, α has been found to be small. On the other hand, in nonideal liquid mixtures, α may be large; it becomes very large in the near-critical region for both near-critical gas and liquid phases. At the critical point, however, it has a limiting value. In this respect, α and the molecular diffusion coefficient, D, have opposite trends. Molecular diffusion is pronounced for low pressure gaseous mixtures and becomes small as the critical region is approached.

In the absence of convection and in the horizontal direction (that is, no gravity effect), Eq. (2.88) simplifies to

$$\frac{dx_1}{dx} = \left(\frac{\partial \ln x_1}{\partial \ln f_1}\right) k_T \frac{d \ln T}{dx}, \tag{2.93}$$

where x_1 is the mole fraction of component 1 and x represents distance in the horizontal direction. Combining Eqs. (2.92) and (2.93) results in

$$\frac{dx_1}{dx} = \alpha^{exp} x_1 x_2 \frac{d \ln T}{dx} \tag{2.94}$$

where
$$\alpha^{exp} = \alpha \left(\frac{\partial \ln x_1}{\partial \ln f_1}\right) \tag{2.95}$$

Note that the factor $(\partial \ln x_1 / \partial \ln f_1)_{T,P}$ is included in the expression for α^{exp}. If α^{exp} is assumed to be constant, then Eq. (2.94) can be integrated between the hot and cold regions to provide

$$\boxed{\alpha^{exp} = \ln[(x_1/x_2)^H/(x_2/x_1)^C]/\ln(T^H/T^C)}, \tag{2.96}$$

where H and C represent, respectively, the hot and cold regions in a two-chamber cell. Equation (2.96) is the expression that has been used to infer experimental thermal diffusion factors from measured concentration and temperature data. In order to use α^{exp}, however, we need to multiply it by $(\partial \ln f_1 / \partial \ln x_1)$ for substitution into Eq. (2.87) or Eq. (2.88).

There are a number of theoretical models for the calculation of α for binary mixtures. These models often rely on equilibrium and nonequilibrium properties of mixtures (Shukla and Firoozabadi, 1998). The general expression for α for a binary mixture is

$$\alpha = \frac{Q_2^* - Q_1^*}{x_1 \left(\dfrac{\partial \mu_1}{\partial x_1}\right)_{T,P}} \tag{2.97}$$

where Q_i^* is the net heat of transport of component i (see Example 1.11 of Chapter 1, and Shukla and Firoozabadi, 1998). In Eq. (2.97), the numerator is in terms of nonequilibrium properties and the denominator is in terms of equilibrium properties. The complexity is the evaluation of the net heat of transport, Q_i^*, in the numerator. Note that α in Eq. (2.97) is also factored with $(\partial \ln x_1 / \partial \ln f_1)$. Therefore the results from Eq. (2.97) can be compared to measured values from Eq. (2.96).

There are very few experimental data for α of binary hydrocarbon mixtures at high pressures; the available data are limited to the C_1/C_3 system from Haase *et al.* (1971) and the C_1/nC_4 system from Rutherford and Roof (1959). Figures (2.15) to (2.19) show the data and the theory. The solid line in these figures is from a new model by Shukla and Firoozabadi (1998). The comparison between theory and data for the C_1/C_3 system suggests that there is a major discrepancy in the critical region. To highlight the critical region, Fig. 2.20 presents the pressure–composition plot of C_1/C_3 at 346 K. The composition $x_1 = 0.34$ (mole fraction) with pressure range $P = 50$ to 70 bar, lies in the critical region. In order to further illustrate the effect of the critical region on α, Figs. 2.21 and 2.22 show the calculated results for $(\partial \ln f_1/\partial \ln x_1)$ and $\ln f_1$; $(\partial \ln f_1/\partial \ln x_1)$ is a measure of nonideality and is unity for an ideal mixture and it is zero at this critical point for a nonideal binary mixture. Apparently, critical condition effects are not properly taken into account by models that are used to calculate α.

The agreement between prediction results and data are better for the C_1/nC_4 mixture. For this system, unlike for the C_1/C_3 system, the fluid is on the left side of the critical point on the $P–T$ plot, implying liquid state. For both the C_1/C_3 and C_1/nC_4 systems, $\alpha > 0$, which implies that methane segregrates towards the hot region.

Next we will discuss thermal convection and then formulate the problem of composition variation in hydrocarbon reservoirs with natural convection and diffusion.

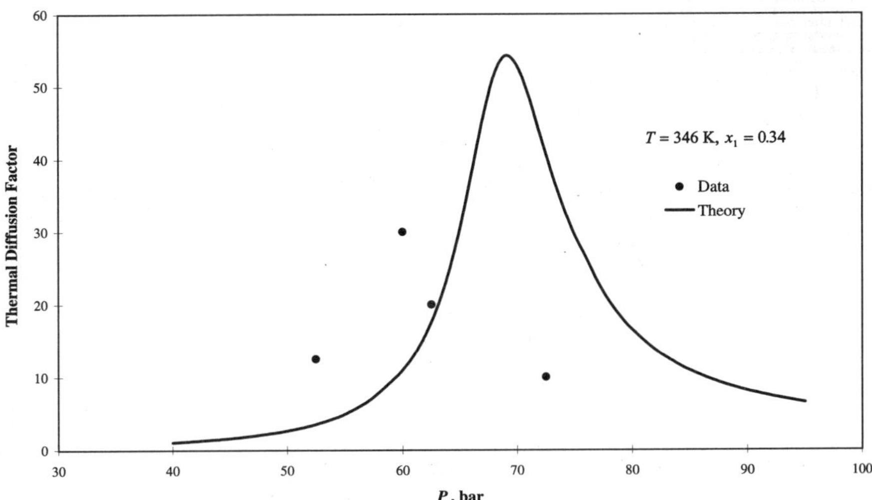

Figure 2.15 Comparison of thermal diffusion factors from theory and experiments for the C_1/C_3 system (adapted from Shukla and Firoozabadi, 1998).

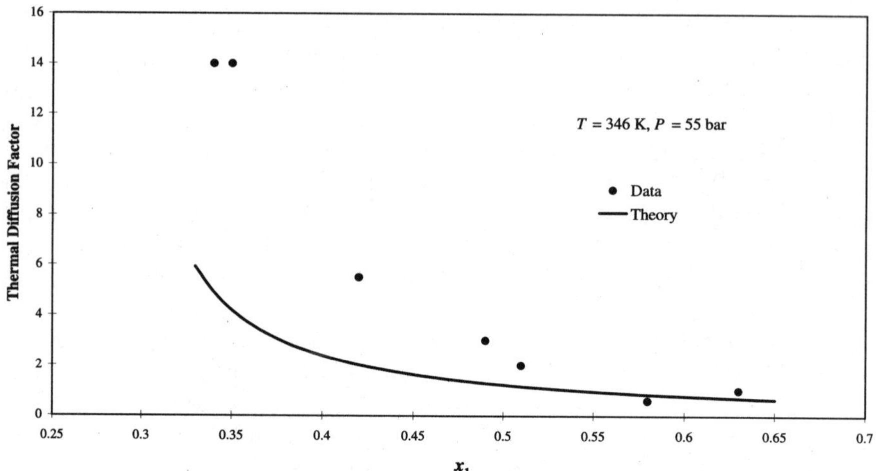

Figure 2.16   Comparison of thermal diffusion factors from theory with experiments for the C_1/C_3 system (adapted from Shukla and Firoozabadi, 1998).

Thermal convection

One major goal of this section is to understand what drives flow in thermal convection at steady state; it may not be the buoyancy as one may believe. The benefit of buoyancy is that it will (if it is large enough) cause instabilities and these instabilities will cause a horizontal density

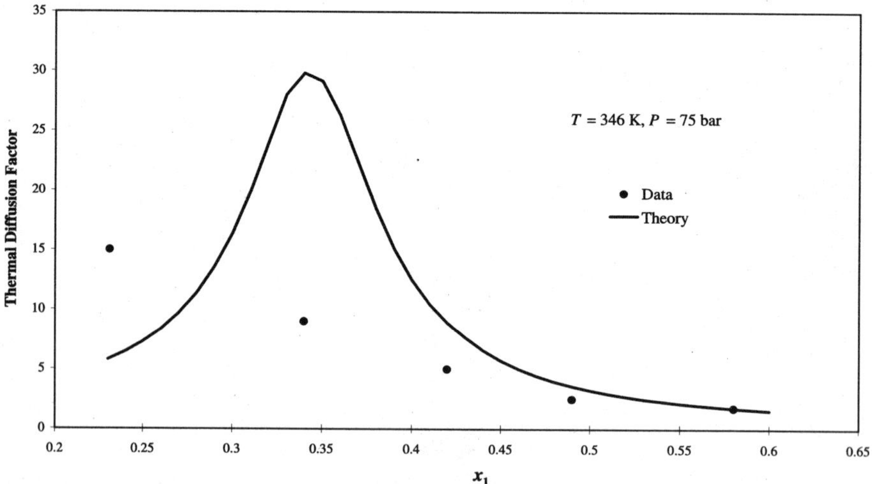

Figure 2.17   Comparison of thermal diffusion factors from theory and experiments for the C_1/C_3 system (adapted from Shukla and Firoozabadi, 1998).

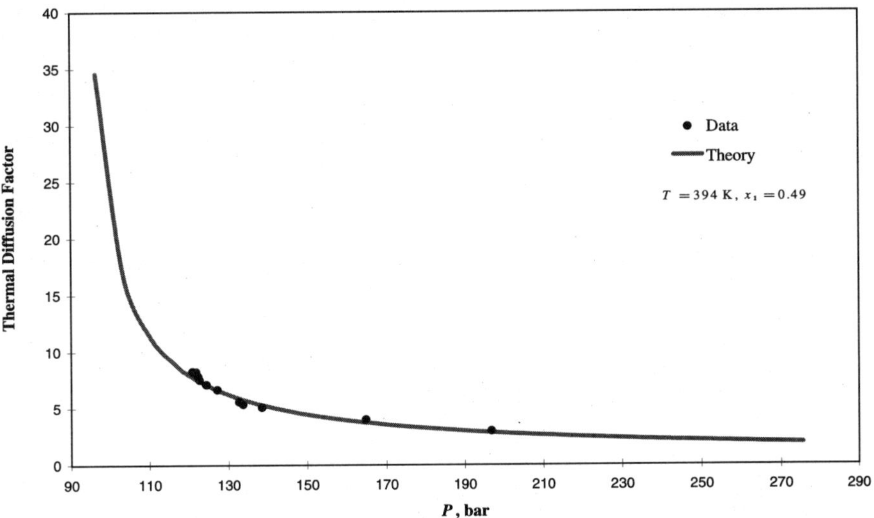

Figure 2.18　Comparison of thermal diffusion factors from theory and experiments for the C_1/nC_4 system (adapted from Shukla and Firoozabadi, 1998).

gradient. Let us consider a single-component fluid in a two-dimensional cross-sectional (x, z) reservoir. We assume that the process has been going on long enough to have reached steady state. The reservoir boundaries are closed to flow and there is a steady-state temperature field imposed on the system.

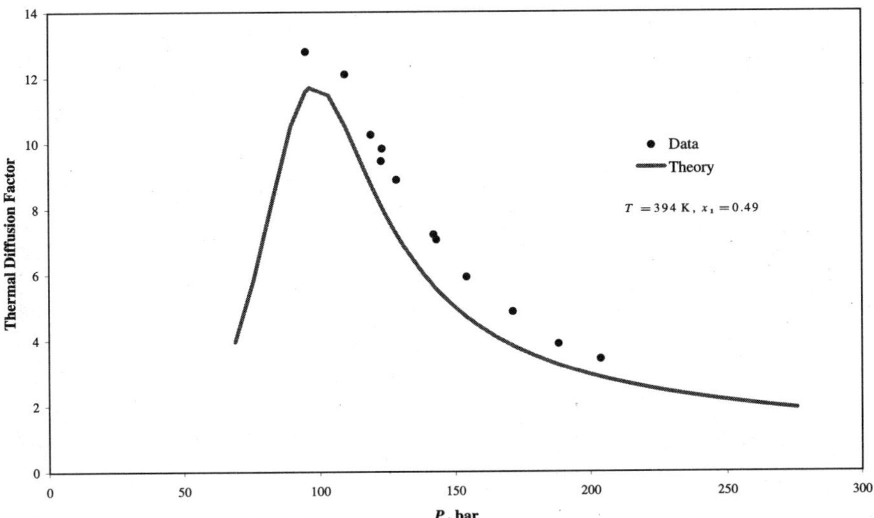

Figure 2.19　Comparison of thermal diffusion factors from theory and experiments for the C_1/nC_4 system (adapted from Shukla and Firoozabadi, 1998).

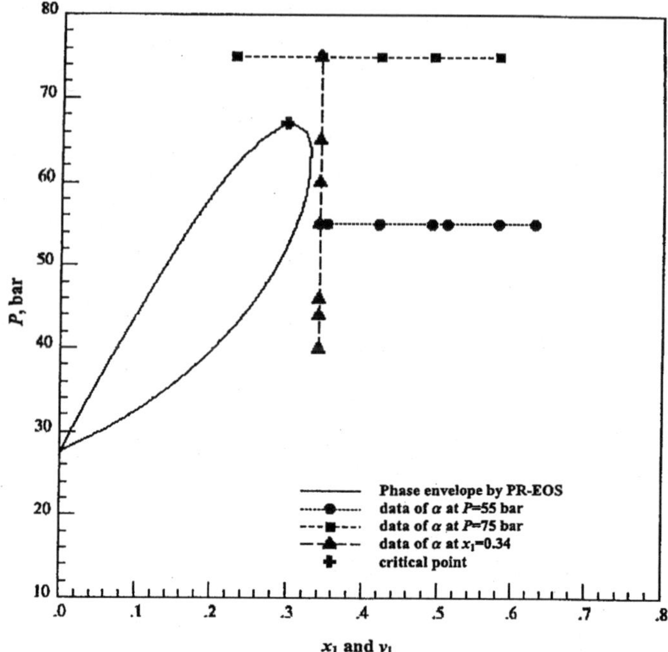

Figure 2.20 Phase diagram of the C_1/C_3 system at $T = 346\,\text{K}$, $\delta_{C_1-C_3} = 0.01$ (adapted from Shukla and Firoozabadi, 1998).

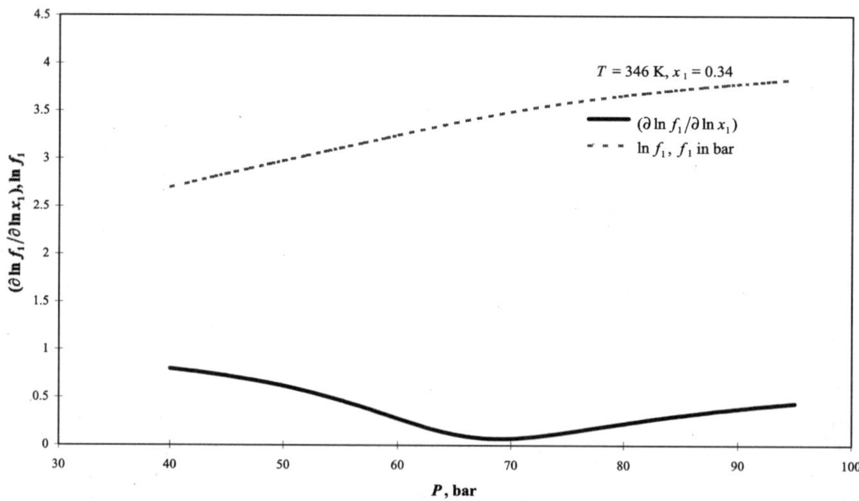

Figure 2.21 Fugacity of C_1 in the C_1/C_3 system (adapted from Shukla and Firoozabadi, 1998).

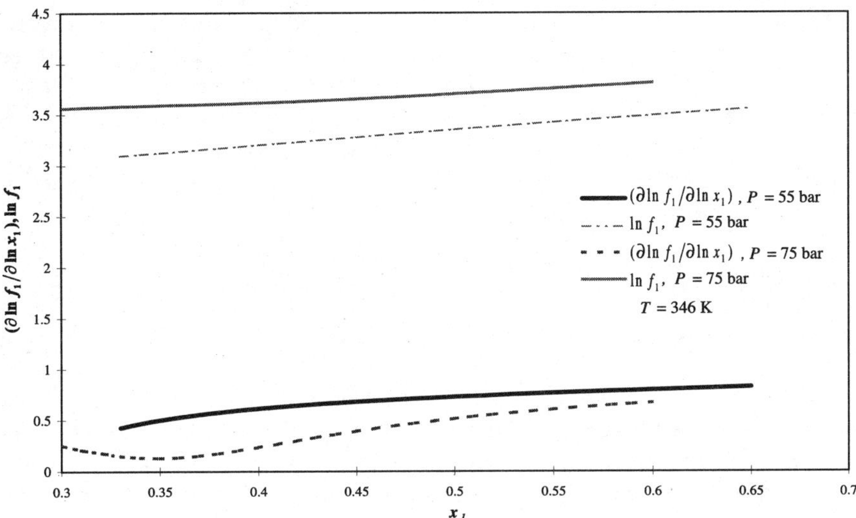

Figure 2.22   Fugacity of C_1 in the C_1/C_3 system (adapted from Shukla and Firoozabadi, 1998).

The flow is governed by the continuity equation:

$$\nabla \cdot (\rho \vec{v}) = 0, \qquad (2.98)$$

where the velocity is given by Darcy's law,

$$\vec{v} = -(k/\mu)(\nabla P + \rho \vec{g}\nabla z). \qquad (2.99)$$

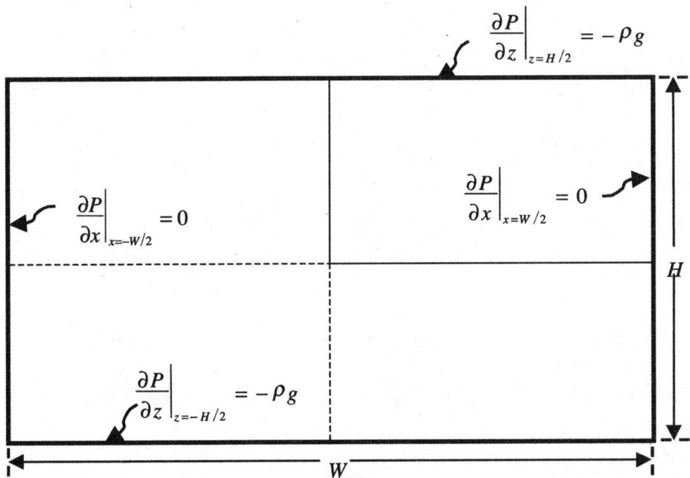

Figure 2.23   Geometry and boundary conditions for the 2D single-component fluid system used in the study of thermal convection.

Note that in Eq. (2.99), $\vec{v}$ is not the true velocity; the true velocity is given by $\vec{v}/\phi$ where ϕ is the fractional porosity. Figure 2.23 provides the boundary conditions of the 2D geometry. The boundary conditions are derived from the assumption that fluid does not cross the boundaries; i.e., the normal component of velocity is zero.

By combining the continuity equation, Eq. (2.98), and Darcy's equation, Eq. (2.99), one obtains,

$$\frac{\partial^2 P}{\partial x^2} + \frac{\partial^2 P}{\partial z^2} + g\frac{\partial \rho}{\partial z} = -\left[\frac{1}{\rho/\mu}\right]\frac{\partial}{\partial x}(\rho/\mu)\frac{\partial P}{\partial x} - \left[\frac{1}{\rho/\mu}\right]\frac{\partial}{\partial z}(\rho/\mu)\left[\frac{\partial P}{\partial x} + \rho g\right].$$

$$(2.100)$$

In the derivation of Eq. (2.100), it is assumed that the permeability is independent of x and z. In order to solve the problem of thermal convection in a simple manner, one may use the Boussinesq approximation (see Chapter 2 of Chandrasekhar, 1961), neglect the variation of ρ/μ in the continuity equation, and retain the density variation in Darcy's equation. Riley and Firoozabadi (1996) have examined the Boussinesq approximation by using the perturbation technique and found that the neglect of the right-side terms of Eq. (2.100) is an excellent approximation. (In the perturbation technique, one can check the solution and then the error. The solution from the neglected terms is inserted in the equation terms. The approximate solution is then improved by solving the equation with values of the neglected terms from the original solution.) Using the Boussinesq approximation, Eq. (2.100) transforms into

$$\frac{\partial^2 P}{\partial x^2} + \frac{\partial^2 P}{\partial z^2} + g\frac{\partial \rho}{\partial z} = 0. \qquad (2.101)$$

Now let us examine the $(\partial \rho/\partial z)$ term. In general, for a single-component fluid, $\rho = \rho(T, P)$, and therefore

$$\frac{\partial \rho}{\partial z} = \frac{\partial \rho}{\partial P}\frac{\partial P}{\partial z} + \frac{\partial \rho}{\partial T}\frac{\partial T}{\partial z}. \qquad (2.102)$$

The effect of pressure in density variation does not contribute to thermal convection (see Problem 2.17), and therefore we write

$$\frac{\partial \rho}{\partial z} \approx \frac{d\rho}{dT}\frac{\partial T}{\partial z}. \qquad (2.103)$$

The temperature dependence of the density can be approximated using the Taylor series,

$$\rho(T) \approx \rho(T^0) + \frac{d\rho}{dT}\bigg|_{T^0}(T - T^0), \qquad (2.104)$$

where T^0 is the reference temperature. From the definition of thermal expansivity, $e = (1/V)(dV/dT)$ and $e = 1/(V/m)d(V/m)/dT = 1/(1/\rho)d(1/\rho)/dT = -(1/\rho)(d\rho/dT)$ (we have divided the numerator and denominator by the mass of volume m), then $d\rho/dT|_{T^0} = -\rho^0 e$, where ρ^0 is the density at temperature T^0. Therefore, Eq. (2.104) can be written as

$$\rho(T) \approx \rho^0[1 - e(T - T^0)]. \tag{2.105}$$

Now let us choose a temperature variation in a hydrocarbon reservoir according to

$$T = T^0 - BW \cos\frac{n\pi(x + W/2)}{W} + Cz, \tag{2.106}$$

where parameter B influences the horizontal temperature variation and parameter C influences the vertical temperature variation. The constant n is an odd number; we will discuss its effect on the shape of the temperature variation in Example 2.11. Instead of Eq. (2.106), one may use the simpler expression,

$$T = T^0 + 2Bx + Cz. \tag{2.107}$$

The factor "2" is included so that the two temperature profiles, from Eqs. (2.106) and (2.107), will have the same overall horizontal temperature change for the same value of B. Equation (2.107) describes very well the temperature variation in hydrocarbon reservoirs; it is also a solution to the energy equation where heat transfer is by pure conduction, i.e., the solution to $\nabla^2 T = 0$ (see Example 2.10). The advantage of using the cosine temperature distribution is that $\cos(n\pi(x + W/2)/W)$ is orthogonal to the eigenfunctions of the problem, $\cos(m\pi(x + W/2)/W)$. This reduces the number of infinite sum cosines to two terms, $m = 0$ and $m = n$.

From Eqs. (2.103), (2.105), and (2.106), $\partial\rho/\partial z = -\rho^0 eC$, and then Eq. (2.101) transforms into

$$\frac{\partial^2 P}{\partial x^2} + \frac{\partial^2 P}{\partial z^2} - \rho^0 geC = 0, \tag{2.108}$$

and the boundary conditions are

$$\frac{\partial P}{\partial x}\Big|_{x=-W/2} = \frac{\partial P}{\partial x}\Big|_{x=W/2} = 0 \tag{2.109}$$

$$\frac{\partial P}{\partial z}\Big|_{z=-H/2} = -\rho g\big|_{z=-H/2}$$

$$= -\rho^0 g\left[1 + eBW\cos\frac{n\pi(x+W/2)}{W} + eCH/2\right] \tag{2.110}$$

$$\frac{\partial P}{\partial z}\Big|_{z=H/2} = -\rho g\big|_{z=H/2}$$

$$= -\rho^0 g\left[1 + eBW\cos\frac{n\pi(x+W/2)}{W} - eCH/2\right]. \tag{2.111}$$

The solution of the above partial differential equation, Eq. (2.108), and the boundary conditions, Eqs. (2.109) to (2.111), are readily obtained:

$$P(x,z) = P^0 - \rho^0 gz + \rho^0 geCz^2/2$$

$$- \frac{\rho^0 geW^2}{n\pi}B\frac{\sinh(n\pi z/W)}{\cosh(n\pi H/2W)}\cos\frac{n\pi(x+W/2)}{W}. \tag{2.112}$$

In the above equation, P^0 is the pressure at the origin $x = 0$ and $z = 0$. T^0 and ρ^0 are also the temperature and density at the origin. The Darcy velocities in the x and z directions are obtained from $v_x = -(k/\mu)(\partial P/\partial x)$ and $v_z = -(k/\mu)(\partial P/\partial z + \rho g)$:

$$v_x = -\frac{k\rho^0 geWB}{\mu}\frac{\sinh(n\pi z/W)}{\cosh(n\pi H/2W)}\sin\frac{n\pi(x+W/2)}{W} \tag{2.113}$$

and

$$v_z = \frac{k\rho^0 geWB}{\mu}\left(\frac{\cosh(n\pi z/W)}{\cosh(n\pi H/2W)} - 1\right)\cos\frac{n\pi(x+W/2)}{W}. \tag{2.114}$$

Note that in Eqs. (2.113) and (2.114), only the parameter of the horizontal temperature variation, B, appears. The vertical temperature variation parameter, C, is absent. In other words, the above two equations reveal that v_x and v_z are independent of the vertical temperature gradient and the vertical density difference; the thermal convection is proportional to the magnitude of the horizontal density gradient.

Now let us write the flow equation in terms of the stream function. In rotational flow in fluid mechanics, the stream function is often used instead of pressure. The stream function, ψ, is the complement of pressure, P. The value of the stream function is constant across a streamline;

there is no flow across a streamline. The stream function is defined by (Lu, 1973):

$$v_x = -\partial\psi/\partial z, \text{ and } v_z = +\partial\psi/\partial x. \tag{2.115}$$

Note that the stream function automatically satisfies the continuity equation, $\nabla \cdot \vec{v} = 0$, for the Boussinesq case where ρ is assumed constant in the continuity equation. The equation that defines the stream function is from the curl of velocity (Lu, 1973); the curl of velocity is called the vorticity. The vorticity provides a measure of rotational flow. The curl of velocity, $\nabla \times \vec{v}$, is given by (see Borisenko and Tarapov, 1968)

$$\nabla \times \vec{v} = \frac{\partial}{\partial z}(v_x) - \frac{\partial}{\partial x}(v_z). \tag{2.116}$$

Substituting for $v_x = -(k/\mu)\partial P/\partial x$ and $v_z = -(k/\mu)(\partial P/\partial z + \rho g)$, assuming μ and k constant, and combining the results with Eq. (2.115),

$$\frac{\partial^2\psi}{\partial x^2} + \frac{\partial^2\psi}{\partial z^2} = -\frac{k}{\mu}g\frac{\partial\rho}{\partial x}. \tag{2.117}$$

The expression for $\partial\rho/\partial x$ is $\partial\rho/\partial x \approx (d\rho/dT)(\partial T/\partial x)$, where the pressure dependence of ρ does not contribute to convection (see Problem 2.17). Using Eqs. (2.105) and (2.106),

$$(\partial\rho/\partial x) = -(\rho^0 e)(n\pi B)\sin\frac{n\pi(x + W/2)}{W}. \tag{2.118}$$

Combining Eqs. (2.117) and (2.118),

$$\frac{\partial^2\psi}{\partial x^2} + \frac{\partial^2\psi}{\partial z^2} = \left(\frac{k}{\mu}\right)(\rho^0 ge)(n\pi B)\sin\frac{n\pi(x + W/2)}{W}. \tag{2.119}$$

The boundary conditions for Eq. (2.119) are $\partial\psi/\partial z = 0$ and $\partial\psi/\partial x = 0$. Therefore $\psi(x = -W/2, z) = \psi(x = W/2, z) = \psi(x, z = -H/2)$, $\psi(x, z = H/2) = \text{constant}$. In other words, the boundary of the system is a streamline. We may assign $\psi = 0$, since one may choose any value for this streamline. Therefore, the boundary conditions of Eq. (2.119) are

$$\psi(x = -W/2, z) = \psi(x = W/2, z) = \psi(x, z = -H/2) = \psi(x, z = H/2) = 0. \tag{2.120}$$

The solution to Eqs. (2.119) and (2.120) is also in the form of Fourier series with only one term,

$$\psi(x, z) = +\frac{k\rho^0 ge W^2}{\mu\pi n}B\left(\frac{\cosh(n\pi z/W)}{\cosh(n\pi H/2W)} - 1\right)\sin\left(\frac{n\pi(x + W/2)}{W}\right). \tag{2.121}$$

The velocities on the streamline can be readily calculated from Eq. (2.115); the results are the same as from Eqs. (2.113) and (2.114). The advantage of stream function formulation is that streamlines are provided.

Example 2.11 at the end of this chapter provides numerical results for vertical and horizontal velocities in a 2D cross-sectional reservoir. Comments on the vertical and horizontal velocity profiles will be made in that example.

The final topic of this chapter is natural convection and diffusion in porous media with the objective of studying composition variation in hydrocarbon reservoirs. The understanding of irreversible phenomena facilitates such a study; the use of the Gibbs sedimentation equation, $d\mu_i = -M_i g dz$, which has been used by some authors in the literature, is not justified because of entropy production.

Natural convection and diffusion in porous media

In the formulation of thermal convection in porous media at steady state, it was demonstrated that the horizontal gradient of temperature drives the thermal convection. In fact, the driving force for both thermal convection (that is, the convection due to thermal gradient) and natural convection (that is, the convection due to both thermal gradient and composition gradient) is governed by $(\partial\rho/\partial x)$ at steady state. The expression for $(\partial\rho/\partial x)$ is given by

$$\frac{\partial\rho}{\partial x} = \left(\frac{\partial\rho}{\partial T}\right)\left(\frac{\partial T}{\partial x}\right) + \sum_{i=1}^{c-1}\left(\frac{\partial\rho}{\partial x_i}\right)\left(\frac{\partial x_i}{\partial x}\right). \tag{2.122}$$

When there is only bulk flow and diffusive fluxes are zero, the temperature gradient $(\partial T/\partial x)$ is the sole contributor to density gradient $(\partial\rho/\partial x)$. With diffusion, the second term on the right side of Eq. (2.122) becomes effective. The two terms on the right side of Eq. (2.122) may have the same sign, or opposite signs and may have different magnitudes relative to each other. Therefore, convection may enhance composition variation due to the effect of the second term. Such a behavior is not in line with the common belief in the literature that convection always reduces composition variation in hydrocarbon reservoirs.

We now present the equations that describe the combined effect of convection and diffusion in porous media. Let us assume that there are two components in the mixture, and that there is a single phase, either gas or liquid; the geometry is a two-dimensional rectangle (that is, $x - z$). It is also assumed that the temperature field is known. In hydrocarbon reservoirs, temperature data can be measured with modern tools. At steady state, one may write the continuity equation for components 1 and 2:

$$\nabla \cdot (\rho w_1 \vec{v} + \vec{J}_1) = 0 \tag{2.123}$$

$$\nabla \cdot (\rho w_2 \vec{v} + \vec{J}_2) = 0, \tag{2.124}$$

where $\vec{J}$ is the total diffusion mass flux. Adding Eqs. (2.123) and (2.124) results in the continuity equation for bulk flow given by Eq. (2.98), since $\vec{J}_1 + \vec{J}_2 = 0$ (see Problem 2.15). Since one of the three equations among Eqs. (2.98), (2.123) and (2.124) is a linear combination of the other two, we may use Eqs. (2.98) and (2.123). The velocity in porous media is given by Darcy's law (see Eq. (2.99)).

The diffusive mass flux of component 1 is given by the expression

$$\vec{J}_1 = C^x(x_1, T, P)\nabla x_1 + C^P(x_1, T, P)\nabla P + C^T(x_1, T, P)\nabla T, \tag{2.125}$$

which is an alternative form of Eq. (2.87). The coefficients of the above expression are

$$C^x = -\rho \frac{dw_1}{dx_1} D_{12} \left(\frac{\partial \ln f_1}{\partial \ln x_1} \right)_{T,P} \tag{2.126}$$

$$C^P = -\rho \frac{dw_1}{dx_1} D_{12} \frac{x_1}{RT} \left(\overline{V}_1 - \frac{M_1}{\rho} \right) \tag{2.127}$$

$$C^T = -\rho \frac{dw_1}{dx_1} D_{12} \left(\frac{k_T}{T} \right). \tag{2.128}$$

The mole fraction x_1 and the mass fraction w_1 are related to each other by

$$w_1 = \frac{x_1 M_1}{x_1 M_1 + (1 - x_1)M_2}. \tag{2.129}$$

The boundary conditions for the 2D problem are

$$J_{1,x} = v_x = 0 \text{ at } x = \pm W/2 \tag{2.130}$$

$$J_{1,z} = v_z = 0 \text{ at } x = \pm H/2, \tag{2.131}$$

where H and W are the height and width of the reservoir (see Fig. 2.23 for geometry). We also need to specify pressure and composition in one point in the reservoir, say at the origin at, $x = 0$ and $z = 0$ because the boundary conditions are of Neumann type (that is, derivatives are provided at the boundary). The problem formulation is then completed. The unknowns are pressure P and composition x_1, similarly to the 1D case, which does not allow for convection. The above system of equations in its general form has been solved by Riley and Firoozabadi (1998) using a method of successive approximation. In that work, the equations are first transformed into Poisson's equations. Riley and Firoozabadi studied the effect of convection on horizontal composition variation in a two-component mixture. Let us review some results from the work of these authors for the binary system C_1/nC_4. The pressure, temperature, and composition at the origin ($x = 0, z = 0$) are fixed: $P° = 110$ atm, $T° = 339$ K, and $x_1° = 0.20$. The dimensions of the reservoir are: $H = 150$ m and $W = 3000$ m. The horizontal and vertical temperature gradients are $\partial T/\partial x = 1$ K/300 m, and $\partial T/\partial z = -2$ K/30 m. Note that the geothermal temperature decreases towards the surface of the earth. The fluid viscosity is assumed constant in the reservoir, $\mu = 0.2$ cp—independent of pressure, temperature, and composition. Throughout the reservoir, the C_1/nC_4 mixture remains a liquid for all cases studied by Riley and Firoozabadi. The fractional porosity of the reservoir $\phi = 0.20$, and the permeability is varied to examine its effect on composition variation. Constant diffusion coefficient $D_{12} = 1.02 \times 10^{-9}$ m^2/s was used in all the calculations. The thermal diffusion ratio $k_T = 82,600[x_1(1 - x_1)/RT]$ was estimated from the work of Rutherford and Roof (1959), with R having units of atm $\cdot$ cm^3/mol $\cdot$ K, and T in K. No adjustments were made for the factor $(\partial \ln f_1/\partial \ln x_1)$ in k_T. The Peng-Robinson equation of state (1976) was used for the estimation of volumetric and thermodynamic properties.

A contour plot of the methane mole fraction is shown for $k = 0$ (that is, no convection) in Fig. 2.24. The contour interval is 0.1 mole % in this and subsequent contour plots. This figure shows that the constant composition contours are essentially straight lines. The horizontal composition variation is 0.91 mole % at $z = 0$. Figure 2.25 shows the composition contours for $k = 0.2$ md. The surprise is that with a small permeability and the introduction of convection, the horizontal composition variation increases. The expectation is that convection would decrease the composition variation. Figure 2.26 shows that by increasing the permeability to 10 md, the trend of compositional gradient is reversed. Figures 2.27 and 2.28 also show that the horizontal compositional gradients are decreasing with increasing permeability. Figures 2.26 to 2.28 also reveal that the curves have a negative slope near $z = 0$, indicating that the vertical concentration gradients are reversing sign in this region. The compositional changes in Figs. 2.26 to 2.28 are 0.55, 0.21, and 0.06 mole% which are roughly proportional to $1/k$. There are other features in all the compositional contours: (1) they become more vertical; (2) except near the side boundaries, the curves seem to have the same shape and spacing, indicating $(\partial x_1/\partial x)$ being constant; (3) the curves develop a subtle "S" shape, which indicates that the verti-

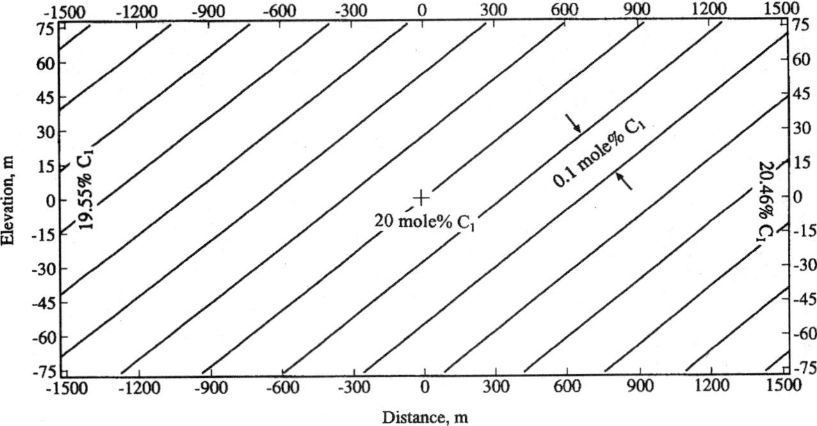

Figure 2.24 Methane mole fraction contours: $k = 0$ md (from Riley and Firoozabadi, 1998).

cal compositional gradient is not monotonic. Fig. 2.29 provides the variation of horizontal composition gradient of methane vs. permeability at the origin ($x = 0$, $z = 0$). The trend was also evident from contour plots in Figs. 2.24 to 2.28

Figures 2.30 and 2.31 show the vertical and horizontal velocities for $k = 10$ md. Figure 2.30 indicates that the vertical velocity is nearly

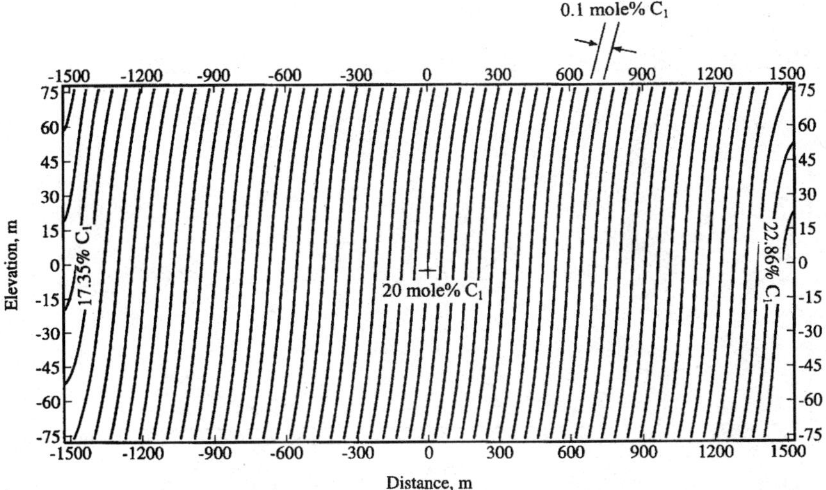

Figure 2.25 Methane mole fraction contours: $k = 0.2$ md (from Riley and Firoozabadi, 1998).

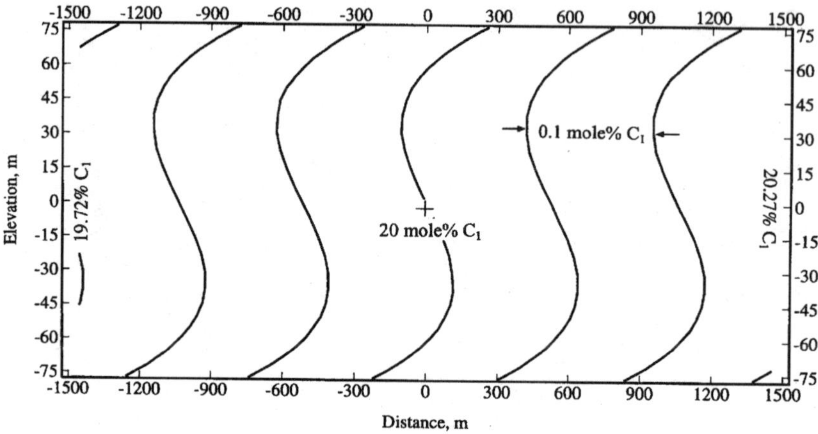

Figure 2.26   Methane mole fraction contours: $k = 10$ md (from Riley and Firoozabadi, 1998).

zero except close to the vertical boundaries. Figure 2.31 reveals that the horizontal velocity v_x varies linearly with z. (An explicit approximate expression for v_x will be derived shortly.) Figure 2.32 depicts the velocity contours where the features are the same as those in Figs. 2.30 and 2.31.

Now that the features of the solution are available, we can make appropriate assumptions to derive an approximate analytical expression for v_x. Let us define a stream function that can accommodate

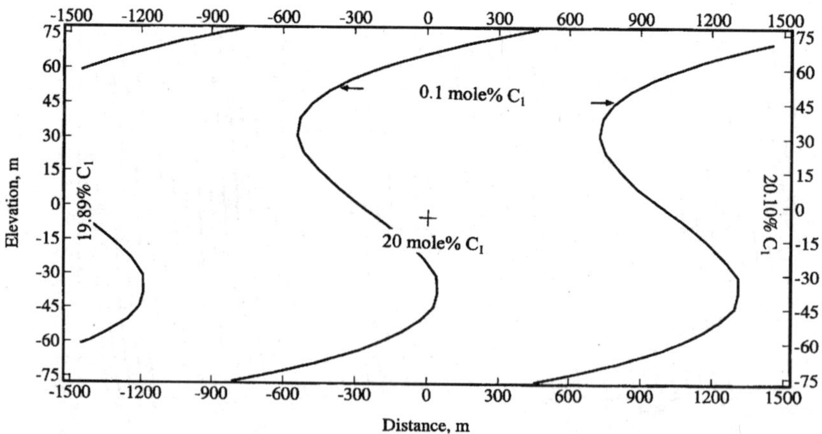

Figure 2.27   Methane mole fraction contours: $k = 30$ md (from Riley and Firoozabadi, 1998).

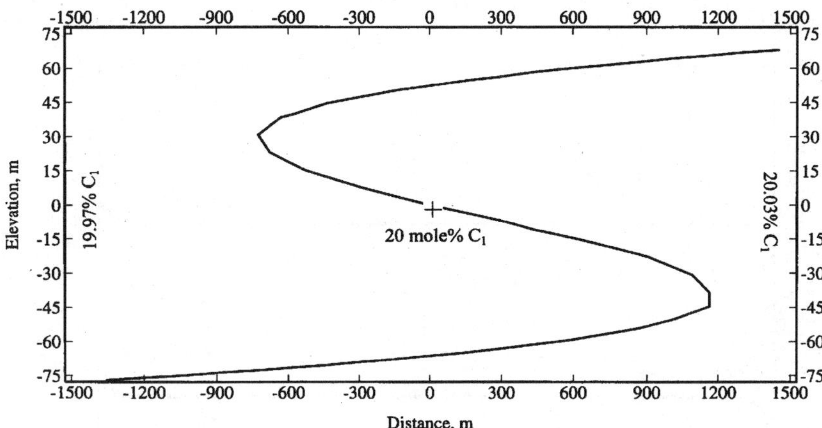

Figure 2.28   Methane mole fraction contours: $k = 100$ md (from Riley and Firoozabadi, 1998).

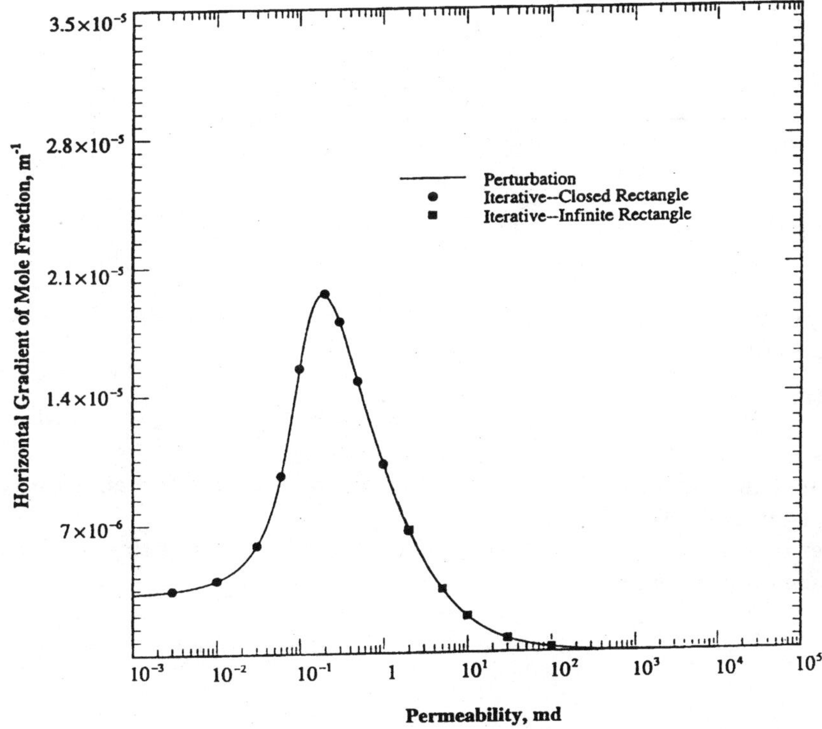

Figure 2.29   Horizontal composition gradient at $x = 0, z = 0$ vs. permeability (adapted from Riley and Firoozabadi, 1998).

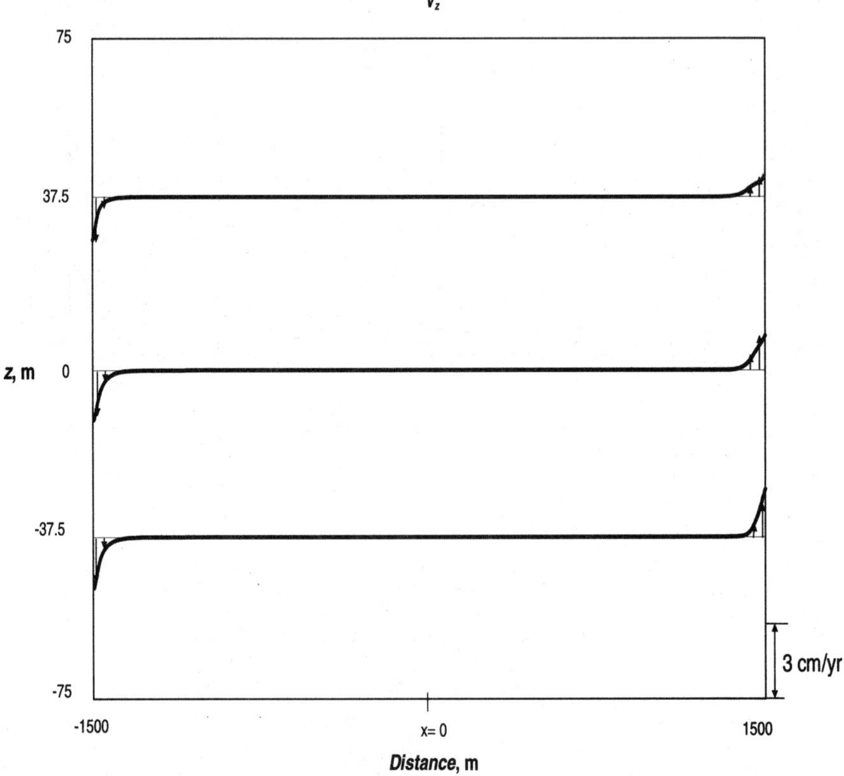

Figure 2.30 v_z vs. z at $x = -37.5, 0$, and 37.5 m: $k = 10$ md.

compressible flow. The modified stream function can be defined in terms of mass flux:

$$\rho v_x = -\partial \psi_m / \partial z, \ \text{ and } \ \rho v_z = + \partial \psi_m / \partial x, \tag{2.132}$$

Defined in this way, the stream function ψ_m satisfies the continuity equation, Eq. (2.98).

Let us take the curl of $\rho \vec{v}$ and write it first in terms of the modified stream function and then in terms of $\vec{v}$:

$$\nabla \times \rho \vec{v} = \frac{\partial}{\partial z}(\rho v_x) - \frac{\partial}{\partial x}(\rho v_z) = -\frac{\partial^2 \psi_m}{\partial z^2} - \frac{\partial^2 \psi_m}{\partial x^2} \tag{2.133}$$

and
$$\nabla \times \rho \vec{v} = \rho \left[\frac{\partial v_x}{\partial z} - \frac{\partial v_z}{\partial x} \right] + v_x \frac{\partial \rho}{\partial z} - v_z \frac{\partial \rho}{\partial x}. \tag{2.134}$$

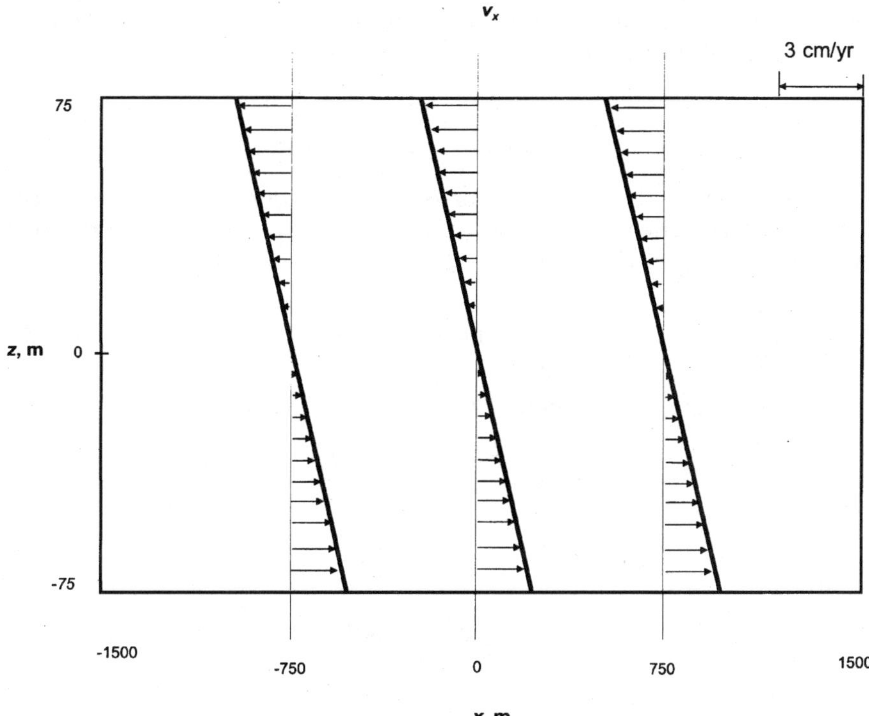

Figure 2.31 v_x vs. z at $x = -750, 0,$ and 750 m: $k = 10$ md.

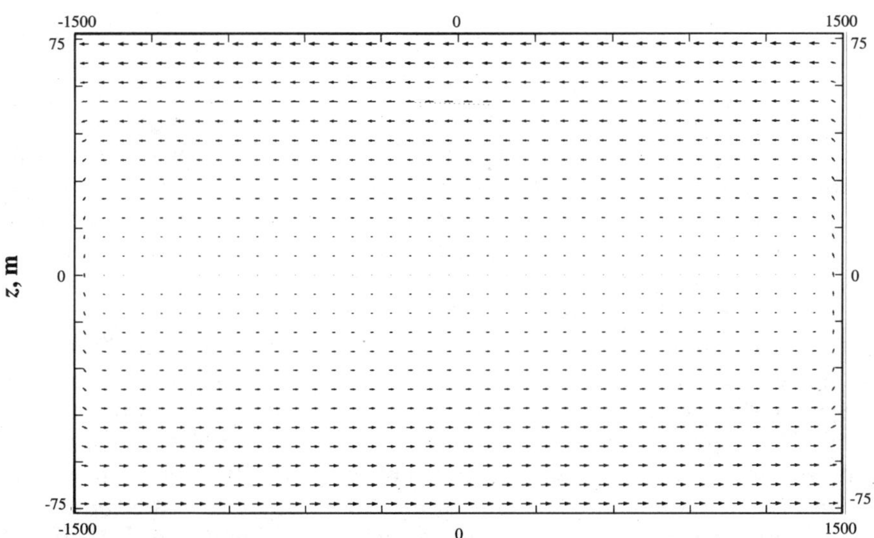

Figure 2.32 Velocity contours: $k = 10$ md.

Substitution of v_x and v_z in the expressions in the brackets from Darcy's law (Eq. (2.99)) and after simplification results in

$$\nabla \times \rho\vec{v} = \frac{kg}{2\mu}\frac{\partial\rho^2}{\partial x} + \frac{\rho v_x}{2\rho^2}\frac{\partial\rho^2}{\partial z} - \frac{\rho v_z}{2\rho^2}\frac{\partial\rho^2}{\partial x}. \tag{2.135}$$

The last term on the right side is very small when compared with the second term on the right; the second term is very small when compared with the first term; therefore, Eq. (2.135) can be simplified to

$$\nabla \times \rho\vec{v} \approx \frac{kg}{2\mu}\frac{\partial\rho^2}{\partial x}. \tag{2.136}$$

Combining Eqs. (2.133) and (2.136),

$$\frac{\partial^2\psi_m}{\partial z^2} + \frac{\partial^2\psi_m}{\partial x^2} \approx -\frac{kg}{2\mu}\frac{\partial\rho^2}{\partial x}. \tag{2.137}$$

Since $\partial^2\psi_m/\partial x^2 = \partial/\partial x(\rho v_z)$ and since $\rho v_z \approx 0$ except around the side boundaries,

$$\frac{\partial^2\psi_m}{\partial z^2} \approx -\frac{kg}{2\mu}\frac{\partial\rho^2}{\partial x}. \tag{2.138}$$

Also note that $v_x = 0$ at $z = 0$ (see Fig. 2.31). Therefore, the integration of Eq. (2.138) provides the following simple expression for horizontal velocity:

$$\boxed{v_x \approx \frac{kg}{\mu\phi}\frac{\partial\rho}{\partial x}\, z}. \tag{2.139}$$

Note that we have divided the Darcy velocity by fractional porosity in the last step to have true velocity. The $(\partial\rho/\partial x)$ term was previously expressed in terms of $(\partial T/\partial x)$ and $(\partial x_i/\partial x)$ (see Eq. (2.122)). Equation (2.139) applies to both thermal convection, where the convection is driven by $(\partial T/\partial x)$ as well as natural convection where flow is driven by $(\partial T/\partial x)$ and $(\partial x_1/\partial x)$. As was stated before, convection may weaken or enhance composition variation. Figure 2.33 provides a simple explanation of the change in composition due to convection. In this figure, the diagram on the right (Fig. 2.33a) shows the composition variation vs. depth with zero convection at $x = 0$ assuming that $C^T = 0$, and that C^P and C^x are not functions of temperature (see Eq. (2.125)). The thin line shows zero vertical compositional grading. Now allow for small values of ρv_x (proportional to z) as shown by thick line B. Assume that ρv_z is identically zero. Because of convection, the composition profile A cannot stay the same, otherwise the material balance for component 1

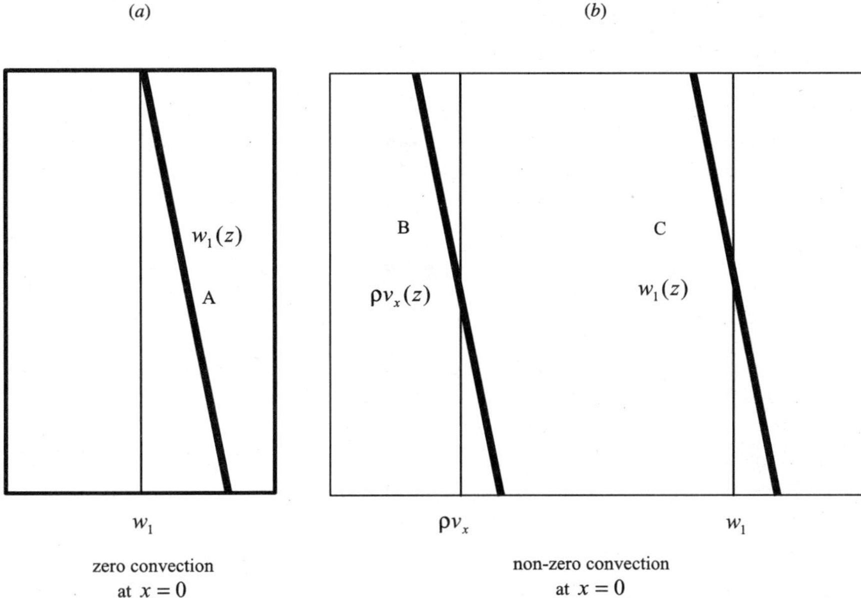

Figure 2.33   The effect of convection on horizontal composition variation (adapted from Riley and Firoozabadi, 1998).

is violated since $\int_{-H/2}^{H/2} \rho w_1 v_x dz$ is not zero. There will be more of component 1 moving to the right than to the left. In order for this not to happen, either the composition must redistribute vertically (see line C) or $\int_{-H/2}^{H/2} \rho w_1 v_x dz + \int_{-H/2}^{H/2} J_{1,x} dz = 0$. The second integral represents the contribution of horizontal diffusion (that is, horizontal composition gradient).

We would like to make the concluding comment that in ternary and multicomponent systems, a different trend in composition is expected to develop. In Figs. 2.24 to 2.28, the amount of C_1 decreases as we move to the top of the reservoir. This behavior, which implies that the heavier component, normal butane, floats on the top, is caused by the sign of k_T being positive; methane segregates to the hot bottom region. In multicomponent hydrocarbon systems, methane may have a higher concentration at the top. The topic of diffusion in multicomponent mixtures and the detailed derivation of diffusion flux expression from the entropy production given by Eq. (2.82) and the Onsager Reciprocal Relations (1931a and b) are presented by Ghorayeb and Firoozabadi (1998, 1999).

Examples and theory extension

Example 2.1

(a) Consider an isothermal column of an ideal liquid mixture of thickness h. Suppose the composition at the bottom of the liquid column is $x^0 = (x_1^0, x_2^0, \ldots, x_{c-1}^0)$. Derive the following expression for the composition at the top of the liquid column:

$$x_i = x_i^0 \exp\left[\frac{(\overline{\rho v_i} - M_i)}{RT} gh\right] \qquad i = 1, \ldots, (c-1),$$

where x_i and x_i^0 are the mole fractions at the top and bottom of the column, respectively, $\overline{\rho v_i}$ is the average of the product of the mass density and the molar volume of component i along the liquid column, and M_i is the molecular weight of component i.

(b) Consider an isothermal column of an ideal gas mixture of thickness h. Suppose the composition at the bottom of the gas column is $y^0 = (y_1^0, y_2^0, \ldots, y_{c-1}^0)$. Derive the following expression for the composition at the top of the gas column:

$$y_i = y_i^0 \exp\left[\frac{(\bar{M} - M_i)}{RT} gh\right] \qquad i = 1, \ldots, (c-1),$$

where $\bar{M}$ is the average molecular weight of the gas mixture along the gas column.

(c) Calculate the mole fraction of C_3 at $h = 1000$ and 5000 ft for a mixture of C_3 and nC_7. At $z = 0$, $x_{C_3}^0 = 0.50$ and $P^0 = 2000$ psia. Assume $T = 100°$F.

(d) Calculate the mole fraction of C_1 at $h = 1000$ and 5000 ft for a mixture of C_1 and N_2. At $z = 0$, $y_{C_1}^0 = 0.50$ and $P^0 = 2000$ psia. Assume $T = 100°$F.

Solution   (a) For an ideal-liquid solution,

$$\mu_i(T, P, x) = \mu_i^0(T, P) + RT \ln x_i. \qquad i = 1, \ldots, c$$

The derivatives of μ_i with respect to x_i and $x_j, j \neq i$, are

$$(\partial \mu_i / \partial x_i) = RT/x_i \qquad i = 1, \ldots, c$$

$$(\partial \mu_i / \partial x_j) = 0 \qquad i \neq j.$$

Now let us write Eq. (2.19a) for an ideal-liquid solution,

$$\sum_{j=1}^{c-1} \left(\frac{\partial \mu_i}{\partial x_j}\right)_{P, x_j} \left(\frac{dx_j}{dz}\right) = (\rho v_i - M_i)g \qquad i = 1, \ldots, c-1.$$

In the above equation, v_i is the molar volume of component i at temperature T and pressure P (that is, $\bar{V}_i = v_i$ for an ideal-liquid solution).

Combining the above three equations,

$$\left(\frac{RT}{x_i}\right)\left(\frac{dx_i}{dz}\right) = (\rho v_i - M_i)g \qquad i = 1, \ldots, c-1.$$

Integrating the above equation from $z = 0$ to $z = h$,

$$\boxed{x_i = x_i^0 \exp\left[\frac{\overline{\rho v_i} - M_i}{RT}gh\right]} \qquad i = 1, \ldots, c-1.$$

(b) For an ideal gas mixture,

$$\mu_i(T, P, y) = \mu_i^0(T, P^0) + RT \ln P y_i - RT \ln P^0.$$

Taking the derivatives of μ_i with respect to y_i and y_j, $i \neq j$,

$$(\partial\mu_i/\partial y_i) = RT/y_i \qquad i = 1, \ldots, c-1$$

$$(\partial\mu_i/\partial y_j) = 0 \qquad i \neq j.$$

For an ideal gas mixture, $\bar{V}_i = RT/P$ and $\rho = (PM/RT)$. Combining Eq. (2.19a) with the above equations,

$$\left(\frac{RT}{y_i}\right)\left(\frac{dy_i}{dz}\right) = (\bar{M} - M_i) \qquad i = 1, \ldots, c-1.$$

Integrating the above equation from $z = 0$ to $z = h$,

$$\boxed{y_i = y_i^0 \exp\left[\frac{(\bar{M} - M_i)}{RT}gh\right]} \qquad i = 1, \ldots, c-1.$$

(c) The specific volumes of C_3 and nC_7 at $100°$F and 2000 psia are (Starling, 1973)

$$v_{C_3}(100°\text{F}, 2000 \text{ psia}) = 0.03157 \text{ ft}^3/\text{lbm}$$

$$v_{nC_7}(100°\text{F}, 2000 \text{ psia}) = 0.02283 \text{ ft}^3/\text{lbm}.$$

The mass density of the mixture of C_3 and $nC_5(x_{C_3} = 0.5)$ is then $\rho_{mix} = 39.212$ lbm/ft^3. Since $R = 1545$ (ft.lbf)/(lbmole.R) and $g/g_c = 1$ (lbf/lbm), then

$$\frac{(\rho v_{C_3} - M_{C_3})}{RT}\left(\frac{g}{g_c}\right)h = 1.2 \times 10^{-5}h(\text{ft}).$$

Therefore, for $h = 1000$ ft, $x_{C_3} = 0.506$ and for $h = 5000$ ft, $x_{C_3} = 0.53$.

(d) The average molecular weight of the gas mixture at the bottom of the column is $\bar{M} = 22$ lbm/lbmole. Therefore, $(\bar{M} - M_{C_1})/RT(g/g_c)h = 6.9 \times 10^{-6}h(\text{ft})$.
At $h = 1000$ ft, $y_{C_1} = 0.503$ and at $h = 5000$ ft, $y_{C_1} = 0.517$.
Note that in the above two examples, there is very little variation of composition with height.

Example 2.2   Consider a mixture of methane and normal butane in a gravitational field. Use the PR-EOS to compute the composition of the C_1/nC_4 system at intervals of 1000, 2000, 3000, 4000, 5000 and 7000 ft below the reference depth for the following cases.

(a) At the reference depth, C_1 and nC_4 compositions are 27.27 and 72.73 mole percent, respectively. Pressure at the reference depth is 1300 psia. Temperature throughout the liquid column is assumed to be $220°\,F$.

(b) At a given reference depth, C_1 and nC_4 compositions are 88.88 and 11.12 mole%, respectively. Pressure at the reference depth is 514 psia. Temperature throughout the gas column is assumed to be $160°\,F$.

Solution   From Eq. (2.15),

$$f_1(P, y_1) = f_1^0(P^0, y_1^0) \exp\left[-\frac{M_1}{RT} gz\right]$$

$$f_2(P, y_1) = f_2^0(P^0, y_1^0) \exp\left[-\frac{M_2}{RT} gz\right],$$

where the subscripts 1 and 2 represent methane and normal butane, respectively. We need to use the expression for fugacity of component i in a mixture from Eq. (3.32) of Chapter 3 for the PR-EOS. In the above equations, the right side is known and the two unknowns P and y_1, on the left are to be found from the two nonlinear equations. One can use Newton's method for the solution. The procedure is to calculate the right side at every position z and then obtain P and y_1. The results of the calculation are shown in Figs. 2.34 and 2.35.

Example 2.3   Consider a binary system of C_1/nC_{10} with the following composition at $100°\,F$:

$$x_{C_1} = 99.894 \text{ mole}\% \text{ and } x_{nC_{10}} = 0.106 \text{ mole}\%$$

(a) Compute the dewpoint pressure of the above system in a PVT cell (measured $P_d \approx 1450$ psia).

(b) What would be the dewpoint pressure of the above fluid system for droplets of mean radii of 10, 1, 0.1, and 0.01 microns?

(c) What are the equilibrium liquid-phase composition of the above cases?

Data

$\sigma_{C_1/nC_{10}}$ at $100°\,F$ and 1500 psia $= 9.76$ dyne/cm. Assume the interfacial tension to be independent of the interface curvature in this problem.

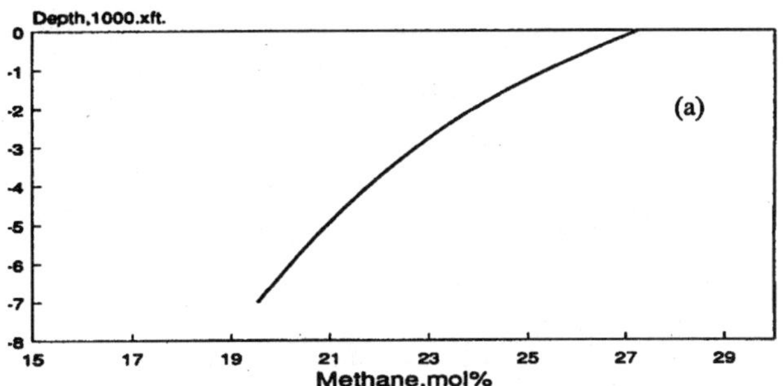

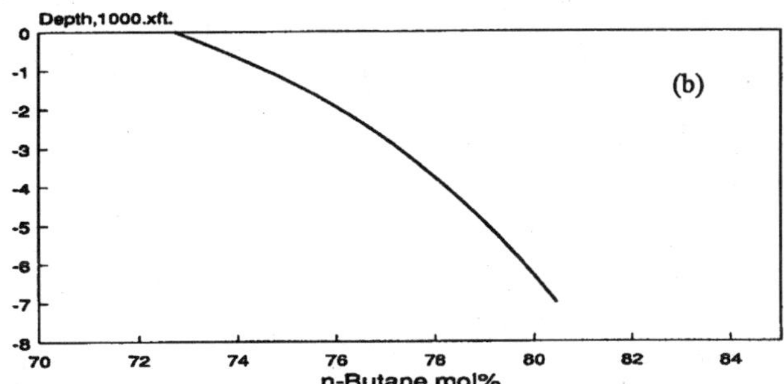

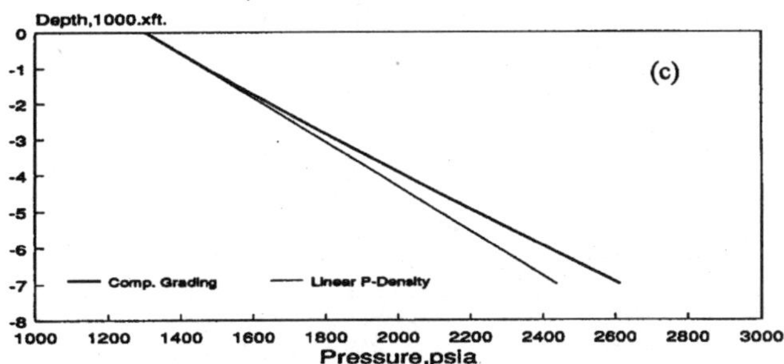

Figure 2.34   Compositional grading: C_1/nC_4 (liquid), $T = 220°F$.

Solution   At the dewpoint,

$$f_i^L(T, P^L, x_1) = f_i^V(T, P_d^V, z_1) \qquad i = 1, 2$$

$$P_d^V - P^L = 2\sigma/r$$

$$\sum_{i=1}^{2} x_i = 1.$$

The fugacity expression (see Eq. (3.32) of Chapter 3) is used to calculate the fugacities. Newton's method can be used to solve the above system of nonlinear

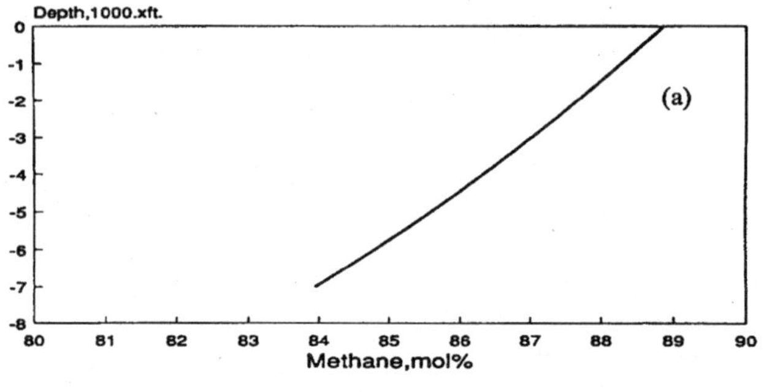

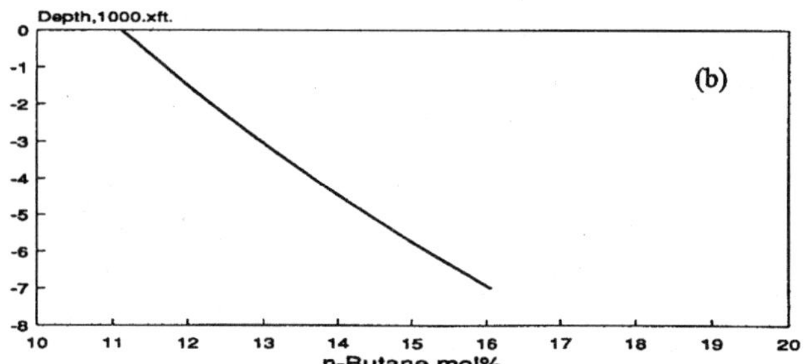

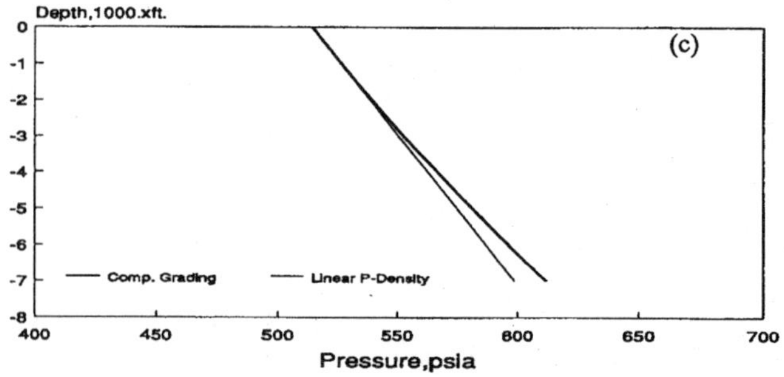

Figure 2.35   Compositional grading: C_1/nC_4 (gas), $T = 160°F$.

equations to obtain P^L, P_d^V, and x_1 at a given T and z_1. Numerical results from the problem are shown in Fig. 2.36. Note that there is hardly any increase in dewpoint pressure when $r > 0.1$ microns. When $r < 0.1$ microns, dewpoint pressure increases with an increase in curvature. Similarly, the effect of interface curvature on the liquid phase composition is negligible when $r > 0.1$ microns. The results presented in Fig. 2.36 suggest that the porous medium unless it is very tight, may not have a significant effect on equilibrium. The implication from this problem is that when condensation takes place, the liquid phase may first form in smaller pores in porous media.

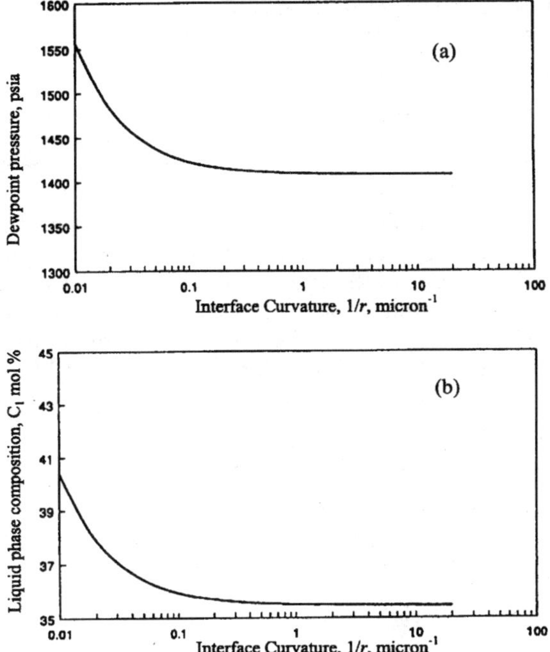

Figure 2.36 Effect of interface curvature on the dewpoint pressure and liquid-phase composition of the C_1/nC_{10} system (99.894 mol% C_1) at 100°F.

Example 2.4 Use Eq. (2.53) to calculate the vapor pressure of a bubble and a droplet of water for radii of $10^{-4}, 10^{-5}, 5 \times 10^{-6}$, and 2×10^{-6} cm at 20°C. Assume the interfacial tension of water-vapor to be 72 dyne/cm.

Solution The expressions for the vapor pressure of a pure substance for bubble and droplet are

$$P^G = P^\infty \exp\left[-\frac{2\sigma}{r}\frac{v^L}{RT}\right]$$

$$P^G = P^\infty \exp\left[+\frac{2\sigma}{r}\frac{v^L}{RT}\right],$$

respectively. Note that the curvature of a bubble or a droplet is related to its diameter by $1/r = \cos\theta/d$ where θ is the contact angle. For a bubble, $\theta = 0$ and $\cos\theta = 1$. For a droplet, $\theta = 180°$ and $\cos\theta = -1$.

Using the above two equations with $v^L = 18$ cm³/gmole and $R = 82.06$ (atm·cm³)/(gmole·K), the following results are obtained:

TABLE 2.1   Vapor pressure ratio, P^G/P^∞, of water at 20°C

r, cm	∞	10^{-4} cm	10^{-5} cm	5×10^{-6} cm	2×10^{-6} cm
Bubble	1	0.999	0.989	0.979	0.948
Droplet	1	1.001	1.010	1.022	1.055

Note that the vapor pressure of a bubble is less than the vapor pressure for a flat interface. On the other hand, the vapor pressure of a droplet is more than the vapor pressure for a flat interface. In the above solution, we have assumed that the interfacial tension is independent of the curvature, even at $r < 10^{-5}$ cm. For a bubble, the interfacial tension increases as the curvature increases, whereas it decreases for a droplet as the curvature increases (Defay and Prigogine, Chapter XV, 1966). However, this trend is only true for pure components; for mixtures, the effect of curvature change on the interfacial tension may be different (see Example 3.6, Chapter 3)

Example 2.5   The work required to create a gas bubble of radius r for a pure substance is given by

$$W = (1/3)\sigma\mathcal{A}^b,$$

where W is the work, σ is the interfacial tension, and $\mathcal{A}^b$ is the surface area of the spherical bubble. Derive the above equation.

Solution   In order to create a gas bubble in a liquid phase, one needs (1) to displace the liquid by the gas bubble and (2) to create the interfacial area of the gas bubble (see Eq. (2.20) and Fig. 2.37a). Therefore, the work required to create a gas bubble consists of two parts:

$$W = -\int^{V^b} \Delta P dV + \sigma\mathcal{A}^b,$$

where ΔP is the difference in pressure in the gas bubble, P^b, and the liquid surrounding the gas bubble, P^L; $\Delta P = P^b - P^L$. Assuming that $P^b - P^L$ is constant,

$$W = -(P^b - P^L)V^b + \sigma\mathcal{A}^b,$$

where V^b is the volume of the gas bubble. Since for a spherical gas bubble, $P^b - P^L = 2\sigma/r$ (see Eq. (2.34)) and $V^b = (1/3)r\mathcal{A}^b$, then

$$\boxed{W = (1/3)\sigma\mathcal{A}^b}.$$

The above fundamental equation is of importance in nucleation theory. Nucleation is a phenomenon of interest in many engineering applications, including metallurgical processes and solution-gas drive in porous media for oil production (Firoozabadi and Kashchiev, 1996).

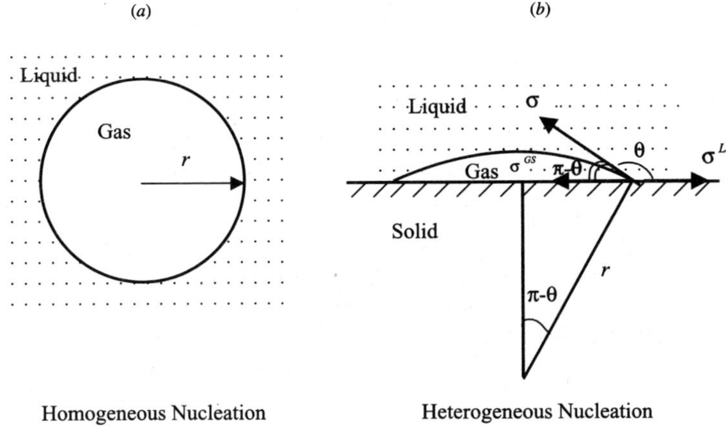

Figure 2.37   Homogeneous and heterogeneous bubble nucleation.

Example 2.6   Consider the creation of a new gas phase that consists of a segment of a spherical bubble as shown in Fig. 2.37b. Derive the expression for the work required to create the new gas phase, given by

$$W_{het} = W\phi_{het}$$

and
$$\phi_{het} = \tfrac{1}{4}(2 - \cos\theta)(1 + \cos\theta)^2,$$

where θ is the contact angle measured through the liquid phase.

Solution   The solution to this problem is similar to Example 2.5, with the difference that the presence of the solid should be considered (see Fig. 2.37).

The work required to displace the liquid is the volume of the spherical segment, V^b, multiplied by $-(P^b - P^L)$. The volume of the spherical segment shown in Fig. 2.37 is

$$V^b = \tfrac{1}{3}\pi r^3(2 + 3\cos\theta - \cos^3\theta).$$

There are now two distinct surface areas created; one is the interface between the gas and the liquid, given by

$$\mathcal{A}^{GL} = 2\pi r^2(1 + \cos\theta),$$

and the other interface is between the gas and the solid, given by

$$\mathcal{A}^{GS} = \pi r^2(1 - \cos^2\theta).$$

The work terms are given by

$$W_{het} = -\tfrac{1}{3}\pi r^3 (2 + 3\cos\theta - \cos^3\theta)(P^b - P^L) + 2\pi r^2 (1 + \cos\theta)\sigma$$
$$+ \pi r^2 (1 - \cos^2\theta)(\sigma^{GS} - \sigma^{LS}).$$

In the above expression, the first term on the right represents the work of liquid displacement by gas, the second term represents the work of creating the surface area between the gas and the liquid, and the last term represents the work of creating the interface between the gas and the solid. σ is the gas–liquid interfacial tension, and σ^{GL} and σ^{LS} are the interfacial tensions between the gas and solid and liquid and solid, respectively. The gas–solid and liquid–solid interfacial tensions from a simple force balance are related through

$$\sigma^{GS} - \sigma^{LS} = \sigma\cos\theta.$$

Combining the above equations and the results from Example 2.5,

$$\boxed{W_{het} = W\phi_{het}},$$

where
$$\boxed{\phi_{het} = \tfrac{1}{4}(2 - \cos\theta)(1 + \cos\theta)^2}.$$

In nucleation theory, W_{het} is the heterogeneous nucleation work as compared with the homogeneous nucleation work W. The heterogeneous nucleation refers to the fact that the gas bubble forms at the interface of a liquid and a solid phase. Note that ϕ_{het} is simply W_{het}/W, which varies in the range of 0 and 1.

Example 2.7   Derive the following relationships for thermodynamic functions U, A, and G of an open system with a curved interface:

$$U = TS - PV + \sigma\mathcal{A} + \sum_{i=1}^{c} n_i\mu_i$$

$$A = -PV + \sigma\mathcal{A} + \sum_{i=1}^{c} n_i\mu_i$$

$$G = \sigma\mathcal{A} + \sum_{i=1}^{c} n_i\mu_i \quad \text{and} \quad \sigma = \left(G - \sum_{i=1}^{c} n_i\mu_i\right)/\mathcal{A}.$$

Show that the interfacial tension, σ, is given by

$$\sigma = (\partial U/\partial\mathcal{A})_{S,V,\underline{n}} = (\partial A/\partial\mathcal{A})_{T,V,\underline{n}} = (\partial G/\partial\mathcal{A})_{T,P,\underline{n}}.$$

Solution The internal energy of an open system with a curved interface is given by (see Eq. (2.21))

$$dU = TdS - PdV + \sum_{i=1}^{c} \mu_i dn_i + \sigma d\mathcal{A}$$

and, therefore, $U = U(S, V, n_1, n_2, \ldots, n_c, \mathcal{A})$. Note that for an open system with a curved interface, the extensive variables S, V, $n_1, \ldots, n_c$ and $\mathcal{A}$ define the system. The bulk phase, which excludes the interface, is still defined by S, V, and $n_1, \ldots, n_c$. The intensive variables T, P, μ_i, and σ are conjugates to the extensive variables S, V, $n_1, n_2, \ldots, n_c$ and $\mathcal{A}$. Writing the total differential of $U(S, V, n_1, n_2, \ldots, n_c, \mathcal{A})$ and comparing the results with the above equation,

$$\sigma = (\partial U / \partial \mathcal{A})_{S, V, \underline{n}},$$

which defines the interfacial tension as the change in internal energy when the interfacial area, $\mathcal{A}$, of a closed system is changed at constant S and V. Since σ is intrinsically positive, the internal energy U increases as the interfacial area increases.

From the property that U is a first-order homogeneous function (see Chapter 1),

$$U(\lambda S, \lambda V, \lambda n_1, \ldots, \lambda n_c, \lambda \mathcal{A}) = \lambda U(S, V, n_1, \ldots, n_c, \mathcal{A}),$$

where λ is a positive parameter (see Eq. (1.33) of Chapter 1). Differentiating the above equation with respect to λ and performing the algebra similarly to Eqs. (1.33) to (1.35) of Chapter 1, one obtains

$$U = TS - PV + \sigma \mathcal{A} + \sum_{i=1}^{c} n_i \mu_i.$$

The expressions for A, G, and the corresponding expressions for σ are obtained in a similar manner. In the Problem section, the expressions for A and G are asked to be obtained using the Legendre transformation.

Example 2.8 Derive the following Gibbs-Duhem equation for an open phase with a curved interface:

$$d\sigma = -(S/\mathcal{A})dT + (V/\mathcal{A})dP - \sum_{i=1}^{c} (n_i/\mathcal{A})d\mu_i. \qquad \text{(E.2.8)}$$

Note that the Gibbs-Duhem equation for a bulk phase does not depend on the interface; it is the same whether the interface is flat or curved. In other words, the Gibbs-Duhem equation of the bulk phase, b, is always

$$-S^b dT + V^b dP - \sum_{i=1}^{c} n_i^b d\mu_i = 0.$$

Solution   The internal energy of an open phase with a curved interface when the interface is counted as part of the phase is given by

$$U = TS - PV + \sigma\mathcal{A} + \sum_{i=1}^{c} n_i\mu_i.$$

Writing the total differential of the above equation and comparing the results with the equation for dU in Example 1.7, one obtains

$$\boxed{d\sigma = -(S/\mathcal{A})dT + (V/\mathcal{A})dP - \sum_{i=1}^{c}(n_i/\mathcal{A})d\mu_i}.$$

The above equation provides the relationship between the intensive variables $\sigma, T, P, \mu_1, \ldots, \mu_c$ of a phase with a curved interface when the interface is counted as part of that phase.

Example 2.9   *Gibbs Phase Rule for curved interfaces*   Derive the phase rule for a composite system of p phases and c components with curved interfaces; $F = c + 1$ where F is the number of degrees of freedom. If some of the interfaces are flat, then $F = c + 1 - I$, where I is the number of flat interfaces between the bulk phases.

Solution   The Gibbs Phase Rule for a flat interface between the phases, which was established in Example 1.9 of Chapter 1, is based on the assumption that the PdV work is the only mode of work. As we have seen in this chapter, the equilibrium conditions for systems with curved interfaces and under the influence of gravity are different from the equilibrium conditions of systems with flat interfaces and negligible gravity. For systems with curved interfaces and also with gravity effect, the Gibbs Phase Rule should be modified. In this example, we will only consider the effect of the curved interface.

The criteria of equilibrium of multicomponent systems with c components and p phases with curved interface between phase 1 and the other phases are (see Fig. 2.38)

$$T^{(1)} = T^{(2)} = \cdots = T^{(p)}$$
$$P^{(1)} - P^{(2)} = \sigma^{(1,2)}J^{(1,2)}$$
$$\vdots$$
$$P^{(1)} - P^{(p)} = \sigma^{(1,p)}J^{(1,p)}$$
$$\mu_1^{(1)}(T^{(1)}, P^{(1)}, x_1^{(1)}, \ldots, x_{c-1}^{(1)}) = \cdots = \mu_1^{(p)}(T^{(p)}, P^{(p)}, x_1^{(p)}, \ldots, x_{c-1}^{(p)})$$
$$\vdots$$
$$\mu_c^{(1)}(T^{(1)}, P^{(1)}, x_1^{(1)}, \ldots, x_{c-1}^{(1)}) = \cdots = \mu_c^{(p)}(T^{(p)}, P^{(p)}, x_1^{(p)}, \ldots, x_{c-1}^{(p)}),$$

where $\sigma^{(1,j)}$ and $J^{(1,j)}$ are the surface tension and the mean curvature of the interfaces between phase 1 and phase j ($j = 2, \ldots, p$), respectively. There are $(p-1)$ equations for temperature, $(p-1)$ equations for pressure relationships, and $c(p-1)$ equations for chemical equilibria. The total number of equations are, therefore, $(p-1)(c+2)$. The intensive variables for phase $j(j = 2, \ldots, p)$

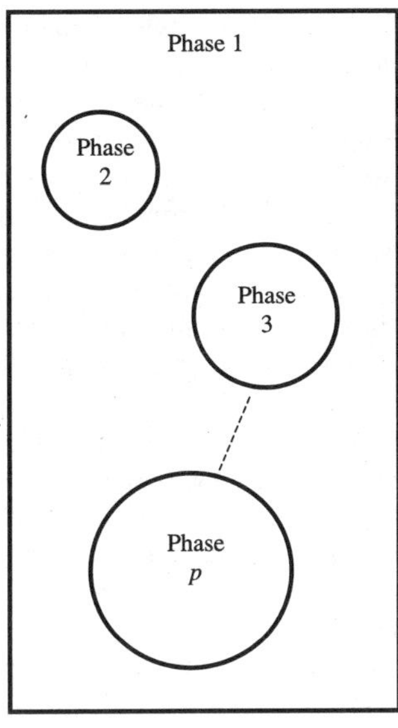

Figure 2.38 Sketch of phases with curved interfaces.

are $T^{(j)}$, $P^{(j)}$, $\mu_i^{(j)}(i = 1, \ldots, c)$, and $\sigma^{(1,j)}$. For phase 1, the intensive variables are $T^{(1)}$, $P^{(1)}$, and $\mu_i^{(1)}(i = 1, \ldots, c)$. Note that we have assigned the interfaces to phases $j = 2, \ldots, p$, which is consistent with our derivation of Eqs. (2.25) and (2.26). In phase 1, only $(c + 1)$ intensive variables among a total of $(c + 2)$ intensive variables are independent (see the Gibbs-Duhem Eq. (1.40) of Chapter 1). Of the $(c + 3)$ intensive variables $T^{(j)}$, $P^{(j)}$, $\mu_i^{(j)}$, and $\sigma^{1,j}$ for phases $j = 2, \ldots, p$, only $(c + 2)$ intensive variables are independent (see Example 2.8 above). Therefore, the total number of independent intensive variables of the composite system is: $(p - 1)(c + 2) + (c + 1)$. The number of degrees of freedom is given by the expression F = number of independent intensive variables − number of independent equations. After substitution in the above expression,

$$\boxed{F = c + 1}.$$

If there are I flat interfaces between the phases, one can readily show that

$$\boxed{F = c + 1 - I}.$$

In the above two equations, the intensive variables are temperature, pressure, chemical potential, and interfacial tension. As we saw in Example 1.9 of Chapter 1, instead of the chemical potentials, one may use the mole fractions as intensive variables. It is also more convenient and meaningful to use the interface curvature instead of the interfacial tension as the intensive variable. For a pure substance, as the interface curvature changes, the interfacial tension also changes. For an isothermal process of a single-component system when the curvature of the bubble in a bulk liquid phase at constant pressure changes, the interfacial tension also changes because of (1) curvature change and (2) gas-phase pressure change. Similarly, for a bubble in a multicomponent system at constant temperature, when the bulk liquid phase is held at constant pressure and composition, if the interface curvature changes, the interfacial tension may also change because of (1) curvature change, (2) gas-phase pressure change, and (3) gas-phase composition change. For a pure substance, the interfacial tension is a weak function of curvature when, say, $r > 10^{-5}$ cm. For mixtures, the effect of curvature change on interfacial-tension change becomes more complicated because of both curvature and composition effects (see Example 3.6 in Chapter 3). In the context of the Phase Rule, if we specify the interface curvature, the interfacial tension is also specified because other intensive variables such as pressure, temperature, and composition are among the variables of the Phase Rule.

For systems in which there is a three-phase contact line between the phases as a result of a solid phase, the concept of contact angle is introduced. For such systems, the Phase Rule remains the same (Li *et al.*, 1989). For highly curved interfaces where the thickness of the heterogeneous region between the phases is not small compared to r, there are other considerations in the derivation of the Phase Rule (Li *et al.*, 1989; Li, 1994).

Let us give two examples for the use of the Phase Rule for curved interfaces. First, consider a single component gas–liquid system with a curved interface between the gas and liquid phases; for this system $F = 2$. Therefore, we can fix two intensive variables, say temperature and pressure of the vapor phase. Then the system is fully defined. We can also specify temperature and curvature. Note that for a single-component gas–liquid system with a flat interface, $F = 1$. If we fix the temperature, the vapor pressure is fixed. In the second example, we consider a two-component two-phase system with a curved interface between the gas and liquid phases, $F = 3$. Unlike the system with a flat interface, specifying the temperature and pressure of the gas phase does not specify the system. We also need to specify the interface curvature.

Example 2.10 Derive the following equation, which describes temperature variation in porous media saturated with a single-phase fluid mixture at steady state. The equation is based on the negligible effect of inter-diffusion on heat transfer.

$$\nabla \cdot (K\nabla T) - \rho \vec{v} c_p \nabla T = 0,$$

where K is the thermal conductivity of the porous medium, and ρ and c_P are the fluid mass density and the heat capacity per unit mass of the fluid, respectively.

Solution We will derive the energy equation in a 2D $x - z$ system and then extend it to 3D. Consider a volume element of Δz and Δx dimensions sketched in Fig. 2.39. At steady state, energy flux in at $z -$ energy flux out at $(z + \Delta z) +$ energy flux in at $x -$ energy flux out at $(x + \Delta x) = 0$. The energy flux consists of two terms: conduction and convection. Note that we have neglected the energy flux due to diffusion.

Therefore,

$$q_x^{cond}\Delta z \times 1|_x - q_x^{cond}\Delta z \times 1|_{x+\Delta x} + q_z^{cond}\Delta x \times 1|_z - q_z^{cond}\Delta x \times 1|_{z+\Delta z}$$
$$+ q_x^{conv}\Delta z \times 1|_x - q_x^{conv}\Delta z \times 1|_{x+\Delta x} + q_z^{conv}\Delta x \times 1|_z - q_z^{conv}\Delta x \times 1|_{z+\Delta z} = 0.$$

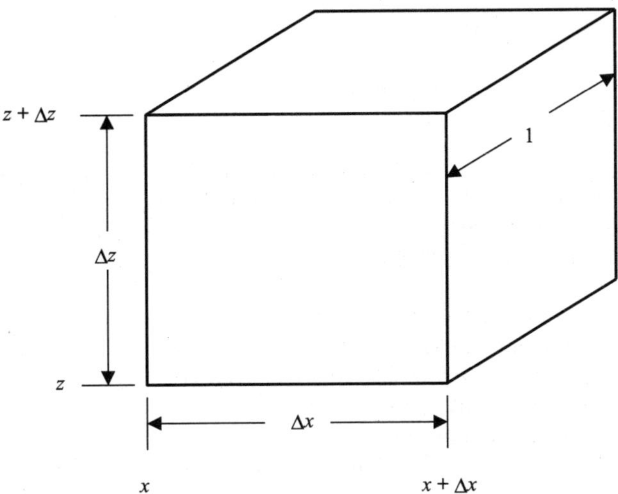

Figure 2.39   Volume element of Δx, Δz and 1.

In the above equation, $\vec{q}^{\,cond}$ represents the heat flux by conduction through the porous medium, and $\vec{q}^{\,conv}$ represents the heat flux by convection of the fluid mixture. The y direction is assumed to be of unit length.

Dividing the above expression by $\Delta x \, \Delta z$ and taking the limit as $\Delta x \to 0$ and $\Delta z \to 0$, one obtains

$$\frac{\partial q_x^{cond}}{\partial z} + \frac{\partial q_z^{cond}}{\partial x} + \frac{\partial q_x^{conv}}{\partial z} + \frac{\partial q_z^{conv}}{\partial x} = 0.$$

But $\vec{q}^{\,cond} = -K\nabla T$ and $\vec{q}^{\,conv} = \rho \vec{v} h$, where h is enthalpy per unit mass. The conduction and convection in the x-direction are $q_x^{cond} = -K_x \partial T / \partial x$ and $q_x^{conv} = \rho v_x h$, respectively. Therefore, the energy-flux expression in 2D is

$$\frac{\partial}{\partial x}\left[-K_x \frac{\partial T}{\partial x}\right] + \frac{\partial}{\partial z}\left[-K_z \frac{\partial T}{\partial z}\right] + \frac{\partial}{\partial x}[\rho v_x h] + \frac{\partial}{\partial z}[\rho v_z h] = 0.$$

The last two terms can be expressed as

$$\frac{\partial}{\partial x}[\rho v_x h] + \frac{\partial}{\partial z}[\rho v_z h] = \rho v_x c_P \frac{\partial T}{\partial x} + \rho v_z c_P \frac{\partial T}{\partial z} + h \frac{\partial}{\partial x}(\rho v_x) + h \frac{\partial}{\partial z}(\rho v_z).$$

Note that from continuity equation, Eq. (2.98),

$$h \frac{\partial}{\partial x}(\rho v_x) + h \frac{\partial}{\partial z}(\rho v_z) = h \nabla \cdot (\rho \vec{v}) = 0.$$

Therefore,

$$\frac{\partial}{\partial x}\left[-K_x \frac{\partial T}{\partial x}\right] + \frac{\partial}{\partial z}\left[-K_z \frac{\partial T}{\partial z}\right] + \rho v_x c_P \frac{\partial T}{\partial x} + \rho v_z c_P \frac{\partial T}{\partial z} = 0.$$

The above equation can be written in vector form as

$$\nabla \cdot (K\nabla T) - \rho \vec{v} c_p \nabla T = 0.$$

Example 2.11   Consider a 2D cross-sectional reservoir of height $H = 1500\,$ft and width $W = 10000$ ft. The values of various parameters are (1) $C = -2°\text{R}/100\,$ft, $B = 5°\text{R}/10000\,$ft   (see   Eqs.   (2.106)   and   (2.107)),   (2) $\mu = 0.2\,$cp, $e = 1.67 \times 10^{-3}/\text{R}$, $\rho^0 = 32.2\,\text{lbm/ft}^3$, $T^0 = 150°\text{F}$, and (3) $k = 1000$ md, and $\phi = 0.25$.

First assume $n = 1$ and $n = 3$ in Eq. (2.106) and compare the temperature profiles from Eqs. (2.106) and (2.107) by plotting T vs. x at $z = 0$. Then use Eqs. (2.113) and (2.114) to calculate v_x and v_z and plot the velocity contours for $n = 1$ in Eq. (2.106) and also plot v_x vs. z at $x = -5000$ ft, 0 ft, and $+5000$ ft, and v_z vs. x at $z = -125$ ft, 0 ft, and 125 ft.

In the next step, assume $n = 3$ in Eq. (2.106) and then calculate v_x and v_z and plot the velocity contours.

Solution   The solution for velocity requires the evaluation of $(k/\mu)(\rho^0 ge\,WB)$ in Eqs. (2.113) and (2.114). The rest is straightforward.

The multiplier $(k/\phi\mu)(\rho^0 ge\,WB) = 90\,$ft/year. Note that with this multiplier one obtains the true velocity, not the Darcy velocity.

Figure 2.40 depicts the temperature plots, indicating that Eqs. (2.107) and (2.106) with $n = 1$ give comparable results. Figure 2.41 depicts the velocity contours. Note that the x- and z-axes have different scales. This figure implies that horizontal velocities are much greater than vertical velocities. Figure 2.42

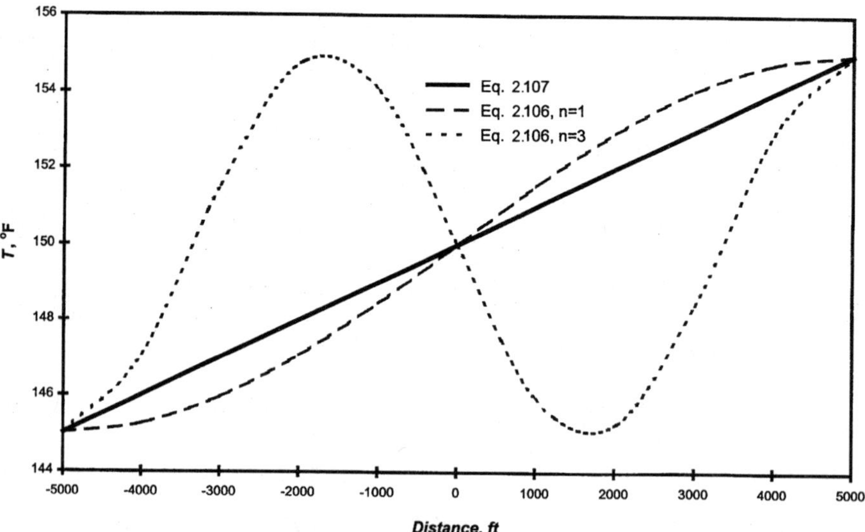

Figure 2.40   Temperature (at $z = 0$) vs. x using the linear and cosine functions.

depicts the horizontal velocity vs. z at $x = -2500$ ft, 0 ft, and $+2500$ ft. An interesting feature of the plot is that v_x varies linearly with z, which was not transparent from Eq. (2.113). Figure 2.43 shows the vertical velocity vs. x at $z = -125$ ft, 0 ft, and 125 ft. Note that vertical velocities are orders of magnitude less than horizontal velocities.

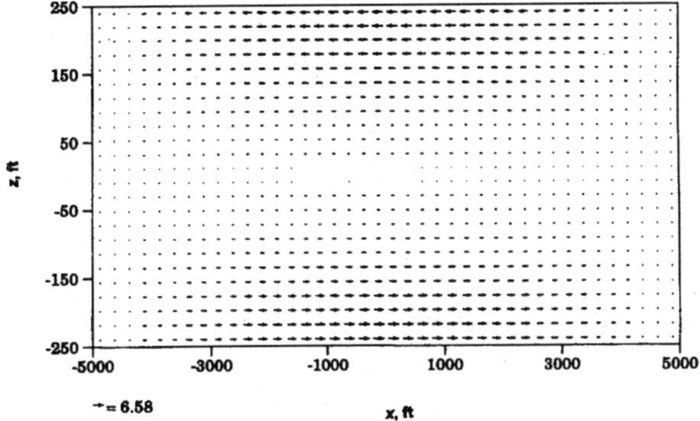

Figure 2.41　Velocity contours due to thermal convection: Example 2.11 ($n = 1$, Eq. (2.106)).

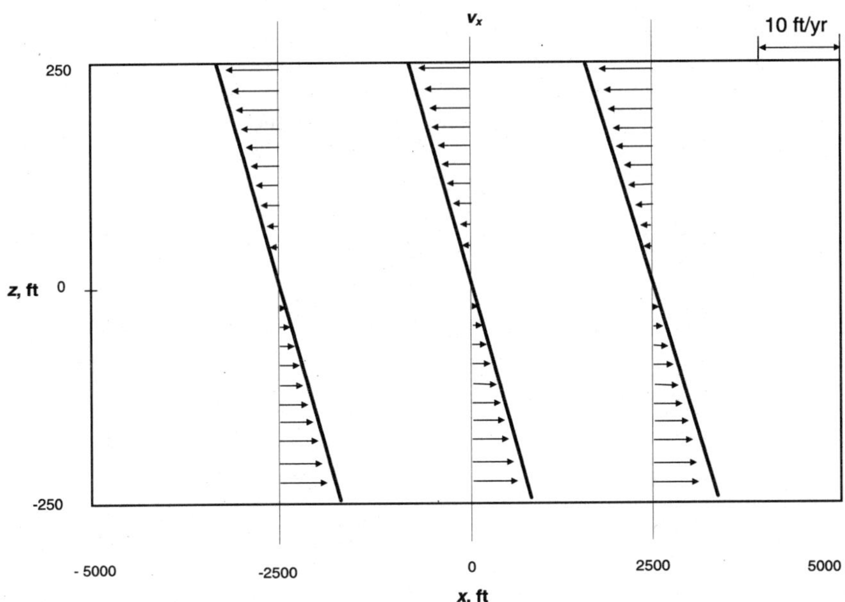

Figure 2.42　v_x vs. z at $x = -2500, 0$, and 2500 ft: Example 2.11 ($n = 1$).

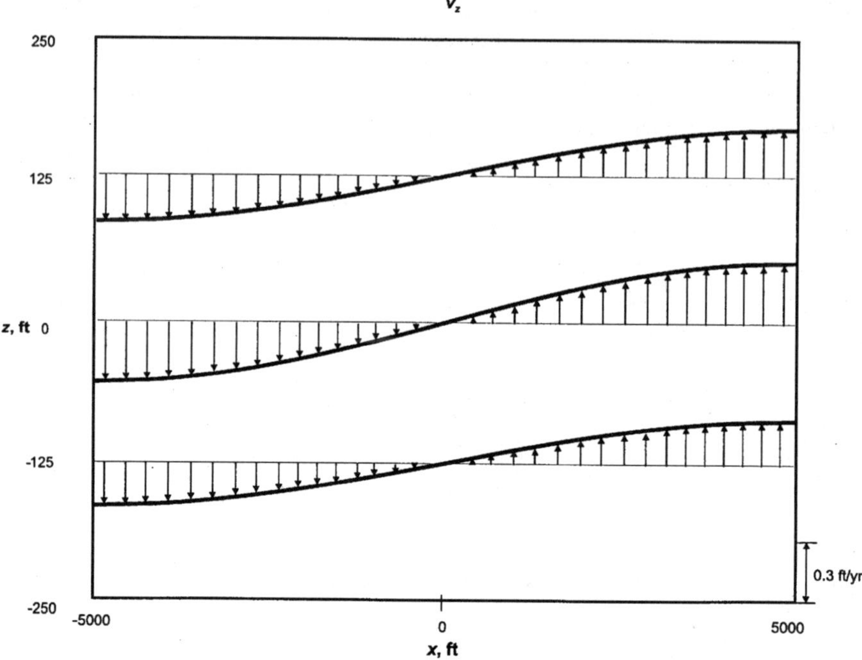

Figure 2.43 v_z vs. x at $z = -125, 0$, and 125 ft: Example 2.11 ($n = 1$).

Figure 2.40 also shows the temperature profiles for $n = 3$. The factor n in the cosine term dictates the number of cells in the x-direction. This is because the flowlines will turn whenever $\partial\rho/\partial x$ is zero. Figure 2.44 provides the velocity contours for $n = 3$, showing that there are indeed three cells. Note the alternation in the velocity directions of the neighboring cells.

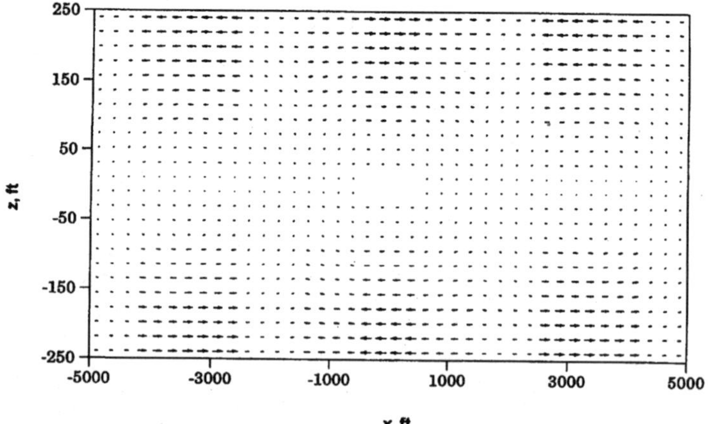

Figure 2.44   Velocity contours due to thermal convection: Example 2.11 ($n = 3$, Eq. (2.106)).

The results from this thermal-convection example and the results from the natural convection–diffusion problem presented in the last section of this chapter have a basic difference: the vertical velocity. For the former, the vertical velocity in the whole distance across the reservoir is effective. For the latter, it is significantly close to the side boundaries and is close to zero across the major part of the reservoir (compare Figs. 2.30 and 2.43).

Problems

2.1   In a gravity field, for a closed reversible process,

$$dH = TdS.$$

When the effect of gravity is neglected,

$$dH = TdS + VdP.$$

Derive the first equation from the second equation.

2.2   Derive the following barometric formula which describes the isothermal change of pressure with increasing altitude z:

$$P = P^0 \exp\left[1 - \frac{Mg}{RT}z\right]$$

P^0 is the pressure at $z = 0$.

2.3   Show that the Phase Rule when the gravity effect is not negligible should be modified to the following form:

$$F = c + 3 - p.$$

Note that in the above equation, the interface between the phases is assumed to be flat.

2.4   Consider capillary condensation in the following configuration where $r_1 = 0.2\,\mu\text{m}$, $r_2 = 10\,\mu\text{m}$, and $r_3 = 100\,\mu\text{m}$. The system is initially saturated with pure nC_5 vapor at 100°F at very low pressure. Calculate the pressure at which condensation occurs in each of the capillary tubes for different contact angles.

(a) $\theta = 0$ (i.e., liquid completely wets the substrate)

(b) $\theta = 30°$

(c) $\theta = 150°$

Pertinent data are

vapor pressure of nC_5 for a flat interface at 100°F $= 15.57$ psia
surface tension for nC_5 at 100°F $= 14$ dynes/cm.

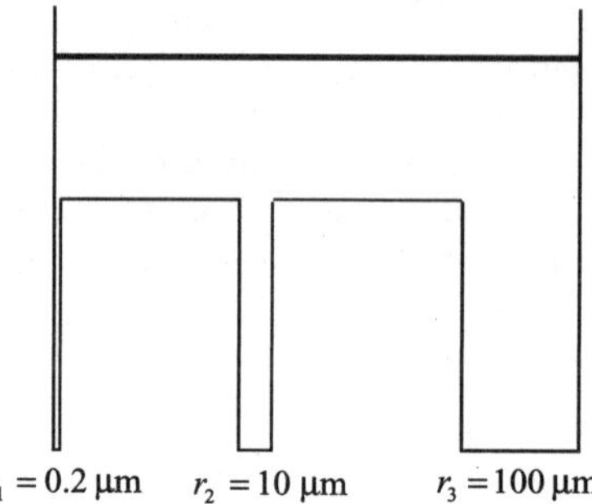

$r_1 = 0.2\,\mu m$ $r_2 = 10\,\mu m$ $r_3 = 100\,\mu m$

2.5   Consider a gas bubble and a liquid droplet sketched in the following. The fluid is nC_5 and the temperature is 100°F. Calculate pressure P' for both systems for the following radii: $r = 0.05, 0.1, 10\ \mu m$.

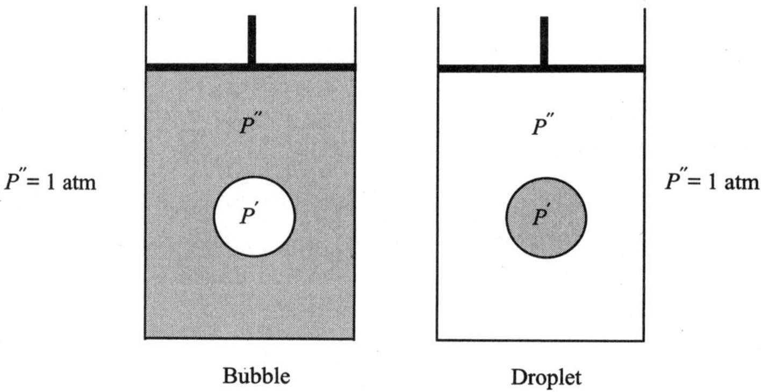

2.6   Using the results from Example 2.5, show that the work required to create a gas bubble can be expressed as

$$W = 16\pi\sigma^3/[3(P^b - P^L)^2].$$

2.7   Consider the P–T plot of a pure substance with a flat interface between the equilibrium gas and liquid phases (see Fig. 2.45). Sketch the P-T plot for the same system when the interface between the equilibrium gas and liquid phases has a constant curvature. Assume liquid is the wetting phase and gas is the nonwetting phase. What would be the results if gas were the wetting phase?

2.8   Consider the gas bubble shown in the cavity in Fig. 2.46. Derive the expression for the work of creating the new gas phase that consists of the top spherical segment and the cone:

$$W_{het} = W\phi_{het},$$

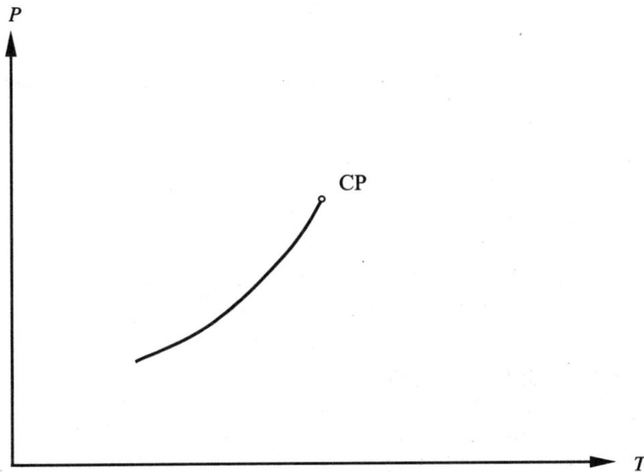

Figure 2.45   Vapor pressure curve for a pure substance: interface curvature $= 0$; $r = \infty$.

where W is the work given in Example 2.5,

$$\phi_{het} = \frac{1}{2}\left[1 - \sin^3 \gamma - \frac{\cos^3 \gamma}{\tan \alpha} + \frac{3}{2}\cos \theta \frac{\cos^3 \gamma}{\sin \alpha}\right]$$

and

$$\gamma = \theta - \alpha.$$

2.9   Derive Eq. (2.49) of the text.

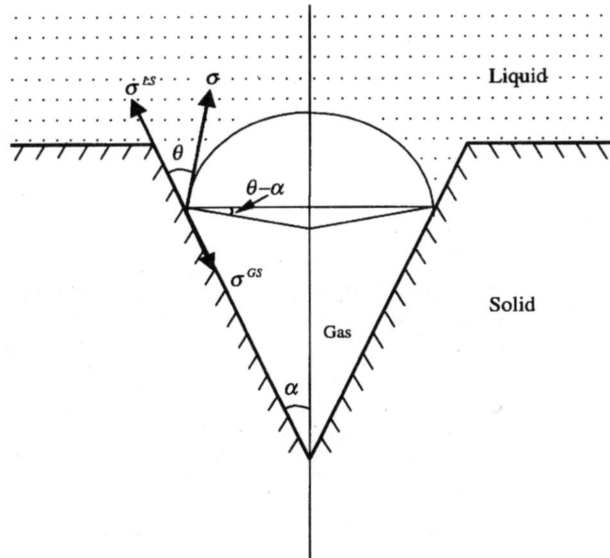

Figure 2.46   Heterogeneous nucleation in a cavity.

2.9   Derive Eq. (2.49) of the text.

2.10   Derive the following simple relations for the gas and liquid pressures corresponding to the bubble and droplet systems:

$$P^G - P^L = 2\sigma/r$$
$$P^G - P^L = -2\sigma/r,$$

where r is the radius of the bubble and the droplet, respectively. Calculate the pressure in the liquid phase for the bubble and droplet of Example 2.4 for $P^G = 0.2$ atm. Note that negative pressure of liquid phase inside the bubble indicates that the liquid is under tension (see Chapter 3).

2.11   Consider the phases sketched in the following figure. In spite of the difference between the arrangement of the phases in the figure below and in Fig. 2.38, show that the Phase Rule remains unchanged: $F = c + 1$.

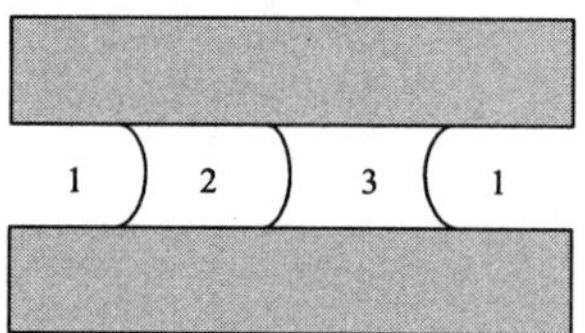

Hint: You may need to establish the following relationship from the criterion of mechanical equilibrium in your derivations:

$$(\sigma^{(1,2)}/r^{(1,2)}) + (\sigma^{(2,3)}/r^{(2,3)}) = (\sigma^{(1,3)}/r^{(1,3)}).$$

2.12   The interface curvature effects the saturation pressure of a pure substance and a multicomponent mixture according to Eqs. (2.48) and (2.47), respectively. The effect of interface curvature on the saturation pressure of a pure substance is given by (see Eq. (2.48))

$$dP^G = \frac{2d(\sigma/r)}{(1 - v^G/v^L)} = \frac{2d(\sigma/r)}{(1 - \rho^L/\rho^G)}.$$

However, as the critical point is approached, both the numerator and denominator of the above equation for a fixed curvature approach zero. Show that for a fixed curvature, approaching the critical point, $dP^G \to 0$ and at the critical point, $dP^G = 0$.

   Hint: At the critical point $\sigma = 0$, but the rates at which σ and $(1 - v^G/v^L)$ approach zero are not the same. From the Macleod-Sugden equation (Macleod, 1923; Sugden, 1924; Fowler, 1937, Firoozabadi *et al.*, 1988),

$$\sigma^{1/4} = (P/M)(\rho^L - \rho^G),$$

the interfacial tension and phase densities are related. In the above equation, P is the parachor of the pure substance (which can be obtained from pure-component interfacial tension measurements), M is the molecular weight, and ρ^L and ρ^G are mass densities of the equilibrium liquid and vapor phases, respectively. Note that in the Macleod-Sugden equation above, the effect of interface curvature on the interfacial tension of a pure substance is neglected.

2.13 Derive the Gibbs-Duhem equation for an open phase with a curved interface (i.e., Eq. (E.2.8)) by taking the Legendre transform of $U = U(S, V, n_1, \ldots, n_c, \mathcal{A})$ with respect to all the extensive variables $S, V, n_1, \ldots, n_c$, and $\mathcal{A}$.

2.14 Prove that for an ideal binary solution, $(\partial \ln f_1 / \partial \ln x_1)_{T,P} = 1$ (see Eqs. (2.87) and (2.88)).

2.15 Show that in a multicomponent system the summation of the diffusive mass flux of all the components is zero:

$$\sum_{i=1}^{c} \vec{J_i} = 0.$$

Hint: Use the diffusive mass-flux expression $\vec{J_i} = \rho_i(\vec{v} - \vec{v})$, where ρ_i is the density of component i (that is, mass of component i per unit volume) and $\vec{v}$ is the velocity of component i, and the definition of mass average velocity is $\vec{v} = \sum_{i=1}^{c} \rho_i \vec{v_i}/\rho$ for the proof above.

2.16 Derive the following expression for the stream function when the temperature variation is represented by Eq. (2.107):

$$\frac{\partial^2 \psi}{\partial x^2} + \frac{\partial^2 \psi}{\partial z^2} = \frac{-2k\rho^0 g\beta B}{\mu}$$

In this case the boundary conditions remain the same as for Eq. (2.120). Derive the solution to the above problem and compare it with Eq. (2.121).

2.17 King Hubert (1956) has shown that so long as the fluid density is constant or is a function of the pressure only, there will be no convective circulation in porous media. However, when ρ is a function of temperature, the fluid may have a convective circulation, which is referred to as thermal convection. In other words, the so-called free convection in porous media cannot occur because of pressure variation of density. For multicomponent systems, it is only the temperature and/or composition variations of density that result in free convection.

Use the rotational and irrotational criteria for a vector field to prove the above.

Hint: Write the Darcy velocity of porous media in the following form: $\vec{v} = -(k/\mu)\rho[(\nabla P/\rho) + \vec{g}]$. Then take the curl of $[(\nabla P/\rho) + \vec{g}]$.

$$\nabla \times [(\nabla P/\rho) + \vec{g}] = \nabla \times (\nabla P/\rho) = \nabla P \times \nabla(1/\rho),$$

since $\nabla \times \vec{g} = 0$. Show that $\nabla P \times \nabla(1/\rho) = 0$ if $\rho = \rho(P)$ and $\nabla P \times \nabla(1/\rho) \neq 0$ if $\rho = \rho(T)$.

References

1. Aavatsmark, I.: *Mathematische Einführug in die Thermodynamik der Gemische*, Akademie Verlag, Berlin, 1995.

2. Bird, R.B., W.E. Stewart and E.N. Lightfoot: *Transport Phenomena*, John Wiley and Sons, New York, 1960.

3. Borisenko, A.I., and I.E. Tarapov: *Vector and Tensor Analysis with Applications*, Dover Publications, Inc., New York, 1968.

4. Chandrasekhar, S.: *Hydrodynamic and Hydrodynamic Stability*, Dover Publications, Inc., New York, 1961.

5. Creek, J.L., and M.L. Schrader: "East Painter Reservoir: An Example of a Compositional Gradient from a Gravitational Field," SPE 14411, paper presented at the 60th Annual Technical Conference and Exhibition of the Society of Petroleum Engineers, Las Vegas, Nevada, Sept. 22–25, 1985.

6. Defay, R., and I. Prigogine: *Surface Tension and Adsorption*, Longmans, London, 1966.

7. de Groot, S.R., and P. Mazur: *Non-Equilibrium Thermodynamics*, Dover Publications, Inc., New York, 1984.

8. Espach, R.H., and J. Fry: "Variable Characteristics in the Oil in the Tensleep Sandstone Reservoir, Elk Basin Field, Wyoming and Montana," *Trans. AIME,* vol. 192, p. 75, 1951.

9. Firoozabadi, A., B. Dindoruk, and E. Chang: "Areal and Vertical Composition Variation in Hydrocarbon Reservoirs: Formulation and One-D Binary Results," *Entropie* No. 198/199, p. 109, 1996.

10. Firoozabadi, A., and D. Kashchiev: "Pressure and Volume Evolution During Gas Phase Formation in Solution Gas Drive," *Soc. Pet. Eng. J.* p. 219, Sept. 1996.

11. Firoozabadi, A., D.L. Katz, H. Soroosh, and V.A. Sajjadian: "Surface Tension of Reservoir Crude Oil/Gas Systems Recognizing the Asphalt in the Heavy Fractions," *SPE Res. Eng.,* p. 265, Feb. 1988.

12. Fowler, R.H.: "A Tentative Statistical Theory of Macleod's Equation for Surface Tension and the Parachor," *Proc, Royal Soc. of London,* Series A, p. 229, 1937.

13. Ghorayeb, K., and Firoozabadi, A.: "Modeling of Multicomponent Diffusion and Convection in Porous Media," SPE 51932, Proceedings of the 15th SPE Reservoir Simulation Symposium, Feb. 14 to 17, 1999, Houston, TX.

14. Ghorayeb, K., and Firoozabadi, A.: Molecular Pressure and Thermal Diffusion in Multicomponent Systems," *J. Chemical Physics* (to be submitted, 1998).

15. Haase, R., H.-W. Borgmann, K.-H. Dücker, and W.-P. Lee; "Thermodiffusion im Kritischen Verdampfungsgebiet Binärer Systeme," *Z. Naturforsch, Teil A.* vol. 26, p. 1224, 1971.

16. Haase, R.: *Thermodynamics of Irreversible Processes,* Dover Publications, Inc., New York, 1969.

17. Hubert, M.K.: "Darcy's Law and the Field Equations of the Flow of Underground Fluids," *Petroleum Transactions,* AIME, vol. 207, p. 222, 1956.

18. Kempers, L.J.T.: "Thermodynamic Theory of the Soret Effect in a Multicomponent Liquid," *J. Chem. Phys.* vol. 90, p. 6541, Jun. 1989.

19. Kingston, P.E., and H. Niko: "Development Planning of the Brent Field," *J. of Pet. Tech.,* p. 1190, Oct. 1975.

20. Li, D.: "Curvature Effects on the Phase Rule," *Fluid Phase Equilibria* vol. 98, p. 13, 1994.

21. Li, D., J. Gaydos, and A.W. Neumann: "The Phase Rule for Systems Containing Surfaces and Lines. 1. Moderate Curvature," *Langmuir,* p. 1133, Sept./Oct. 1989.

22. Lu, P.-C. *Introduction to the Mechanics of Viscous Fluids*, Holt, Rinehart, and Winston, New York, 1973.

23. Macleod, D.B.: "On a Relation Between Surface Tension and Density," *Trans., Faraday Soc.* vol. 19, p. 38, 1923.

24. Neveux, A.R., and S. Sathikumar: "Delineation and Evaluation of a North Sea Reservoir Containing Near-Critical Fluids," *SPE Res. Eng.,* p. 842 Aug. 1988.

25. Onsager, L.: "Reciprocal Relations in Irreversible Processes, I," *Physical Review* (Feb. 1931a) Vol. 37, p. 405, Feb. 1931a.

26. Onsager, L.: "Reciprocal Relations in Irreversible Processes. II." *Physical Review*, vol. 38, p. 2265, Dec. 1931b.

27. Peng, D.Y., and D.B. Robinson: "A New Two-Constant Equation of State," *Ind. Eng. Chem. Fund.* vol. 15, no. 1, p. 59, 1976.

28. Riley, M., and A. Firoozabadi: "Natural Convection in Hydrocarbon Reservoirs," Quarterly Report of Reservoir Engineering Research Institute, Palo Alto, 1996.

29. Riley, M., and A. Firoozabadi: "Compositional Variation in Hydrocarbon Reservoirs with Natural Convection and Diffusion," *AIChE J.*, p. 452, Feb. 1998.

30. Rutherford, W.M., and J.G. Roof: "Thermal Diffusion in Methane n-Butane Mixtures in the Critical Region," *J. Phys. Chem.*, vol. 63, p. 1506, Sept. 1959.

31. Rutherford, W.M.: "Calculation of Thermal Diffusion Factors for the Methane-n-Butane System in the Critical and Liquid Regions." *AIChE J.*, vol. 9, p. 841, 1963.

32. Shukla, K., and A. Firoozabadi: "A New Model of Thermal Diffusion Coefficients in Binary Hydrocarbon Mixtures," *Ind. Eng. Chem. Research*, vol. 37, p. 3331, Aug. 1998.

33. Sigmund, P.M., Dranchuk, P.M., Morrow, N.R., Purvis, R.A.: "Retrograde Condensation in Porous Media," *Soc. Pet. Eng. J.*, p. 93, April 1973.

34. Sugden, S.: "The Variation of Surface Tension with Temperature and some Related Functions," *J. Chem. Soc.*, vol. 125, p. 32, 1924.

35. Tindy, R., and M. Raynal: "Are Test-Cell Saturation Pressures Accurate Enough?" *Oil and Gas Journal*, p. 128, Dec. 1966.

36. Trebin, F.A., and G.I. Zadora: "Experimental Study of the Effect of Porous Media on Phase Changes in Gas Condensate Systems," *Neft'i Gaz*, vol. 81, p. 37, 1968.

37. Wheaton, R., "Treatment of Variations of Composition with Depth in Gas-Condensate Reservoirs," *SPE Res. Eng.*, p. 239, May 1991.

3

Equation-of-state representation of reservoir-fluids phase behavior and properties

Reservoir fluids contain a variety of substances of diverse chemical nature that include hydrocarbons and nonhydrocarbons. Hydrocarbons range from methane to substances that may contain 100 carbon atoms, even when these substances are in the form of singly dispersed molecules (i.e., monomers). Nonhydrocarbons include substances such as N_2, CO_2, H_2S, S, H_2O, He, and even traces of Hg. The chemistry of hydrocarbon-reservoir fluids is very complex. Methane, often a predominant component of natural gases and petroleum-reservoir fluids, is a gas, nC_5 and hydrocarbons as heavy as nC_{15} may be in the liquid state, and normal paraffins heavier than nC_{15} may be in the solid state at room temperatures. However, the mixture of these hydrocarbons may be in a gaseous or liquid state at the pressures and temperatures often encountered in hydrocarbon reservoirs. The mixture may also be a solid as will be seen in Chapter 5. The majority of reservoirs fall within the temperature range of 80 to 350°F, and the pressure range of 50 to 20,000 psia. When steam is injected into hydrocarbon reservoirs, the temperature may exceed 550°F and for in-situ combustion, the temperature may be even higher.

In spite of the complexity of hydrocarbon fluids found in underground reservoirs, simple cubic equations of state have shown surprising performance in the phase-behavior calculations, for both vapor–liquid and vapor–liquid–liquid equilibria of these complex fluids. This chapter is structured on a simple presentation of cubic equations of state and their use based on our own experience with the remarkable accuracy of these equations. For the phase behavior of water–hydrocarbon fluids at high temperatures, the concept of association of water molecules may become necessary and should be integrated into the cubic equations. This integration, however, does not result in complexity; the simplicity of the calculations remains intact.

The calculation of two-phase isothermal and isentropic compressibilities, two-phase sonic velocity, single-phase sonic velocity, and cooling and heating due to expansions are presented in the second part of this chapter. Cubic equations of state facilitate all of these calculations. One basic assumption in the formulation for the two-phase compressibilities and two-phase sonic velocities is the equilibrium state. In the transition from single-phase to two-phase state, compressibilities and sonic velocity may have a sharp discontinuity, which implies lack of validity of averaging procedures.

EOS representation of volumetric and phase behavior

Consider the plot of pressure versus total volume of a pure substance shown in Fig. 3.1. An equation of state (EOS) is desired to represent the volumetric behavior of the pure substance in the entire range of volume—both in the liquid and in the gaseous state. It is also desired that the EOS be a continuous function of volume.

Let us first examine the phase diagram of Fig. 3.1 at T_1, where $T_1 < T_c$; T_c is the critical temperature. At point S, the liquid is compressed. (In petroleum engineering terminology, this state is referred to as undersaturated liquid because more gas could be dissolved in it.) As the pressure is lowered, the volume increases. At point A, the liquid is in the saturated and stable state. Toward the right of point A, as the pressure is lowered, the substance might follow one of two routes. It might follow the line AD, in which case point D represents the saturated and stable vapor, or it might follow curve AB, for which the fluid will be in a metastable condition. In this case, the limit of stability is determined by the condition that $(\partial P/\partial V)_{T_1}$ vanishes (i.e., point B). Similarly, one can start from point R and observe that as the pressure increases, condensation may not occur up to point C, where again $(\partial P/\partial V)_{T_1}$ will vanish. Curve DC represents the locus of metastability and point C is the limit of metastability at T_1 for the vapor. The dashed line in Fig. 3.1 is the locus of the limit of metastability (i.e., the spinodal curve). Any point on this curve or curve AB, below zero pressure, represents the liquid under the state of tension. The envelope shown by the solid thick line is the binodal curve representing the saturated equilibrium liquid and vapor loci.

The region between the spinodal curve and the saturated liquid curve may represent liquid in a superheated state. Similarly, the region

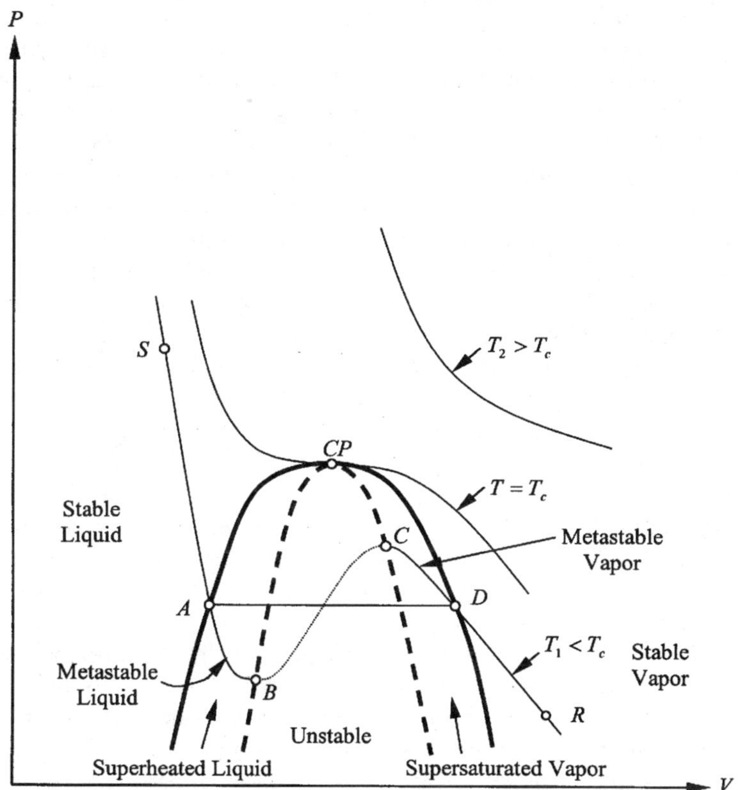

Figure 3.1   Pressure *vs.* volume of a pure substance.

between the saturated vapor curve and the spinodal curve represents the supersaturated vapor state. Inside the binodal curve is the unstable region; there is no physical meaning to the curve between points B and C. In other words, it is not physically possible to have a fluid on the curve between points B and C at temperature T_1, while a fluid may exist on the curves between A and B and between C and D. In the next chapter, we will discuss the stability concept in detail and derive the corresponding mathematical expressions from the first principles.

Kenrick, Gilbert, and Wismer (1924) studied the superheating of a number of liquids at atmospheric pressure and found that there was a limiting temperature for each liquid, at or below which it always exploded. For example, the maximum value for the superheating of water at atmospheric pressure is around 270°C. Debenedetti (1996) presents metastability of water and other liquids in detail.

It was mentioned that a point on curve AB may be at negative pressure. In other words, a liquid could have a negative pressure without vaporizing. Consider a simple experiment that was performed to examine the effect of negative pressure in a tube (Skuse, Firoozabadi, and Ramey, 1992). The apparatus of Fig. 3.2 was used to create a negative pressure in a glass capillary. First, the bottle was partially filled to 2-cm depth with vacuum pump oil and a partial vacuum of 0.00038 psia (2.6 Pa) was created to remove air from the oil, then the vacuum was released. A capillary tube of 0.025-cm radius was placed in a vertical position inside the bottle, dipping into the vacuum pump oil. A capillary rise of 2.5 cm was observed. The capillary rise remained the same when the bottle was pumped to a vacuum of 0.00038 psia (2.6 Pa). For an oil density of 0.9 g/cm^3, the hydrostatic pressure at the bottom of the column was, therefore, about -0.0315 psia (-217 Pa). The entire column was at negative pressure. No vaporization was observed. Briggs (1950) measured the limiting negative pressure of water as a function of temperature using a centrifuge to generate negative pressure in a capillary tube. Figure. 3.3 is a reproduction of Briggs' data. The maximum negative pressure of water from Fig. 3.3 is about 280 bar at a temperature of 9°C. The equation of state and the stability criteria can be used to estimate the limiting negative pressure of pure substances and mixtures (see Chapter 4).

As will be seen in Problem 3.11 toward the end of this chapter, when the interface is curved, it is possible to be in a stable state instead of a metastable state inside the binodal curve. The binodal curve of a curved interface for the same fluid and the same temperature is different from the binodal curve when the interface between phases is flat. We could have negative pressures while the fluid is stable; this can occur in a porous medium (Skuse, Firoozabadi, Ramey, 1992).

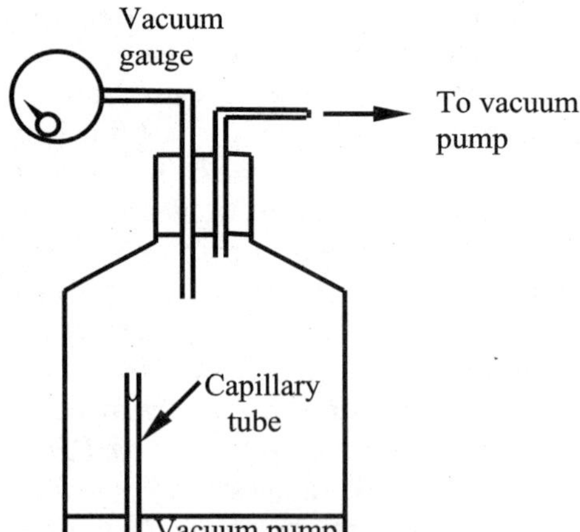

Figure 3.2 Apparatus used to establish negative pressure in a capillary tube (adapted from Skuse, Firoozabadi, and Ramey, 1992).

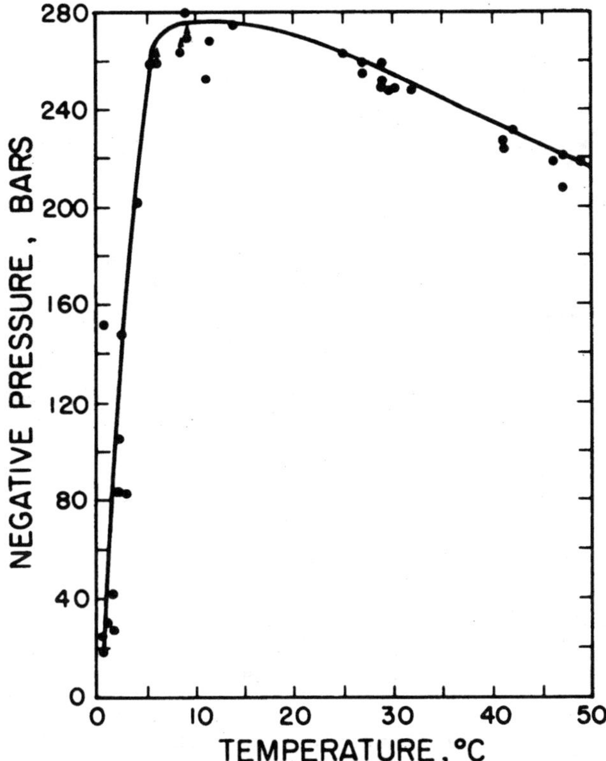

Figure 3.3   Limiting negative pressure of water between 0 and 50°C (adapted from Briggs, 1950).

It is also desired that an equation of state predict the limits of stability. Another feature of an EOS is that it can describe the volumetric behavior even in the negative pressure region. Now consider the pressure versus total volume plot shown in Fig. 3.4.

An equation of state is an algebraic expression that can represent the phase behavior of the fluid, both in the two-phase envelope (i.e., inside the binodal curve), as was demanded above, on the two-phase envelope, and outside the binodal curve. The equations of state are divided into two main groups: cubic and noncubic. Cubic equations have three roots when $T \leqslant T_c$ and only one root when $T > T_c$. At $T = T_c$, there are three equal roots. Figure. 3.4 portrays the deficiency with most of the cubic equations of state. In this plot, the solid circles show measured data and the solid line represents the predictions from an EOS. The flatness around the critical point can not be adequately described by most cubic equations. The liquid phase description is also not so good as the description of the gas phase. Later, we will discuss how the volume-

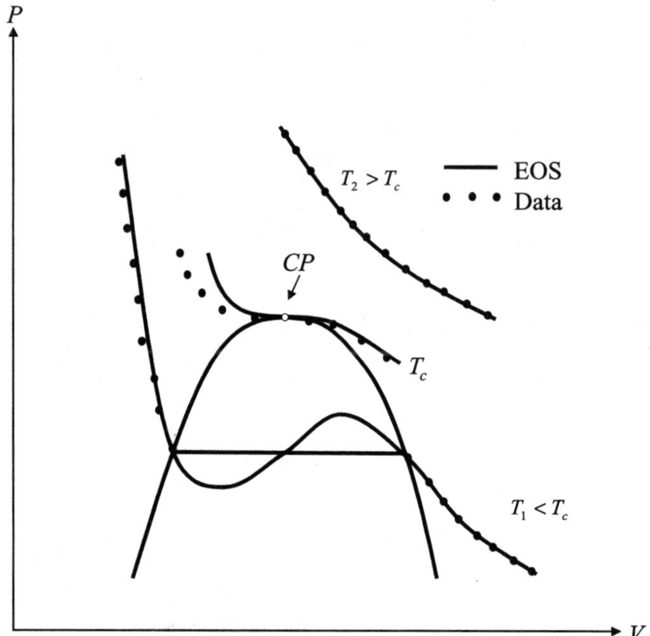

Figure 3.4   Equation-of-state representation of P $vs.$ V for a pure substance.

translation technique can alleviate this deficiency. Noncubic equations can better describe the volumetric behavior of pure substances but may not be suitable for complex hydrocarbon mixtures.

For mixtures, the pressure–volume plot differs from that of pure substances; a schematic of a P–V plot for mixtures is shown in Fig. 3.5. The main differences between Figs. 3.4 and 3.5 are (1) V^L and V^G of Fig. 3.5 do not represent the equilibrium states and (2) the critical points have different features. For a pure substance, $(\partial P/\partial V) = (\partial^2 P/\partial V^2) = 0$ at the critical point. For a mixture, these two equations do not hold (see Figs. 3.1 and 3.5). The Z-factors in Fig. 3.4 of the equilibrium gas and liquid phases always meet the condition $Z^L < Z^G$. However, for mixtures, when gas and liquid phases are at equilibrium, Z^L might be smaller or larger than Z^G. At equilibrium, the mass density of the liquid phase is higher than the mass density of the gas phase, $\rho^L > \rho^G$. Then from

$$\rho = \frac{PM}{ZRT} \tag{3.1}$$

it follows that

$$\frac{M^L}{Z^L} > \frac{M^G}{Z^G}. \tag{3.2}$$

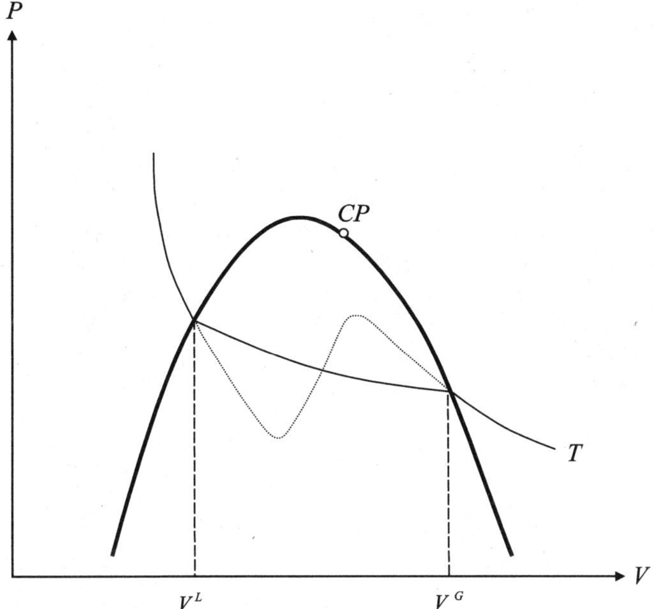

Figure 3.5   Pressure vs. volume of a mixture.

When Z-factors are less than one, then Z^L could be smaller or larger than Z^G.

Similarly to pure substances, an equation of state should also represent the volumetric behavior of multicomponent fluids. The volumetric representation of mixtures is a more difficult task, especially around the critical point.

There are a large number of equations of state that have been proposed to represent the phase behavior of pure substances and mixtures in the gas and liquid states. In 1873, van der Waals introduced an EOS, which is known as the van der Waals equation of state:

$$\left(P + \frac{a}{v^2}\right)(v - b) = RT \tag{3.3}$$

In the above equation, $v = V/n$ is the molar volume. The parameters a and b have clear physical meaning. As an example, b is the so-called hard-core parameter or co-volume parameter, which is the state at which the fluid is completely packed with the molecules at infinite pressure. Therefore, v should be larger than b. Parameter a has a more difficult meaning and will simply be referred to as the attraction parameter.

Equation (3.3) may be written as

$$\left[P_r + \frac{3}{v_r^2}\right](3v_r - 1) = 8T_r,\tag{3.4}$$

where $P_r = P/P_c$, $T_r = T/T_c$, and $v_r = v/v_c$. Therefore Eq. (3.4) is a statement of "the principle of corresponding states," which means that at the same reduced pressure and reduced temperature, all substances have the same reduced volume. Generalized compressibility-factor charts for natural gases are based on the corresponding-states principle; that concept was also introduced by van der Waals in 1873.

In 1949, Redlich and Kwong (RK) made an important modification to the van der Waals EOS. They proposed

$$\left[P + \frac{a}{T^{1/2}v(v + b)}\right](v - b) = RT.\tag{3.5}$$

Later Soave (1972) improved on the RK-EOS by replacing the term $a/T^{1/2}$ with a more general temperature-dependent term $a(T)$ and proposed a simple form for $a \equiv a(T_r, \omega)$ for all pure substances, taking advantage of the concept of the acentric factor of Pitzer, Pitzer (1939), and Pitzer, *et al.* (1955). Pitzer's acentric factor, ω, was intended as an additional parameter for the improvement of the corresponding-states principle. The acentric factor is a measure of the difference in molecular structure between a given component and a gas with spherically symmetric molecules with $\omega = 0$ (such as argon).

Another important variation of the van der Waals EOS was introduced in 1976 by Peng and Robinson:

$$\left[P + \frac{a(T)}{v(v + b) + b(v - b)}\right](v - b)] = RT.\tag{3.6}$$

This equation improves the liquid density prediction, but still cannot describe volumetric behavior around the critical point because of a fundamental reason that will be discussed later. There are thousands of cubic equations of states, and many noncubic equations. The noncubic equations such as the Benedict-Webb-Rubin equation (1942) and its modification by Starling (1973) have a large number of constants; they describe accurately the volumetric behavior of pure substances. But for hydrocarbon mixtures and crude oils, because of mixing rule complexities, they may not be suitable (Katz and Firoozabadi, 1978). Cubic equations with more than two constants also may not improve the volumetric behavior prediction of complex reservoir fluids. In fact, most of the cubic equations have the same accuracy for phase-behavior prediction of complex hydrocarbon systems; the simpler equations often do better. Therefore, the discussion will be limited to the Peng-

Robinson EOS, which according to the author's experience enjoys more simplicity and reliability than many other equations.

Before turning to the specifics of the PR-EOS, Maxwell's equal-area rule for pure substances will be derived for the van der Waals family of equations and the mathematical structure of these equations will be discussed. Maxwell's equal-area rule, which applies to the subcritical isotherm ($T < T_c$), is shown schematically in Fig. 3.6.

From the equality of chemical potentials or the Gibbs free energy at the saturation points A and E,

$$G^L = G^G \tag{3.7}$$

$$A^L + P^{sat}V^L = A^G + P^{sat}V^G \tag{3.8}$$

and, therefore, $$A^G - A^L = -P^{sat}(V^G - V^L), \tag{3.9}$$

which is the area of the rectangle AEFG. An alternative expression for $A^G - A^L$ can be obtained by evaluating the integral of $\int dA$ along the path $ABCDE$ using $dA = -PdV$, since $dT = 0$,

$$A^G - A^L = -\int PdV, \tag{3.10}$$

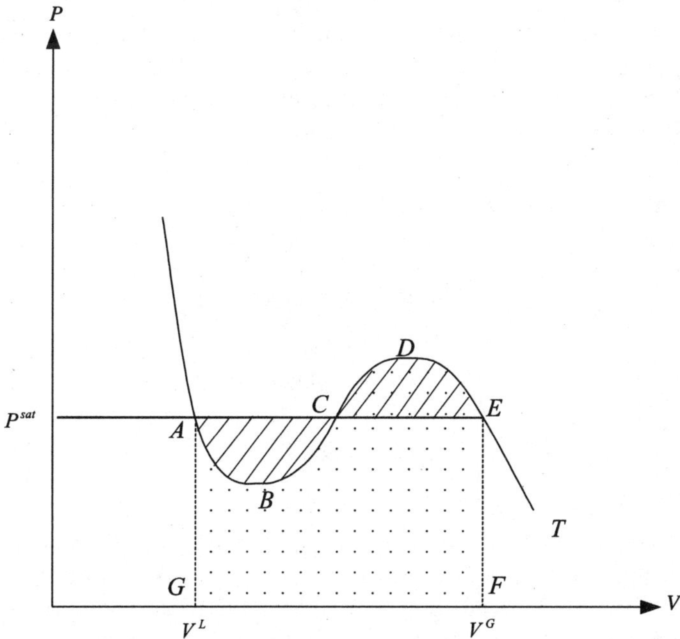

Figure 3.6   Maxwell's equal-area rule.

which is the area shown by the dotted region. It then follows that the two shaded areas are equal. Note that Maxwell's equal-area rule on a $P - V$ plot applies to pure substances only. In Chapter 4, we will address the form of this concept for mixtures.

Algebraic form of cubic equations

The cubic equations of state may have four branches in P–v space. Examine the PR-EOS given by Eq. (3.6.) A plot of Eq. (3.6) shown in Fig. 3.7 exhibits three vertical asymptotes:

$$v = b \tag{3.11}$$

$$v = +(\sqrt{2} - 1)b \tag{3.12}$$

$$v = -(\sqrt{2} + 1)b. \tag{3.13}$$

The branches for which $v < b$ have no physical meaning. When $P > P_1$, there may be three roots: $-(\sqrt{2}+1)b < v < 0$, $-(\sqrt{2}+1)b < v < (\sqrt{2}-1)b$, and $v > b$. Only the root that is larger than b is identified as the molar volume of the liquid phase. The other two roots have no physical meaning. When $P_2 < P < P_1$, the only real root corresponds to the liquid phase. For $P_3 < P < P_2$, three real roots are obtained (see Fig. 3.7a). The largest root corresponds to the vapor phase and the smallest root corresponds to the liquid phase. The intermediate root has no physical meaning. When $P < P_3$, two situations may arise for $v > b$: one shown in Fig. 3.7a and the other shown in Fig. 3.7b. In Fig. 3.7a, the root corresponds to the vapor phase, and in Fig. 3.7b, the smaller root corresponds to the liquid phase (which might have a negative pressure), and the larger root has no physical meaning.

Peng-Robinson equation of state (PR-EOS)

The PR-EOS, Eq. (3.6,) can be written as

$$Z^3 - (1 - B)Z^2 + (A - 3B^2 - 2B)Z - (AB - B^2 - B^3) = 0, \tag{3.14}$$

where

$$A = \frac{aP}{R^2 T^2} \tag{3.15}$$

$$B = \frac{bP}{RT}. \tag{3.16}$$

For pure substances, the first and the second derivatives of pressure with respect to volume at the critical point are equal to zero. (The condition of criticality will be derived in the next chapter.) Note that these derivatives may not be zero for mixtures. Using the criteria of criticality

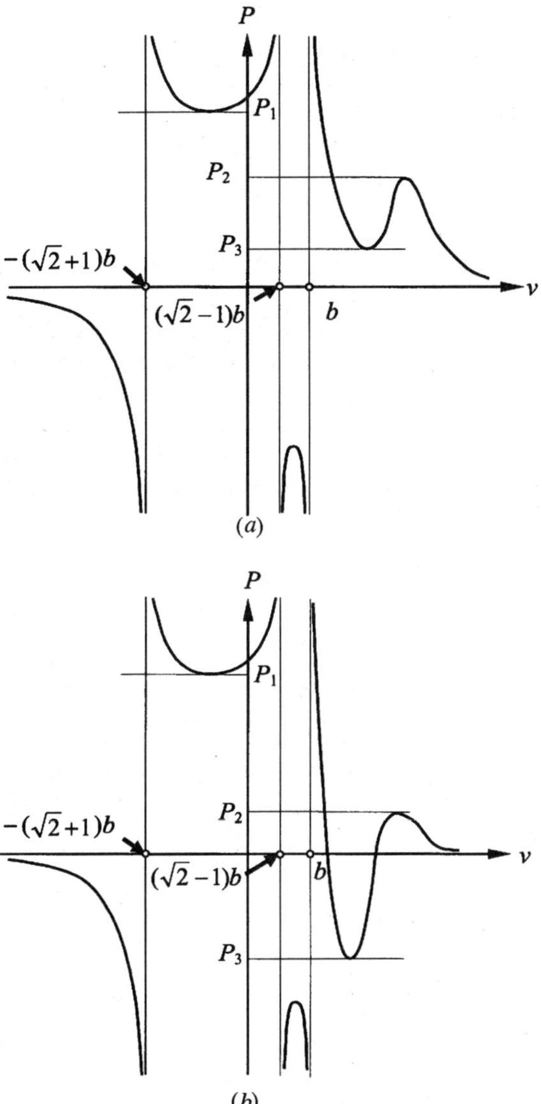

Figure 3.7   Various branches of the PR-EOS.

(that is, the first and second derivatives of P with respect to v being zero at the critical point), and Eq. (3.14),

$$a(T_c) = 0.45724 \frac{R^2 T_c^2}{P_c} \tag{3.17}$$

$$b(T_c) = 0.07780 \frac{RT_c}{P_c} \tag{3.18}$$

$$Z_c = 0.307. \tag{3.19}$$

Equations (3.17) to (3.19) imply that the PR-EOS will pass through the critical pressure and critical temperature, but not through the critical volume. Alternatively, if true T_c and v_c are honored by the equation of state, it cannot pass through the critical pressure. Figure 3.8 highlights the implementation of the criticality criteria on equations of state. In this figure, the pressure is plotted versus the density at $T = T_c$. The solid curve shows the observed data. The dotted curve shows the results when an EOS is forced to pass through the true P_c and T_c; the dashed curve represents the EOS results when it is forced to pass through the

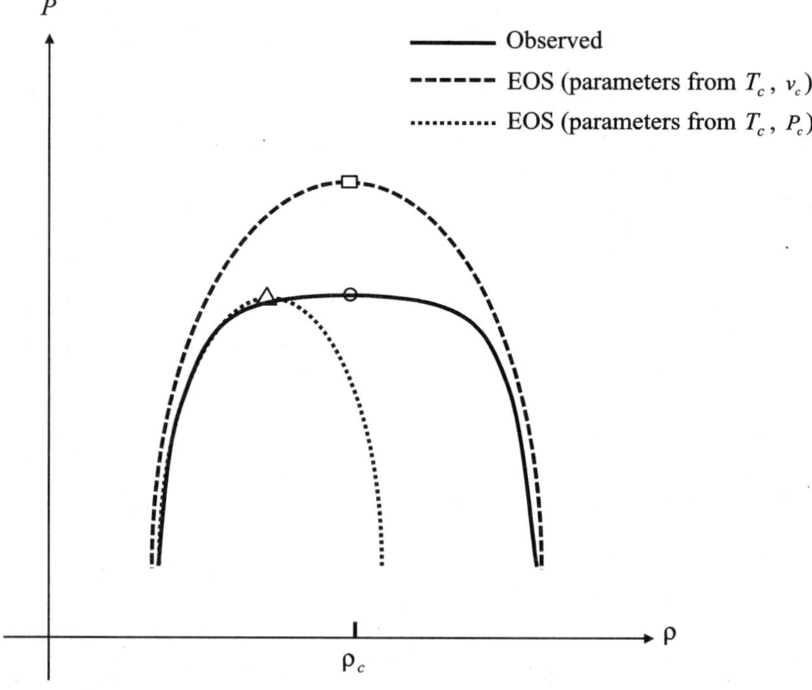

Figure 3.8   Coexistence curve in the critical region: EOS *vs.* observed (adapted from Chu and Prausnitz, 1989).

true T_c and v_c. In either case, the shortcoming of the EOS in the critical region needs improvement. As will soon be seen, the deficiency could be easily alleviated for pure substances. For mixtures, a satisfactory solution is not yet available.

It is well known that the critical compressibility factor depends on the substance. Table 3.1 gives the critical compressibility factors of n-alkanes and some nonhydrocarbons. This table shows a wide variation in Z_c of various substances. All the values listed in Table 3.1 are less than $Z_c = 0.307$ predicted from the PR-EOS. The SRK-EOS gives a critical compressibility factor of 0.333. On the basis of Z_c predictions, one expects the PR-EOS to predict pure component densities better than the SRK-EOS. Density predictions will be further discussed later.

At temperatures other than the critical temperature, the parameter a in Eq. (3.6) is given by

$$a(T) = a(T_c)\alpha(T_r, \omega). \tag{3.20}$$

The dimensionless parameter α is a function of T_r and the acentric factor, ω. Vapor pressure data are used to obtain α in the following manner.

TABLE 3.1 Critical properties and normal boiling points of n-alkanes* and some selected nonhydrocarbons[†]

Comp.	T_c,K	P_c,MPa	ρ_c, g/cm^3	T_b,K	Z_c
C_1	190.58	4.604	0.162	111.63	0.288
C_2	305.42	4.880	0.203	184.55	0.285
C_3	369.82	4.250	0.217	231.05	0.281
nC_4	425.18	3.797	0.228	272.64	0.274
nC_5	469.7	3.369	0.230	309.22	0.271
nC_6	507.3	3.014	0.233	341.88	0.264
nC_7	540.1	2.734	0.233	371.57	0.262
nC_8	568.7	2.495	0.232	398.82	0.260
nC_9	594.6	2.280	0.231	423.97	0.256
nC_{10}	617.7	2.099	0.228	447.30	0.255
nC_{11}	638.8	1.948	0.227	469.08	0.253
nC_{12}	658.4	1.810	0.226	489.47	0.249
nC_{13}	675.9	1.679	0.224	508.62	0.246
nC_{14}	692.3	1.573	0.222	526.73	0.244
nC_{15}	707.8	1.479	0.220	543.84	0.242
nC_{16}	722.6	1.401	0.219	560.01	0.241
nC_{17}	735.6	1.342	0.218	575.17	0.242
nC_{18}	774.2	1.292	0.214	589.50	0.247
H_2O	647.4	22.104	0.400	373.30	0.229
CO_2	304.4	7.398	0.461	194.80	0.274
N_2	126.2	3.392	0.311	77.50	0.290

*n-Alkanes data from Teja *et al.* (1990).
[†]Nonhydrocarbons data from GPA Handbook (1972).

At the boiling point (vapor-pressure condition),

$$f^L = f^G. \tag{3.21}$$

Using Eqs. (3.6), and (1.109) of Chapter 1,

$$\ln(f/P) = Z - 1 - \ln(Z - B) - \frac{A}{2\sqrt{2}B}\ln\frac{Z + 2.414B}{Z - 0.414B}. \tag{3.22}$$

Vapor-pressure data and Eqs. (3.14), (3.21), and (3.22) are used to estimate α for pure substances. The parameter α can be correlated by a simple expression:

$$\alpha = [1 + m(1 - T_r^{1/2})]^2. \tag{3.23}$$

In Eq. (3.23), m is given by

$$m = 0.37464 + 1.54226\omega - 0.26992\omega^2. \tag{3.24}$$

Equation (3.24) is apparently based on vapor pressure data of hydrocarbons with $0 < \omega < 0.5$. The correlation was later expanded for ω in the range of $0.1 < \omega < 2.0$ (Robinson *et al.*, 1985):

$$m = 0.3796 + 1.485\omega - 0.1644\omega^2 + 0.01667\omega^3 \tag{3.25}$$

For water and other polar substances, Mathias and Copeman (1983) suggest a different correlation.

The acentric factor in Eqs. (3.24) and (3.25) is defined as

$$\omega = -\log\frac{P^{sat}}{P_c}\bigg|_{T_r=0.7} -1. \tag{3.26}$$

If we assume the vapor-pressure data to be represented by

$$\log P^{sat} = \beta + \frac{\gamma}{T}, \tag{3.27}$$

where β and γ are constants and P^{sat} is the saturation pressure at absolute temperature T, then

$$\omega = \frac{3}{7}\left[\frac{\log(P_c/14.695)}{(T_c/T_b - 1)}\right] - 1. \tag{3.28}$$

where P_c is in psia, and T_b is the normal boiling point with the same absolute units as T_c. For mixtures, the a and b parameters are defined according to certain mixing rules. The following mixing rules for petroleum fluids have proved useful:

$$a = \sum_{i=1}^{c} \sum_{j=1}^{c} x_i x_j (a_{ij})^{1/2} \tag{3.29}$$

$$a_{ij} = (1 - \delta_{ij}) a_i^{1/2} a_j^{1/2}, \tag{3.30}$$

$$b = \sum_{i=1}^{c} x_i b_i \tag{3.31}$$

where δ_{ij} is the interaction parameter between component i and component j, and $\delta_{ij} = \delta_{ji}$. The interaction parameter is assumed to be independent of pressure and composition and generally independent of temperature. In the above equations, x_i represents the mole fraction and a_i and b_i represent the parameters of pure substance i.

From Eq. (1.109) of Chapter 1 and Eq. (3.6) of this chapter,

$$\ln \frac{f_i}{y_i P} = \frac{b_i}{b}(Z - 1) - \ln(Z - B)$$

$$- \frac{A}{2\sqrt{2}B} \left[\frac{2\sum_{j=1}^{c} y_j a_{ij}}{a} - \frac{b_i}{b} \right] \ln \frac{Z + 2.414B}{Z - 0.414B}. \tag{3.32}$$

This equation is very important in the thermodynamics of phase equilibria; the condition of the equality of the fugacity of equilibrium phases and the above equation provide the phase composition.

Next we will present the use of the cubic equations in predicting (1) the volumetric properties of pure components, (2) the phase behavior of multicomponent mixtures, and (3) the phase behavior of reservoir fluid systems.

Pure substances.   Generally, the PR-EOS and other similar cubic equations reliably represent the vapor pressure of pure substances since vapor pressure data are used to obtain the parameter α. The density prediction is the weak point and may need a modification. An exception is the ZJRK-EOS (Zudkevitch and Jaffe, 1970). In this equation the constants of the a and b parameters of the SRK-EOS, Ω_a^0 and Ω_b^0, are assumed to be temperature-dependent. These two parameters are given by

$$a = \Omega_a^0 \frac{R^2 T_c^{2.5}}{P_c} \tag{3.33}$$

$$b = \Omega_b^0 \frac{R T_c}{P_c}. \tag{3.34}$$

For the SRK-EOS, the dimensionless constants Ω_a^0 and Ω_b^0 are 0.4274 and 0.0867, respectively. However, in the ZJRK-EOS, these two parameters are determined from saturated liquid density and the equality of the saturated liquid and vapor phase fugacities. At the critical temperature and above, these two parameters are assigned values of 0.4274, and 0.0867, respectively. In the PR and SRK equations, no parameter is adjusted for density. As a result, these two equations have a density-prediction deficiency. Figure 3.9 shows the deviation in liquid molar volume of selected substances at $T_r = 0.7$ versus ω. The SRK-EOS underestimates the liquid density of all substances that are shown in the figure. The PR-EOS overestimates the density to $\omega = 0.35$, and then underestimates the density of n-alkanes heavier than nC_8. This figure clearly shows that at $T_r = 0.7$, the SRK-EOS is best suited for density prediction of pure hydrocarbons with $\omega \approx 0$, while the PR-EOS performs best for n-heptane and other hydrocarbons with

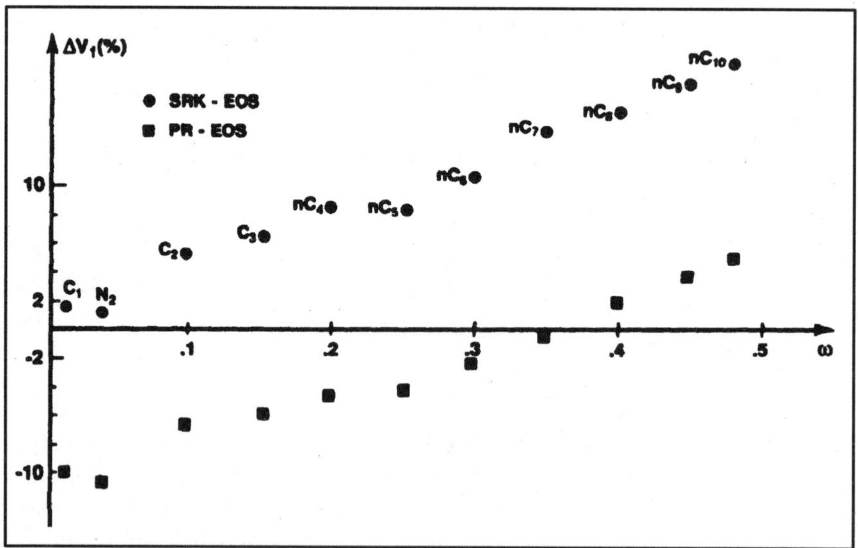

Figure 3.9 Percent deviation in liquid molar volume at $T_r = 0.7$ as a function of acentric factor (adapted from Firoozabadi, 1988).

$\omega \approx 0.35$. Figure 3.10 shows the deviation for the molar liquid volume of *n*-hexane as a function of reduced temperature. This figure reveals that the deviation is nearly constant up to a reduced temperature of 0.75. The volume-translation concept was introduced to take advantage of this feature (Peneloux *et al.*, 1982; Jhaveri and Youngren, 1988). The volume-translation technique separates vapor-liquid-equilibria (VLE) from density calculations (see Example 3.3). The translation along the volume axis is given by

$$v^{true} = v^{EOS} + c, \tag{3.35}$$

where *c* is the volume-translation parameter. Figure 3.10 shows that the volume translation may not improve volumetric prediction above $T_r = 0.7$; for the PR-EOS, it will make the predicted density worse. An additional correction term has been suggested by Mathias, Naheiri, and Oh (1989):

$$v^{true} = v^{EOS} + c + f_c \left[\frac{\lambda}{\lambda - \dfrac{v^2}{RT}\left(\dfrac{\partial P}{\partial v}\right)_T} \right]. \tag{3.36}$$

In Eq. (3.36), $\left(-v^2/RT\right)(\partial P/\partial v)_T$ is a dimensionless quantity related to the inverse of the compressibility. This dimensionless quantity is zero at the critical point, and its value is relatively high at low reduced

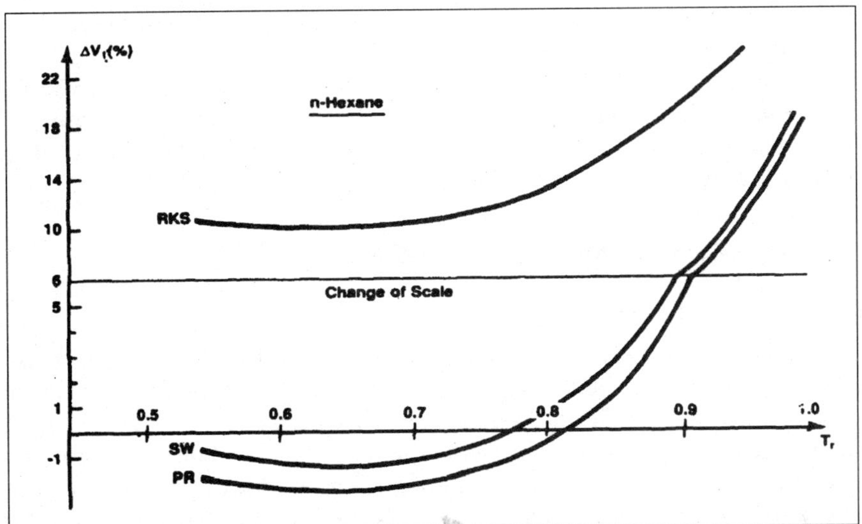

Figure 3.10 Percent deviation in liquid molar volume for *n*-hexane as a function of reduced temperature (adapted from Firoozabadi, 1988).

temperatures. Therefore, at conditions away from the critical point, the third term on the right side of Eq. (3.36) becomes negligible. At the critical point,

$$v_c^{true} = v_c^{EOS} + c + f_c,$$ (3.37)

where f_c from Eq. (3.37) is

$$f_c = v_c^{true} - (c + v_c^{EOS}).$$ (3.38)

The modified volume translation forces the EOS to pass through T_c, P_c, and v_c. There is a universal value for λ for the PR-EOS, 0.41. This value was determined by regressing data for many substances.

Figure 3.11 shows the predicted liquid densities for nitrogen and water when the modified volume translation is used. It is evident that the term on the right side of Eq. (3.36) provides an important description for pure substances in the critical region.

Multicomponent mixtures.   The volume-translation concept is extended to mixtures according to the following mixing rule:

$$c = \sum_{i=1}^{c} x_i c_i$$ (3.39)

$$v_c = \sum_{i=1}^{c} x_i v_{ci}.$$ (3.40)

The application of the conventional volume-translation technique generally improves the volume predictions away from the critical region, although the improvement is not guaranteed. In the retrograde and near-critical regions, both volumetric and modified volumetric translations may fail to predict volumetric behavior accurately (Bjorlykke and Firoozabadi, 1992).

Phase behavior of mixtures with well-defined components

The vapor pressure data for pure components are used to obtain the α parameter of the EOS; the density is then predicted. For mixtures, the EOS can be used to calculate not only the mixture density but also the phase behavior. In a limited sense, the phase behavior means the compositions and amounts of the equilibrium phases. The next chapter presents the equations for phase-behavior calculations. Essentially, phase-behavior calculations rely on the use of the expression for the fugacity of component i in the mixture given by Eq. (3.32) for the PR-EOS.

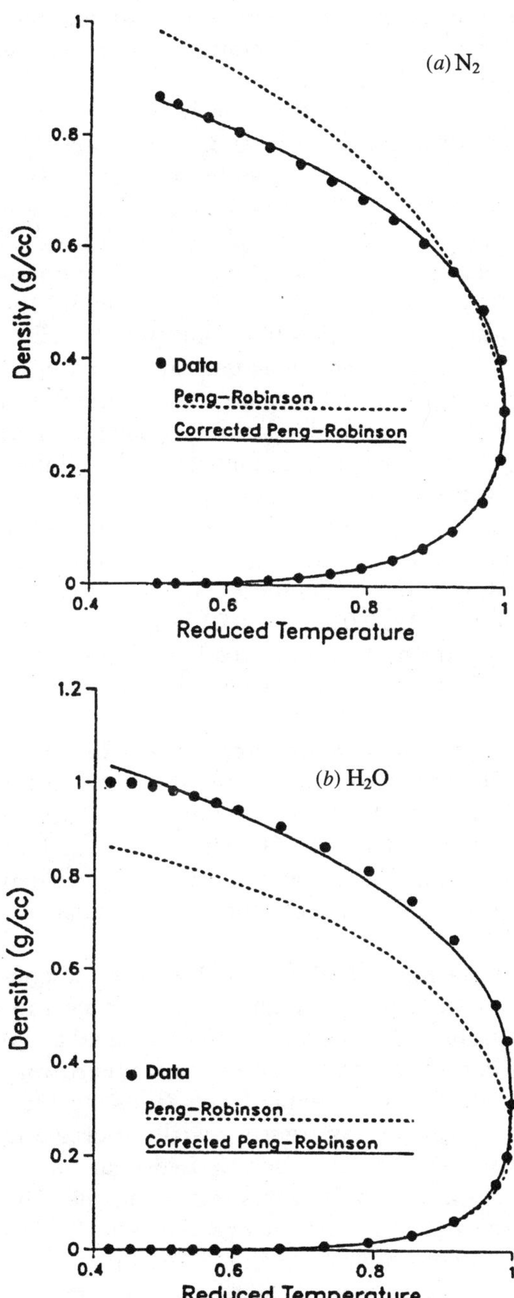

Figure 3.11 Comparison of the measured and predicted saturated densities (adapted from Mathias *et al.* 1989).

Mixtures with well-defined components are comprised of substances with a known boiling point. One may restrict the definition to mixtures with components of known critical properties.

The critical properties (T_c and P_c) and either the boiling point or acentric factor, ω, are needed to describe pure components by an EOS. For mixtures, the only additional parameter is the binary interaction parameter. This coefficient is often empirically determined from one or more data points of binary mixtures in the form of bubblepoint or K-values. Once these coefficients are available, VLE for multicomponent systems can be predicted. Figure 3.12 shows a comparison of the measured and computed K-values of a well-defined mixture (Yarborough, 1979; Firoozabadi, 1988). The SRK-EOS was used in the calculations. Note that the agreement between the measured and calculated values is good. For mixtures of different hydrocarbons where critical properties and normal boiling points or acentric factors are available, the VLE predictions from the PR-EOS and other similar equations are surprisingly good, except in the critical region. Figure 3.13 depicts the measured and predicted phase behavior of the $C_1/nC_4/nC_{10}$ system at 3000 psia and 160°F. The interaction coefficients between C_1/nC_4, C_1/nC_{10}, and C_4/nC_{10} were set at 0.012, 0.044, and 0.01 respectively, in the calculations. This ternary diagram demonstrates the limitation of the PR-EOS and other similar equations in the critical region.

Reservoir-fluids phase behavior and volumetric properties. Reservoir fluids are a complex mixture of thousands of components that exhibit very complex phase behavior. It is, however, surprising that a simple two-constant equation of state such as the PR-EOS can do an excellent job for vapor–liquid equilibria calculations away from the critical region. A discussion of the manner in which the calculations are performed is presented.

As was stated earlier, one needs T_c, P_c, and T_b or ω for every component in a mixture. However, critical property data beyond a certain carbon number are not yet available. The measurement of critical properties of heavy hydrocarbons is difficult and is subject to uncertainty because of thermal decomposition and chemical reactions at high temperatures. Thermal decomposition increases with increasing carbon number and temperature. When a heavy hydrocarbon is heated, it reaches a temperature at which it starts to decompose. The product of decomposition increases with temperature and time. Therefore, the measured critical point corresponds to the critical point of a mixture. In order to determine the critical point of the original substance, certain assumptions have to be made. For mildly unstable substances, the critical locus of the mixture (original substance and the products of decomposition) is assumed to be a linear function of

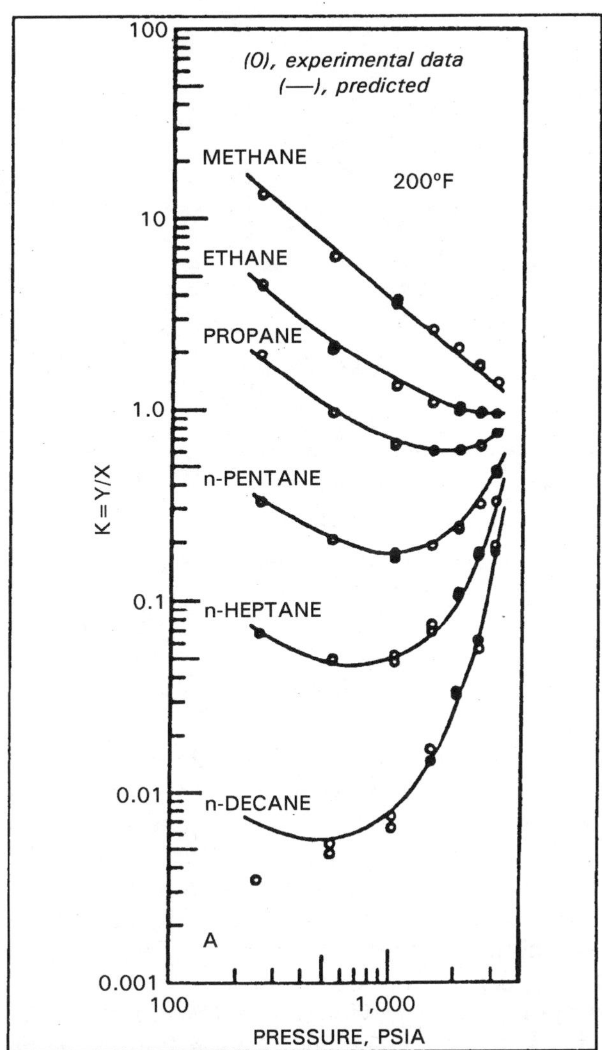

Figure 3.12 Comparison of the predicted and measured K-values of a well-defined hydrocarbon mixture (from Yarborough, 1979).

the mole fractions of the decomposed products. The criteria of stability and criticality from Chapter 4 can be used to determine the critical properties of pure heavy components (Anselme, 1988; Teja *et al*, 1990). This approach has been used to measure T_c and P_c of normal alkanes to nC_{18} by Anselme (1988). The pulse-heating method has also been used to measure the critical pressure and temperature (Nikitin, Pavlov, and Skripov, 1993). Using this method, Nikitin, Pavlov, and

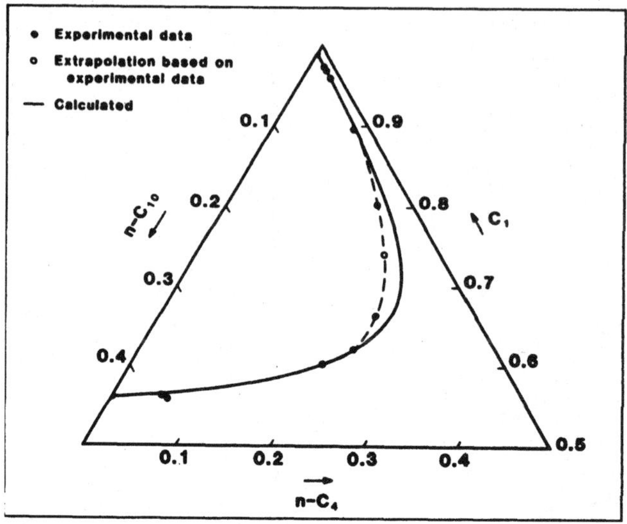

Figure 3.13   Phase diagram showing the predicted and measured phase envelope of $C_1/nC_4/nC_{10}$ at 160°F (from Firoozabadi and Aziz, 1986).

Popov (1997) measured critical pressure and temperatures of several normal alkanes including nC_{36}.

For heavier n-alkanes, one could use various correlations. Figures 3.14 and 3.15 plot critical P_c and T_c from various correlations. These two figures show that the uncertainty in T_c estimation is less than that in P_c. Extrapolation of the correlation to nC_{100} may result in unrealistic values of T_c and P_c. Also note that the trend of P_c data from Anselme (1988) and Nikitin *et al.* (1997) are not the same. Until the data of Nikitin *et al.* (1997) is verified, we should use them with caution. The subject of critical properties is further discussed in Chapter 5.

For crude oils, it is not practical to provide the analysis to thousands of components. Instead, for the C_{6+} residue, average boiling point, liquid density, and molecular weight of groups or cuts are measured. When only the amount and the density and the molecular weight of the C_{6+} or a heavier plus fraction is available, then it may become necessary for some calculations to estimate an extended analysis. The two-parameter gamma distribution function (Johnson and Kotz, 1970; Whitson, 1983) can be used for generating the molar distributions for the plus fractions. Figures 3.16 and 3.17 give average density and molecular weight for groups of compounds boiling between 0.5°C above the previous n-paraffin and 0.5°C above the normal paraffin carbon number used to identify the group. Table 3.2 shows the same data in tabular form. Crude oils, with few exceptions, follow these two graphs. The

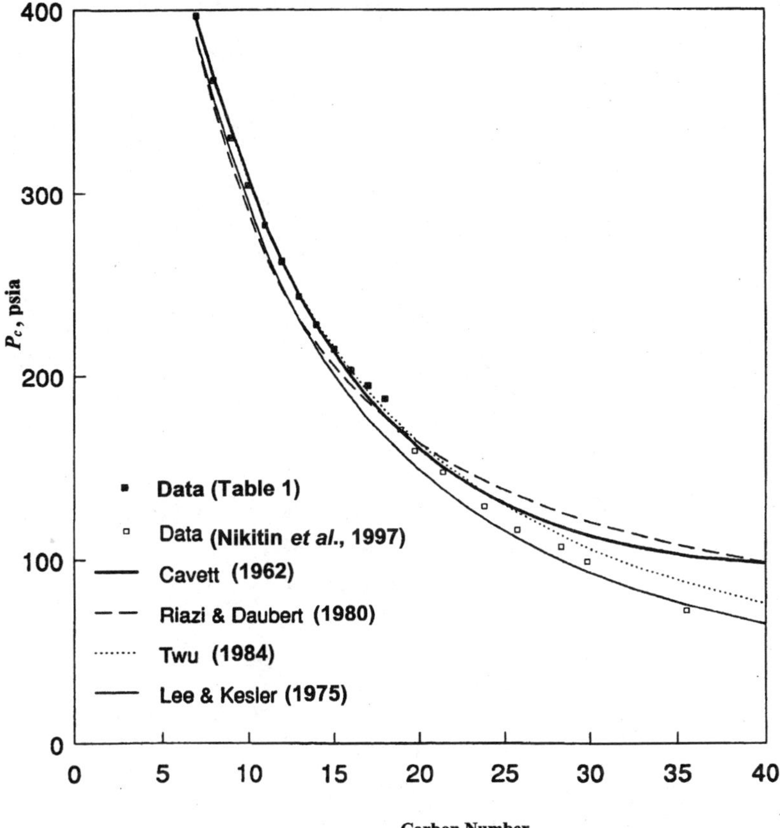

Figure 3.14 Comparison of n-alkanes P_c estimated from various correlations with data.

practice is to use (from measurements or estimation from Figs. 3.16 and 3.17) average normal boiling point and liquid density instead of measured properties available for pure components. From the density and boiling point, then correlations such as those of Cavett (1962) or Lee and Kesler (1975) are used to estimate critical properties. In addition to critical properties and normal boiling point, one needs binary interaction coefficients for phase-behavior calculation of petroleum fluids. It has been shown (Katz and Firoozabadi, 1978) that interaction coefficients for methane–heavy component binaries significantly improve the phase-behavior calculation of reservoir fluids. Measured phase behavior of binary mixtures provides interaction coefficients between hydrocarbon–hydrocarbon, hydrocarbon-nonhydrocarbon, and nonhydrocarbon-nonhydrocarbon pairs. In crude oils only one data point, such as a bubblepoint pressure, which is often available,

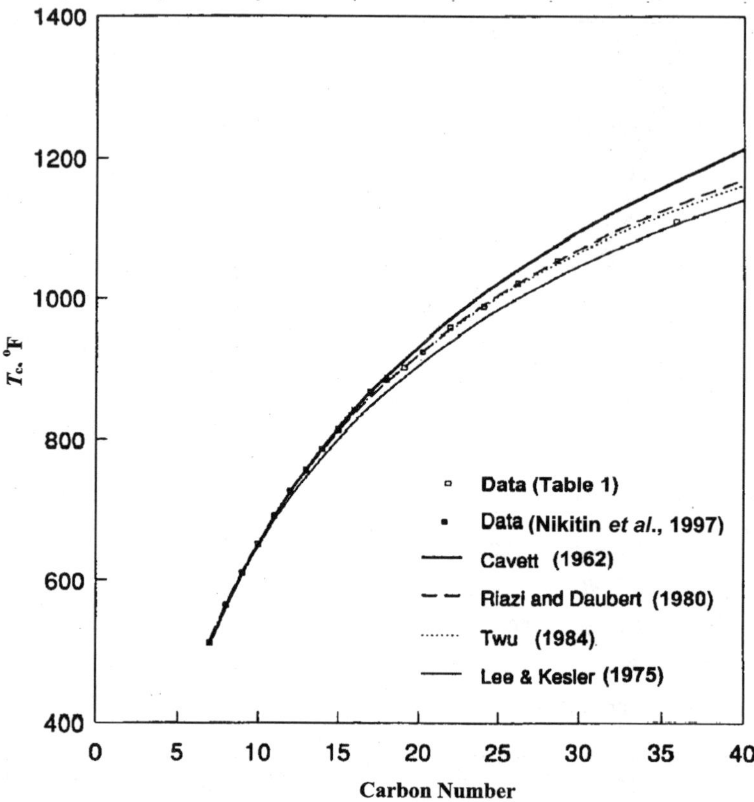

Figure 3.15   Comparison of n-alkanes T_c estimated from various correlations with data.

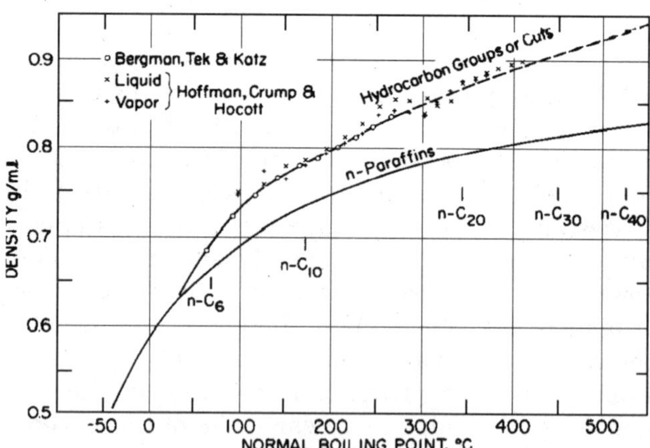

Figure 3.16   Density at 15.5°C $vs.$ boiling points of hydrocarbon groups in crude-oil and condensate systems (from Katz and Firoozabadi, 1978).

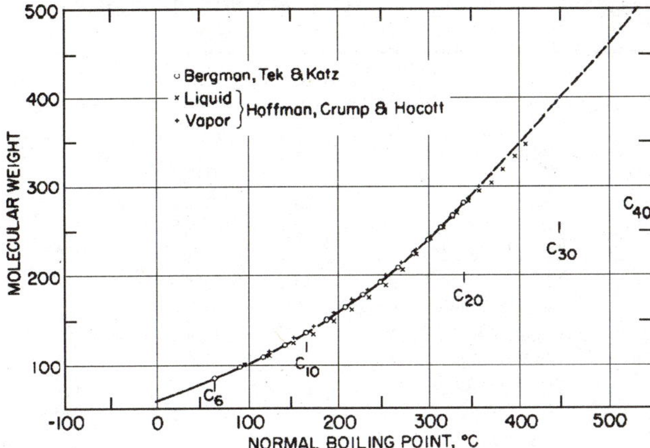

Figure 3.17 Molecular weight *vs.* boiling points of hydrocarbon groups in crude-oil and condensate systems (from Katz and Firoozabadi, 1978).

can provide the interaction coefficient between methane and the heavy end. Various interaction coefficients given by Arbabi and Firoozabadi (1995) are summarized in Table 3.3. Figure 3.18 compares predicted and measured equilibrium ratios for the lean natural gas condensate of Hoffman *et al.* (1953). The data of Hoffman *et al.* (1953) are perhaps the most complete condensate measurement in the literature. The data and prediction results are in good agreement. Figure 3.19 compares the predicted and experimental values for a crude oil from Roland (1945). From two data points at 120°F, the interaction coefficient between C_1 and C_{28+} residue was estimated (Katz and Firoozabadi, 1978). The estimated interaction coefficient was used to predict the K-values at 200°F. The figure shows excellent results from the PR-EOS.

Figure 3.20 shows the computed results and data for a near-critical condensate. The figure shows that as the K-values approach one (i.e., critical point), the prediction results deviate from the measured data.

Both the usefulness and promise of the cubic equations to describe gas and liquid phases and the problems of the description in the critical region and density deviation have forced some users to adjust several parameters of the EOS to match measured and calculated phase and volumetric behavior. These adjusted parameters, in essence, include critical properties and interaction coefficients.

Phase behavior of water and water-hydrocarbon mixtures

Water is an integral part of fluids in the reservoir. In addition to connate water which often covers the rock surface, H_2O at high temperatures, either in the form of liquid or in the form of steam, is injected into the reservoir to enhance the recovery of oil. At high

TABLE 3.2   Generalized properties of petroleum hexane-plus groups (From Katz and Firoozabadi, 1978)

Hydrocarbon Group	Boiling range °C	Boiling range °F	Average Boiling point °C	Average Boiling point °F	Density at 60 °F g/cm³	Molecular weight
C_6	36.5 to 69.2	97.9 to 156.7	63.9	147	0.685	84
C_7	69.2 to 98.9	156.7 to 210.1	91.9	197.5	0.722	96
C_8	98.9 to 126.1	210.1 to 259.1	116.7	242	0.745	107
C_9	126.1 to 151.3	259.1 to 304.4	142.2	288	0.764	121
C_{10}	151.3 to 174.6	304.4 to 346.4	165.8	330.5	0.778	134
C_{11}	174.6 to 196.4	346.3 to 385.5	187.2	369	0.789	147
C_{12}	196.4 to 216.8	385.5 to 422.2	208.3	407	0.800	161
C_{13}	216.8 to 235.9	422.2 to 456.7	227.2	441	0.811	175
C_{14}	235.9 to 253.9	456.7 to 489.2	246.4	475.5	0.822	190
C_{15}	253.9 to 271.1	489.2 to 520	266	511	0.832	206
C_{16}	271.1 to 287.3	520 to 547	283	542	0.839	222
C_{17}	287 to 303	547 to 577	300	572	0.847	237
C_{18}	303 to 317	577 to 603	313	595	0.852	251
C_{19}	317 to 331	603 to 628	325	617	0.857	263
C_{20}	331 to 344	628 to 652	338	640.5	0.862	275
C_{21}	344 to 357	652 to 675	351	664	0.867	291
C_{22}	357 to 369	675 to 696	363	686	0.872	305
C_{23}	369 to 381	696 to 717	375	707	0.877	318
C_{24}	381 to 392	717 to 737	386	727	0.881	331
C_{25}	392 to 402	737 to 756	397	747	0.885	345
C_{26}	402 to 413	756 to 775	408	766	0.889	359
C_{27}	413 to 423	775 to 793	419	784	0.893	374
C_{28}	423 to 432	793 to 810	429	802	0.896	388
C_{29}	432 to 441	810 to 825	438	817	0.899	402
C_{30}	441 to 450	826 to 842	446	834	0.902	416
C_{31}	450 to 459	842 to 857	455	850	0.906	430
C_{32}	459 to 468	857 to 874	463	866	0.909	444
C_{33}	468 to 476	874 to 888	471	881	0.912	458
C_{34}	476 to 483	888 to 901	478	895	0.914	472
C_{35}	483 to 491	901 to 915	486	908	0.917	486
C_{36}			493	922	0.919	500
C_{37}			500	934	0.922	514
C_{38}			508	947	0.924	528
C_{39}			515	959	0.926	542
C_{40}			522	972	0.928	556
C_{41}			528	982	0.930	570
C_{42}			534	993	0.931	584
C_{43}			540	1004	0.933	598
C_{44}			547	1017	0.935	612
C_{45}			553	1027	0.937	625

temperatures, i.e., above 300°F, the solubility of water in the crude oil increases substantially. An equation of state is an ideal tool to study water/reservoir fluid phase equilibria. However, phase behavior of water–hydrocarbon mixtures is complicated by the association between water molecules. Association means the tendency of the molecules to

TABLE 3.3   Binary interaction coefficients for the PR-EOS

(1) $\delta_{C_1-C_i} = 0.0289 + 1.633 \times 10^{-4} MW_{C_i}$ (Arbabi and Firoozabadi, 1995), where C_i is the hydrocarbon heavier than C_1

(2) $\delta_{C_1-CO_2} = 0.15$

(3) $\delta_{C_1-N_2} = 0.10$

(4) $\delta_{C_1-H_2S} = 0.1$

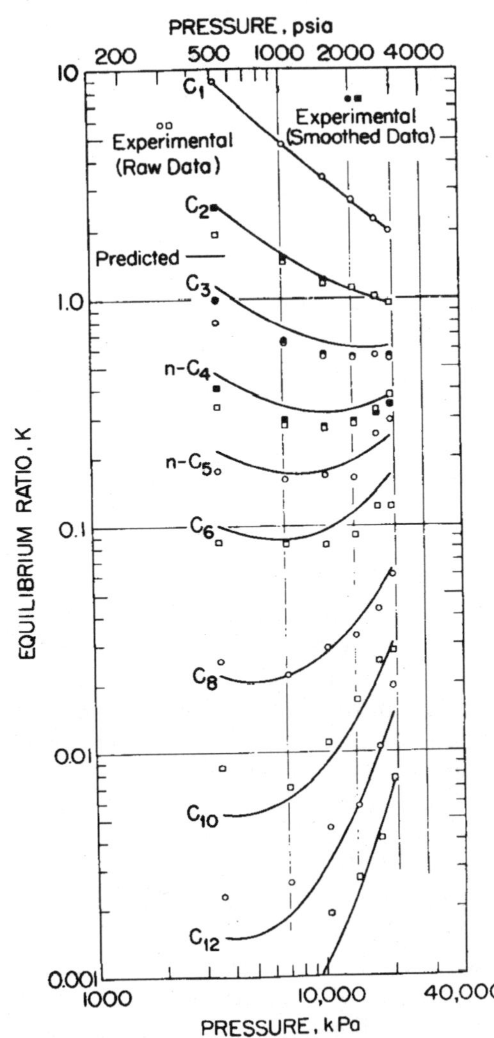

Figure 3.18 Comparison of predicted with experimental K-values of Hoffman *et al.* (1953) condensate at 201°F (from Katz and Firoozabadi, 1978).

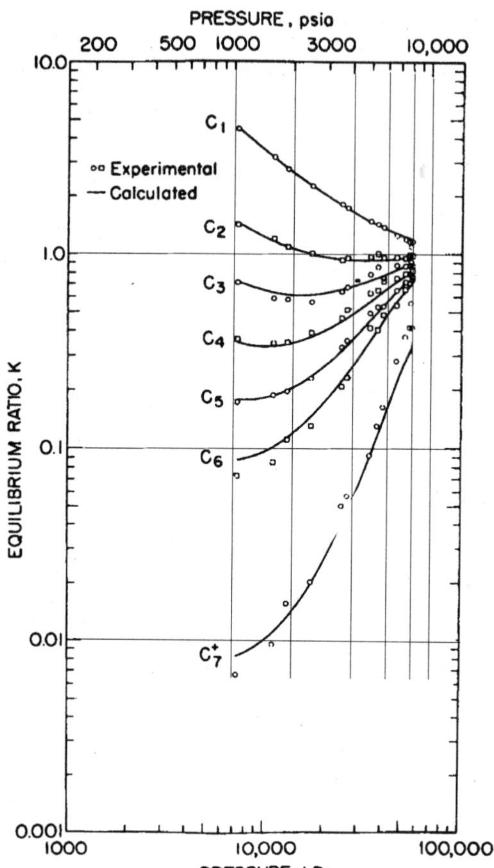

Figure 3.19 Comparison of predicted with experimental K-values for Roland crude oil at 200°F (from Katz and Firoozabadi, 1978).

aggregate and to form clusters. The ability of molecules to associate is mainly controlled by polarity. Cubic equations such as the PR-EOS are not suitable for calculation of the phase behavior of systems in which their molecules associate; only nonpolar substances can be described by cubic equations. Water molecules form a three-dimensional structure of hydrogen bonds (Michel *et al.*, 1989). This structure changes rapidly as the temperature rises. Conventional use of the cubic equations of state cannot, therefore, provide reliable results for water and water/reservoir-fluid phase behavior. While there are many suggestions, the explicit association suggested by Heidemann and Prausnitz (1976 *a, b*) seems simple for the use of the association concept. In this approach a predetermined association model is assumed and then introduced explicitly into the equation of state. Different explicit association models have been proposed by various authors. The following discusses

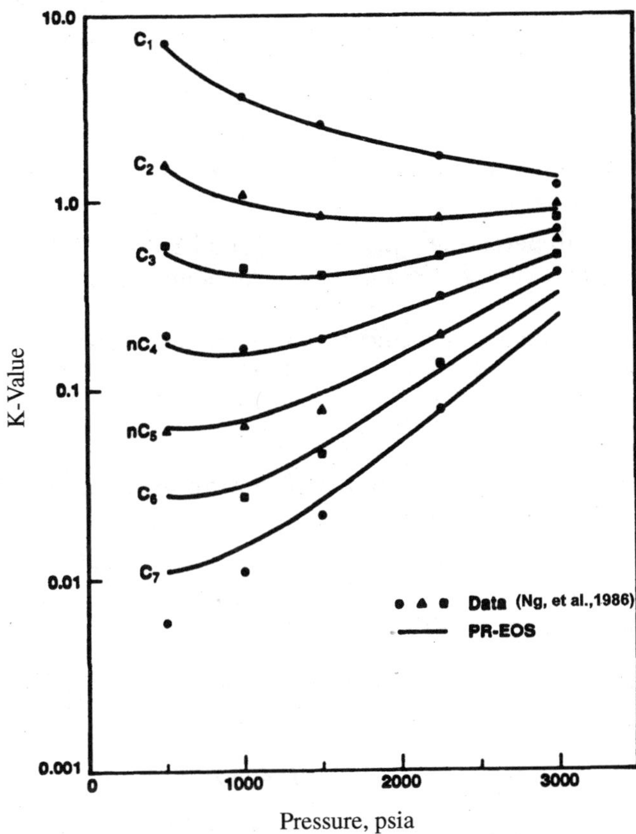

Figure 3.20   Comparison of the measured and computed K-values of the GPA gas-condensate fluid at 100°F (from Firoozabadi, 1988).

how to combine an association model with a cubic equation of state, such as the PR-EOS, to describe phase behavior of water–hydrocarbon mixtures.

Association equation of state (AEOS).   To first introduce the concept of chemical compressibility factor Z^{ch} and the association constant K, assume there are n_0 moles of a substance in which all the molecules are in the form of monomers, i.e., singly dispersed molecules. Then, if the fluid behaves as an ideal gas, $PV_0 = n_0RT$, where V_0 is the volume at pressure P. Let us now assume that some of the molecules will associate so that the true number of moles is n_{true}; then, the ideal gas PVT relationship takes the form

$$PV_{true} = n_{true}RT. \tag{3.41}$$

Equation (3.41) can also be written as

$$PV_{true} = Z^{ch} n_0 RT. \qquad (3.42)$$

From Eqs. (3.41) and (3.42),

$$Z^{ch} = n_{true}/n_0. \qquad (3.43)$$

If all the molecules form dimers, then $n_{true} = (1/2)n_o$ and $Z^{ch} = 1/2$. If all the molecules stay as monomers, $Z^{ch} = 1$. Next the association constant, K, is defined.

Consider a mixture of monomers A and dimers A_2 of a single-component system at temperature T and pressure P. Then

$$\mu_{A_2} = 2\mu_A. \qquad (3.44)$$

Assume that the mixture of species A and A_2 forms an ideal gas. The chemical potentials of species A and A_2 are

$$\mu_A(T, P) = \mu_A^0(T) + RT \ln P_A \qquad (3.45)$$

$$\mu_{A_2}(T, P) = \mu_{A_2}^0(T) + RT \ln P_{A_2}, \qquad (3.46)$$

where P_A and P_{A_2} are the partial pressures of A and A_2, respectively. Combining Eqs. (3.44) to (3.46),

$$\mu_{A_2}^0(T) + RT \ln P_{A_2} = 2\mu_A^0(T) + 2RT \ln P_A. \qquad (3.47)$$

Equation (3.47) can be written as

$$2\mu_A^0(T) - \mu_{A_2}^0(T) = RT \ln P_{A_2}/P_A^2. \qquad (3.48)$$

The association constant is defined by

$$RT \ln K = 2\mu_A^0(T) - \mu_{A_2}^0(T) = RT \ln P_{A_2}/P_A^2. \qquad (3.49)$$

But $\mu = h - Ts$, and therefore

$$RT \ln K = -\left[h_{A_2}^0(T) - 2h_A^0(T) \right] + T\left[s_{A_2}^0(T) - s_A^0(T) \right]. \qquad (3.50)$$

Let us define $\Delta h^o(T) =$ enthalpy of dimer formation $= h_{A_2}^0(T) - 2h_A^0(T)$, and $\Delta s^0(T) =$ entropy of dimer formation $= s_{A_2}^0(T) - 2s_A^0(T)$. From these definitions and Eq. (3.50),

$$\boxed{-\ln K = (\Delta h^0/RT) - \Delta s^0/R}. \qquad (3.51)$$

Similarly, enthalpy and entropy of trimer formation, etc. can be defined. For water, the experimentally determined Δh^0 and Δs^0 are available.

The virial equation of state truncated after the second term can be written as

$$Z = Pv/RT = 1 + B/v, \tag{3.52}$$

where B is the second virial coefficient. Coefficient B is separable into physical and chemical parts (Prausnitz *et al.* 1986):

$$B = B^{ch} + B^{ph}. \tag{3.53}$$

Combining Eqs. (3.52) and (3.53),

$$Z = 1 + (B^{ch} + B^{ph})/v = (1 + B^{ch}/v) + (1 + B^{ph}/v) - 1 \tag{3.54}$$

or

$$\boxed{Z = Z^{ch} + Z^{ph} - 1}. \tag{3.55}$$

In Eq. (3.55), Z^{ch} and Z^{ph} are the contributions from the chemical and physical interactions, respectively. Z^{ph} can be obtained from a cubic EOS, such as the PR-EOS. There are theoretically derived Z^{ch} expressions for different mixtures of associating substances, such as alcohols and phenols. However, because of high dimensionality in its aggregate structure, a sound association model may be difficult to derive for water. One may use theoretical analysis as a guide to find an empirical functional form for the water association constant (Shinta and Firoozabadi, 1995). Anderko (1991) has proposed the following empirical association model for aqueous systems:

$$Z^{ch} = \frac{x_a}{1 + (RTKx_a/v) + 8.2(RTKx_a/v)^2} + 1 - x_a, \tag{3.56}$$

where x_a is the mole fraction of H_2O and K is the association constant. Shinta and Firoozabadi (1995) modified Eq. (3.56) to

$$Z^{ch} = \frac{\xi x_a}{\sqrt{\xi} + (RTKx_a/v) + \beta(RTKx_a/v)^2} + 1 - x_a, \tag{3.57}$$

where $\beta = 0.00005$ and $\xi = 1.06$.

Solution of the association equation of state (AEOS). The incorporation of the association in the PR-EOS results in the following equation:

$$Z = (Pv/RT) = v/(v - n) - av/\left[RT(v^2 + 2bv - b^2)\right]$$
$$+ \xi x_a/\left[\sqrt{\xi} + (RTKx_a/v) + \beta(RTKx_a/v)^2\right] - x_a + 1. \tag{3.58}$$

Equation (3.58) is obtained from combining Eqs. (3.55) and (3.57) and the PR-EOS for Z^{ph}. This equation is a polynomial of degree six in molar volume, and therefore is no longer cubic. One has to find the roots of this polynomial and to relate the appropriate root to the vapor and liquid phases. Fortunately, Eq. (3.58) has a maximum of three significant roots. The remaining roots are either negative, complex, or less than the covolume parameter b (Shinta and Firoozabadi, 1995). The smallest root ($v > b$) is assigned to the liquid phase and the largest root ($v < RT/P$, i.e., less than ideal gas molar volume) is assigned to the gas phase. The Brent method (1973) can be used to conduct the search that satisfies $(\partial P/\partial v)_T < 0$ in the interval b to RT/P to obtain v.

There are six parameters in Eq. (3.58) that have to be evaluated. These parameters are related to

$$\Delta h^0 = \Delta h^{00} + \Delta c_P^0 (T - T^0), \tag{3.59}$$

$$\Delta s^0 = \Delta s^{00} + \Delta c_P^0 \ln(T/T^0), \tag{3.60}$$

and apparent critical properties of water. Parameters Δh^{00} and Δs^{00} are values of enthalpy and entropy of association at the reference temperature of T^0 (say 273.15 K), and Δc_P^0 is the heat capacity of association at the reference temperature. The apparent critical properties T_c' and P_c' and acentric factor ω' relate the physical contribution of water to the cubic equation of state. Apparent properties are equivalent to the properties of an associating monomer that has not aggregated. Therefore, T_c', P_c', and ω' are different from the true measured values. All six parameters can be evaluated by fitting the vapor pressure and saturation liquid density data of water from the triple point to the critical point. Once those parameters are obtained, they can be used for calculation of the aqueous-mixture phase behavior and properties. The association parameters for the PR-EOS for H_2O are (Shinta and Firoozabadi, 1995) $P_c' = 23.305\,\text{MPa}, \quad T_c' = 391\,\text{K}, \quad \omega' = 0.04, \quad \Delta h^{00} = -26.641\,\text{KJ/gmole}, \quad \Delta s^{00} = -93.49\,\text{J/(gmole.K)}$, and $\Delta c_P^0 = 0$.

The next step is the calculation of the fugacity coefficient of the association substance (that is, water), φ_i^{ch}. The physical fugacity coefficient for the PR-EOS is readily available (see Eq. (3.32)). In an AEOS, the fugacity coefficient of an associating component in each phase is the sum of both the chemical and physical contributions:

$$\ln \varphi_i Z = \ln \varphi_i^{ch} Z^{ch} + \ln \varphi_i^{ph} Z^{ph}, \quad i = H_2O \tag{3.61}$$

(see Example 3.5b). Z^{ch} is calculated from Eq. (3.57) and then used to calculate φ_i^{ch}. For the nonassociating components (see Example 3.5a),

$$\varphi_i = \varphi_i^{ph} Z^{ph} / Z \qquad i = 1, \dots, c, \text{ except } H_2O. \tag{3.62}$$

Equation (3.55) can lead to the following pressure expression:

$$P = P^{ph} + P^{ch} - RT/v, \tag{3.63}$$

where P^{ph} and P^{ch} are the physical and chemical contributions to the total pressure. Therefore, Z^{ph} and φ^{ph} in Eqs. (3.61) and (3.62) should be evaluated at P^{ph}. One uses Eq. (3.58) first to calculate the molar volume v which is then used to calculate P^{ph}, Z^{ph}, and φ^{ph} from the PR-EOS.

Figure 3.21 shows a plot of the calculated Z, Z^{ch}, and Z^{ph} for water $vs.$ T_r at saturation pressure. Figure 3.21b shows the contribution of Z^{ph} and Z^{ch} to the liquid compressibility factor. At low reduced tempera-

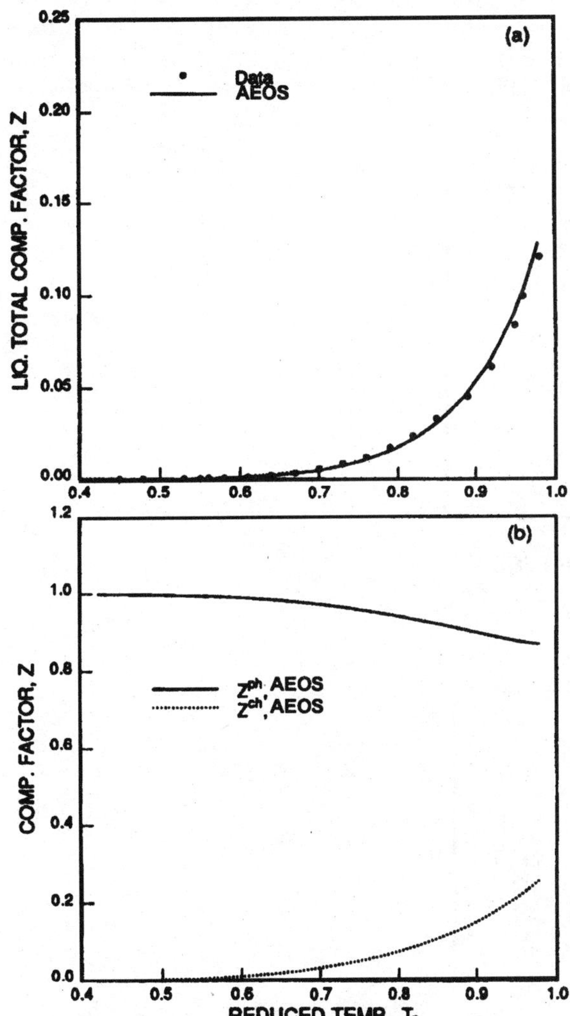

Figure 3.21 Plot of Z, Z^{ph}, and Z^{ch} for H_2O liquid $vs.$ reduced temperature (adapted from Shinta and Firoozabadi, 1995).

tures below 0.55 (also at low pressures), there is strong association (low Z^{ch}). At high pressures and temperatures, the association in the liquid decreases and liquid converges to nonassociating liquid as the pressure increases; Z^{ch} increases and Z^{ph} decreases to keep the calculated total Z from Eq. (3.55) close to the true value.

The application of the AEOS to hydrocarbon–water mixtures requires temperature-dependent binary interaction coefficients. These data can be obtained from binary water-hydrocarbon data. Once such data are available, then the AEOS can be used to predict the phase behavior of H_2O–crude oil systems. Figure 3.22 shows the binary interaction coefficients between C_1 and heavier hydrocarbons from Shinta and Firoozabadi (1997). Binary interaction coefficients of

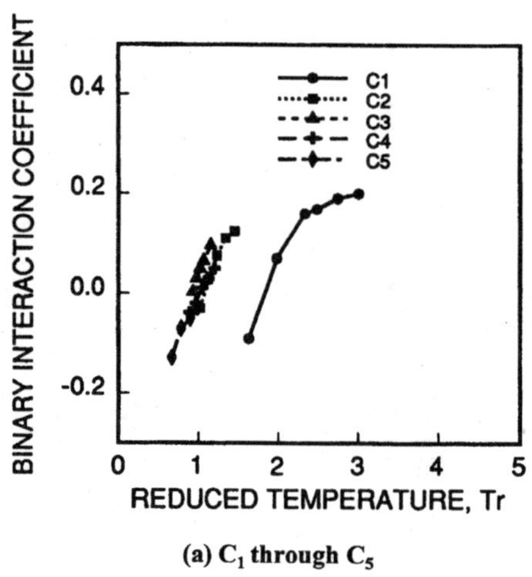

(a) C_1 through C_5

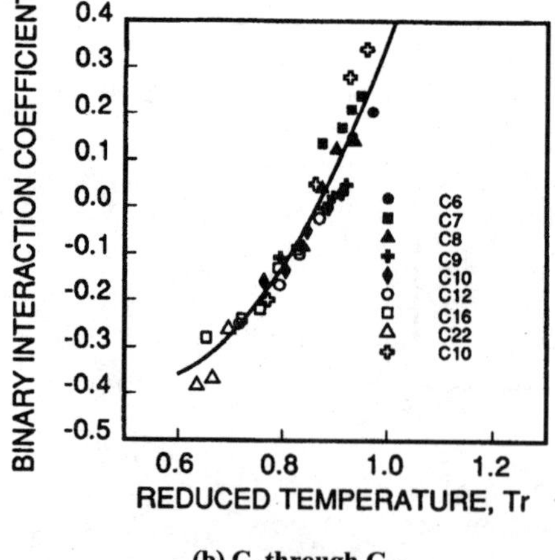

(b) C_6 through C_{22}

Figure 3.22 Binary interaction coefficients of C_1 through C_{22} with water (adapted from Shinta and Firoozabadi, 1997).

TABLE 3.4   Binary interaction coefficients of H_2O/CO_2 and H_2O/H_2S systems for the PR-EOS (from Shinta and Firoozabadi, 1995)

System	Component Reduced Temperature[*,†]	Binary Interaction Coefficient
	1.0000	−0.028
	1.1445	0.002
H_2O/CO_2*	1.2960	0.026
	1.3910	0.055
	1.5700	0.070
	0.9811	0.0450
H_2O/H_2S[†]	1.1298	0.0500
	1.2786	0.0550
	1.4273	0.0570

*Reduced temperature of CO_2.
[†]Reduced temperature of H_2S.

H_2O/CO_2, H_2O/H_2S are given in Table 3.4. Figure 3.23 compares the predicted results for the $C_1/nC_4/H_2O$ from the AEOS with data. Figure 3.24 shows the steam distillation yield of two crude oils from Richardson *et al.* (1992). The API gravities of crudes A and D are 33° and 12.6°, respectively. The steam distillation yield is defined as the volume ratio of the distilled oil to the original oil. The results correspond to a pressure of 500 psig and a temperature of 467°F. The binary interaction coefficients between components other than H_2O remain the same as the ones described earlier. The results presented in Figs. 3.23 and 3.24 demonstrate the usefulness and accuracy of the AEOS.

Two-phase isothermal compressibility

The isothermal compressibility of a single-phase fluid is defined in a straightforward fashion:

$$C_T = -\frac{1}{V}\left(\frac{\partial V}{\partial P}\right)_T. \tag{3.64}$$

Similarly, the isothermal compressibility of a two-phase multicomponent system is defined by

$$C_T^{tp} = -\frac{1}{V_t}\left(\frac{\partial V_t}{\partial P}\right)_{T,\underline{n}}. \tag{3.65}$$

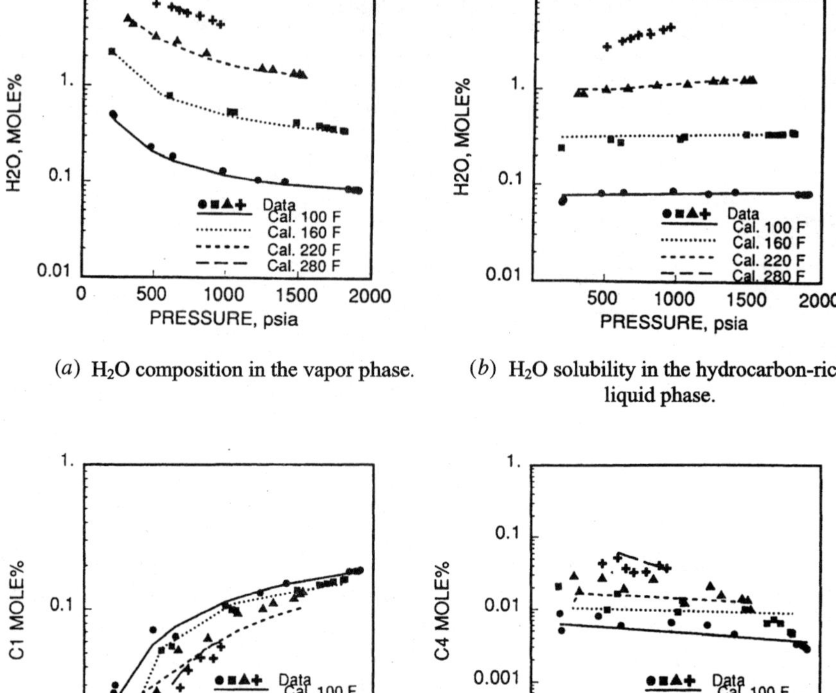

(*a*) H₂O composition in the vapor phase.

(*b*) H₂O solubility in the hydrocarbon-rich liquid phase.

(*c*) Solubility of C₁ in the water-rich liquid phase.

(*d*) Solubility of *n*C₄ in the water-rich liquid phase.

Figure 3.23   Comparison of the calculated and measured (McKetta and Katz, 1948) phase composition of $C_1/nC_4/H_2O$ system (adapted from Shinta and Firoozabadi, 1997).

In the above equation, $\underline{n} = (n_1, n_2, \ldots, n_c)$ represents the total moles of the components in both phases, and the derivative term on the right side represents the change in the total fluid volume, V_t, caused by a small change in the pressure of the closed system at constant temperature. Isothermal two-phase compressibility is important in solving certain well-test (Macais, 1985) and compositional reservoir simulation equations (Acs, Doleschall, and Farkas, 1985; Watts, 1986).

Figures 3.25 and 3.26 highlight isothermal two-phase compressibility in the two-phase region. Figure 3.25 shows the isothermal $P\text{–}V$ relationship for a single-component fluid; the compressibility is proportional

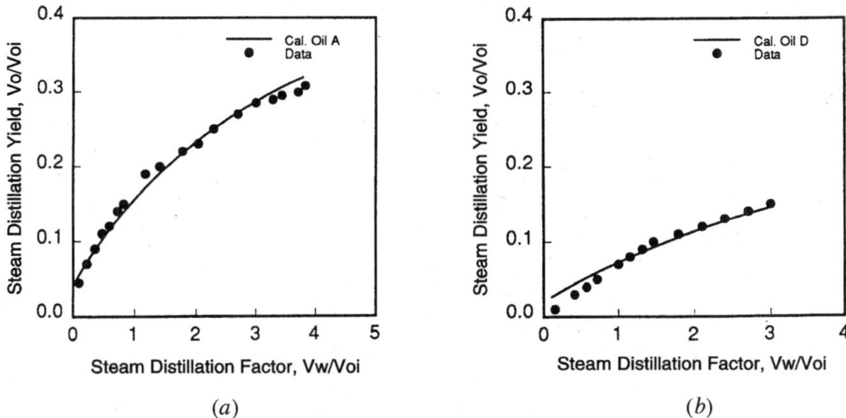

Figure 3.24 Cumulative steam-distillation yield of two crude oils (adapted from Shinta and Firoozabadi, 1997).

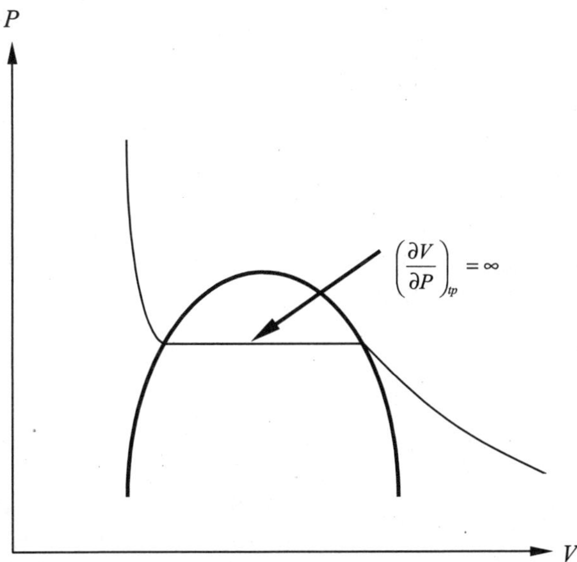

Figure 3.25 Compressibility of a pure-component fluid in the two-phase region.

to $(\partial V / \partial P)_T$. This figure shows that the $(\partial V / \partial P)_T$ of the gas phase is higher than that of the liquid phase, as expected. However, in the two-phase region, $(\partial V / \partial P)_T = \infty$, implying that the isothermal two-phase compressibility for a single-component fluid is infinite.

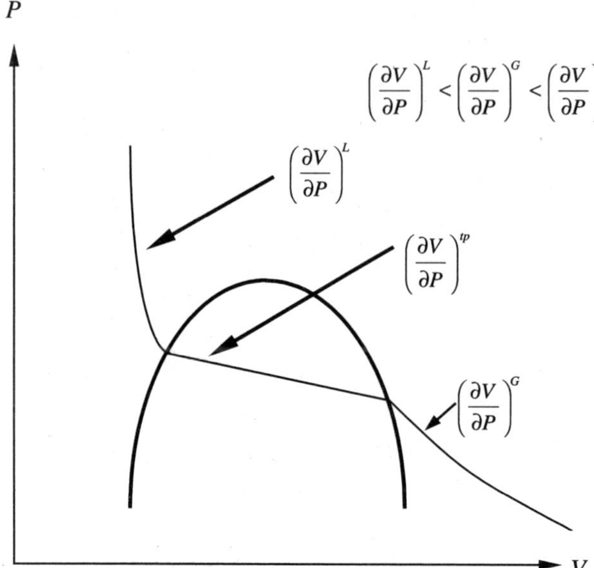

Figure 3.26 Compressibility of a mixture in the two-phase region.

Figure 3.26 reveals that $(\partial V/\partial P)_{T,\boldsymbol{n}}$ for a mixture in the two-phase region might be higher than in the single-phase gas region. The implication is that the isothermal two-phase compressibility might be higher than the isothermal gas-phase compressibility. Practicing engineers often use the following relationship to obtain the two-phase compressibility:

$$C_T^{tp} = C_T^G S^G + C_T^L S^L, \tag{3.66}$$

where C_T^{tp} is the two-phase compressibility and S^G and S^L are the gas and liquid saturations. This equation is invalid when there is mass transfer between the phases; at the bubblepoint, for example, $S^G = 0$ and $C_T^{tp} = C_T^L$ according to Eq. (3.66). As we will see shortly, when we get even one bubble of gas, C_T^{tp} may become several times higher than the compressibility of the compressed liquid just before gas-phase appearance.

The equations for the calculation of two-phase isothermal compressibility (Firoozabadi *et al.*, 1988b) follow. The key term in the evaluation of two-phase compressibility is the derivative of volume with respect to pressure at constant temperature and overall composition. To calculate this term, one writes

$$V_t = V_1 + V_2, \tag{3.67}$$

where V_1 and V_2 are the gas- and liquid-phase volumes, respectively. From the relation $PV_j = Z_j n_{j,t} RT$, where $n_{j,t}$ is the total number of moles in phase j,

$$V_t = \frac{RT}{P} \sum_{j=1}^{2} Z_j n_{j,t}. \tag{3.68}$$

Taking the derivative of Eq. (3.68) with respect to pressure and holding temperature and overall moles constant yields

$$\left(\frac{\partial V_t}{\partial P}\right)_{T,\underline{n}} = -\frac{RT}{P^2} \sum_{j=1}^{2} Z_j n_{j,t} + \frac{RT}{P} \sum_{j=1}^{2} \left[n_{j,t}\left(\frac{\partial Z_j}{\partial P}\right)_{T,\underline{n}} + Z_j\left(\frac{\partial n_{j,t}}{\partial P}\right)_{T,\underline{n}} \right]. \tag{3.69}$$

The unknown derivatives in the right side of Eq. (3.69) are $(\partial n_{j,t}/\partial P)_{T,\underline{n}}$ and $(\partial Z_j/\partial P)_{T,\underline{n}}$. The change in the total number of moles of phase j with pressure can be expressed as the sum of the changes in moles of each component in phase j with pressure:

$$\left(\frac{\partial n_{j,t}}{\partial P}\right)_{T,\underline{n}} = \sum_{i=1}^{c} \left(\frac{\partial n_{j,i}}{\partial P}\right)_{T,\underline{n}} \quad j = 1, 2, \tag{3.70}$$

where $n_{j,i}$ is the number of moles of component i in phase j. There are $2c$ unknown derivatives $(\partial n_{j,i}/\partial P)_{T,\underline{n}}$ in Eq. (3.70); $2c$ equations are needed. These equations come from c material balance equations,

$$n_i = n_{1,i} + n_{2,i} \quad i = 1, \ldots, c, \tag{3.71}$$

which yield $\quad \left(\frac{\partial n_{1,i}}{\partial P}\right)_{T,\underline{n}} + \left(\frac{\partial n_{2,i}}{\partial P}\right)_{T,\underline{n}} = 0 \quad i = 1, \ldots, c, \tag{3.72}$

and the remaining c equations result from the equilibrium conditions before and after the change in pressure of the system:

$$\left(\frac{\partial f_{1,i}}{\partial P} - \frac{\partial f_{2,i}}{\partial P}\right)_{T,\underline{n}} = \left[\sum_{k=1}^{c} \left(\frac{\partial f_{1,i}}{\partial n_{1,k}}\right)_{n_{1,k}} \left(\frac{\partial n_{1,k}}{\partial P}\right)_{\underline{n}} - \sum_{k=1}^{c} \left(\frac{\partial f_{2,i}}{\partial n_{2,k}}\right)_{n_{2,k}} \left(\frac{\partial n_{2,k}}{\partial P}\right)_{\underline{n}} \right]$$

$$+ \left[\left(\frac{\partial f_{1,i}}{\partial P}\right)_{\underline{n}_1} - \left(\frac{\partial f_{2,i}}{\partial P}\right)_{\underline{n}_2} \right] = 0 \quad i = 1, \ldots, c. \tag{3.73}$$

The subscripts $n_{1,k} = (n_{1,1}, \ldots, n_{1,k-1}, n_{1,k+1}, \ldots n_{1,c})$, and $\underline{n}_1 = (n_{1,1}, \ldots, n_{1,c})$. A similar definition applies to $n_{2,k}$ and $\underline{n}_2$. The subscript T has been dropped from the terms in the right side of the above equation. The derivatives of fugacity in the right side of Eq. (3.73) for the PR-EOS can be calculated readily by using software such as Mathematica. For the sake of brevity, we will drop some of the subscripts in the following equations.

Combining Eqs. (3.72) and (3.73) yields

$$\sum_{k=1}^{c}\left(\frac{\partial f_{1,i}}{\partial n_{1,k}}+\frac{\partial f_{2,j}}{\partial n_{2,k}}\right)\left(\frac{\partial n_{2,k}}{\partial P}\right)=\left[\left(\frac{\partial f_{1,j}}{\partial P}\right)_{T,\underline{n}_1}-\left(\frac{\partial f_{2,j}}{\partial P}\right)_{T,\underline{n}_2}\right]\qquad i=1,\ldots,c$$

$$(3.74)$$

After solving for $(\partial n_{j,k}/\partial P)_{T,\underline{n}},\, j=1,2$, the following expression gives the various terms contributing to $(\partial Z_j/\partial P)_{T,\underline{n}}$:

$$\left(\frac{\partial Z_j}{\partial P}\right)_{T,\underline{n}}=\left(\frac{\partial Z_j}{\partial P}\right)_{T,\underline{n}_j}+\sum_{k=1}^{c}\left(\frac{\partial Z_j}{n_{j,k}}\right)\left(\frac{\partial n_{j,k}}{\partial P}\right)\qquad j=1,2.\qquad(3.75)$$

An EOS can be used to calculate the $(\partial Z_j/\partial P)$ and $(\partial Z_j/\partial n_{j,i})$ terms on the right side of Eq. (3.75).

The following example illustrates the use of the above equations in computing the isothermal two-phase compressibility of multicomponent systems. Consider a three-component system in the two-phase gas–liquid region. The calculation of the $(\partial n_{j,i}/\partial P)_{T,\underline{n}}$ terms is the major task of computing two-phase compressibility. The matrix representation of Eq. (3.74) for the three-component system is

$$\begin{bmatrix}\dfrac{\partial f_{1,1}}{\partial n_{1,1}}+\dfrac{\partial f_{2,1}}{\partial n_{2,1}} & \dfrac{\partial f_{1,1}}{\partial n_{1,2}}+\dfrac{\partial f_{2,1}}{\partial n_{2,2}} & \dfrac{\partial f_{1,1}}{\partial n_{1,3}}+\dfrac{\partial f_{2,1}}{\partial n_{2,3}} \\[2ex] \dfrac{\partial f_{1,2}}{\partial n_{1,1}}+\dfrac{\partial f_{2,2}}{\partial n_{2,1}} & \dfrac{\partial f_{1,2}}{\partial n_{1,2}}+\dfrac{\partial f_{2,2}}{\partial n_{2,2}} & \dfrac{\partial f_{1,2}}{\partial n_{1,3}}+\dfrac{\partial f_{2,2}}{\partial n_{2,3}} \\[2ex] \dfrac{\partial f_{1,3}}{\partial n_{1,1}}+\dfrac{\partial f_{2,3}}{\partial n_{2,1}} & \dfrac{\partial f_{1,3}}{\partial n_{1,2}}+\dfrac{\partial f_{2,3}}{\partial n_{2,2}} & \dfrac{\partial f_{1,3}}{\partial n_{1,3}}+\dfrac{\partial f_{2,3}}{\partial n_{2,3}} \end{bmatrix}$$

$$\times\begin{bmatrix}\dfrac{\partial n_{2,1}}{\partial P} \\[2ex] \dfrac{\partial n_{2,2}}{\partial P} \\[2ex] \dfrac{\partial n_{2,3}}{\partial P}\end{bmatrix}=\begin{bmatrix}\dfrac{\partial f_{1,1}}{\partial P}-\dfrac{\partial f_{2,1}}{\partial P} \\[2ex] \dfrac{\partial f_{1,2}}{\partial P}-\dfrac{\partial f_{2,2}}{\partial P} \\[2ex] \dfrac{\partial f_{1,3}}{\partial P}-\dfrac{\partial f_{2,3}}{\partial P}\end{bmatrix}.\qquad(3.76)$$

A similar procedure can be derived to compute two-phase volume expansivity for use in thermal models. The defining equation for volume expansivity is

$$e = \frac{1}{V_t}\left(\frac{\partial V_t}{\partial T}\right)_{P,\underline{n}}.$$
(3.77)

The procedure is virtually identical to the calculation of isothermal compressibility.

Taking the derivative of Eq. (3.77) with respect to temperature, holding pressure and overall mole fractions constant, yields

$$\left(\frac{\partial V_t}{\partial T}\right)_{P,\underline{n}} = \frac{R}{P}\sum_{j=1}^{2} Z_j n_{j,t} + \frac{RT}{P}\sum_{j=1}^{2}\left[n_{j,t}\left(\frac{\partial Z_j}{\partial T}\right)_{P,\underline{n}} + Z_j\left(\frac{\partial n_{j,t}}{\partial T}\right)_{P,\underline{n}}\right].$$
(3.78)

The $(\partial n_{j,t}/\partial T)_{P,\underline{n}}$ term is expressed as

$$(\partial n_{j,t}/\partial T)_{P,\underline{n}} = \sum_{i=1}^{c}\left(\frac{\partial n_{j,i}}{\partial T}\right)_{P,\underline{n}} \qquad j = 1, 2.$$
(3.79)

The evaluation of $(\partial n_{j,t}/\partial T)_{P,\underline{n}}$ and $(\partial Z_j/\partial T)_{P,\underline{n}}$ terms are straightforward (see Example 3.9).

Two-phase isentropic compressibility and two-phase sonic velocity

In the previous section, the two-phase isothermal compressibility of multicomponent systems was formulated using the equilibrium assumption that there is no gradient of chemical potentials in the systems. In this section, the two-phase isentropic compressibility and the two-phase sonic velocity for multicomponent systems will be formulated. Again, we will make the assumption of equilibrium, which implies the gradients of chemical potentials are zero. The equilibrium assumption regarding the two-phase compressibilities depends on the problem and may or may not be justified. As long as there is no supersaturation, and adequate time is allowed to reach the state of equilibrium, then the equilibrium criterion can be invoked.

The isentropic compressibility and the thermodynamic sonic velocity are related to each other both in the single-phase and two-phase states. They are used in problems in the exploration and production of hydrocarbon reservoirs and in different disciplines. We should also make a comment on the thermodynamic sonic velocity, which is a purely thermodynamic property. The thermodynamic sonic velocity is equal to the true sonic velocity over a wide range of frequencies and amplitudes.

However, at high frequencies where the thermal properties depend on the rate of heating, then the true sonic velocity and the thermodynamic sonic velocity are not the same (see Chapters 2 and 3 of Rowlinson and Swinton, 1982; Voronel, 1976).

In the following, we will first derive the expressions that can be used to calculate the isentropic two-phase compressibility. The thermodynamic sonic velocity then can be readily calculated from the isentropic compressibility. We could have combined the derivations for the isothermal and isentropic compressibilities, but have decided on separate derivations for the sake of simplicity.

Compressibility is defined on the basis of the thermodynamic path. For isothermal compressibility, the path is constant temperature; for isentropic compressibility C_S, the thermodynamic path is constant entropy and is defined by

$$C_S = -\frac{1}{V}\left(\frac{\partial V}{\partial P}\right)_{S,\underline{n}}. \tag{3.80}$$

The isothermal and isentropic compressibilities are related by a simple expression in the single-phase state,

$$C_T = \frac{c_P}{c_V}C_S \tag{3.81}$$

where c_P and c_V are heat capacities at constant pressure and volume, respectively. The derivation of Eq. (3.81) is provided in Example 3.8. Since $c_P \geqslant c_V$ (see Example 3.8), then $C_T \geq C_S$.

The difference between C_T and C_S depends on pressure, temperature, and composition and may vary from 10 to several hundred percent in the single-phase state.

The two-phase isentropic compressibility C_S^{tp} is defined as,

$$C_S^{tp} = -\frac{1}{V_t}\left(\frac{\partial V_t}{\partial P_2}\right)_{S,\underline{n}}. \tag{3.82}$$

Note that in Eq. (3.82), we define C_S using pressure of phase 2, say the liquid phase to take into account the effect of interface curvature. Figure 3.27 shows the schematics of the process for the estimation of C_S^{tp} by increasing the volume a small amount from V_t to $V_t + \Delta V_t$. The volume of each phase j is given by

$$V_j = RT\frac{Z_j n_{j,t}}{P_j}, \tag{3.83}$$

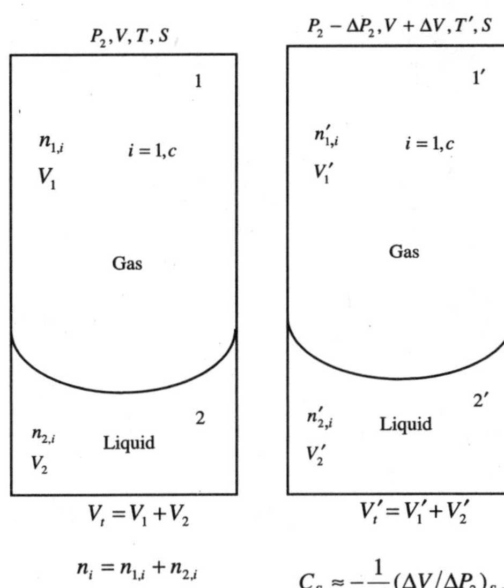

Figure 3.27 Schematics of changes associated with the calculation of two-phase C_S (adapted from Firoozabadi and Pan, 1997).

where P_j is the pressure of phase j. The total volume V_t is then

$$V_t = RT \sum_{j=1}^{2}\left(\frac{Z_j n_{j,t}}{P_j}\right). \tag{3.84}$$

The variation of V_t with respect to P_2 at constant S and $\underline{n}$ can be obtained from the use of Eq. (3.83):

$$
\begin{aligned}
(\partial V_t/\partial P_2)_{S,\underline{n}} = {} & RT \sum_{j=1}^{2}[(Z_j/P_j)(\partial n_{j,t}/\partial P_2)_{S,\underline{n}} + (n_{j,t}/P_j)(\partial Z_j/\partial P_2)_{S,\underline{n}} \\
& - (Z_j n_{j,t}/P_j^2)(\partial P_j/\partial P_2)_{S,\underline{n}}] \\
& + \left[R \sum_{j=1}^{2}(Z_j n_{j,t}/P_j)(\partial T/\partial P_2)_{S,\underline{n}} \right]
\end{aligned}
\tag{3.85}
$$

Various derivative expressions on the right side of Eq. (3.85) are evaluated next.

$(\partial Z_j/\partial P_2)_{S,\underline{n}}$ — One can write the differential of

$$dZ_j = (\partial Z_j/\partial T)_{P,\underline{n}_j} dT + (\partial Z_j/\partial P_j)_{T,\underline{n}_j} dP_j + \sum_{i=1}^{c}(\partial Z_j/\partial n_{j,i})_{T,P_j,n_{j,i}} dn_{j,i},$$

$$\tag{3.86}$$

where $\underline{n}_j = (n_{j,1}, \ldots, n_{j,c})$ and $\underline{n}_{j,i} = (n_{j,1}, \ldots, n_{j,i-1}, n_{j,i+1}, \ldots, n_{j,c})$; $n_{j,i}$ is the number of moles of component i in phase j. By dividing Eq. (3.86) by ∂P_2 at constant $\underline{n}$ and S,

$$(\partial Z_j/\partial P_2)_{S,\underline{n}} = (\partial Z_j/\partial T)_{P_j,\underline{n}_j}(\partial T/\partial P_2)_{S,\underline{n}} + (\partial Z_j/\partial P_j)_{T,\underline{n}_j}$$

$$\cdot(\partial P_j/\partial P_2)_{S,\underline{n}} + \sum_{i=1}^{c}(\partial Z_j/\partial n_{j,i})_{T,P_j,\underline{n}_{j,i}}(\partial n_{j,i}/\partial P_2)_{S,\underline{n}} \qquad j = 1, 2, \qquad (3.87)$$

where coefficients $(\partial Z_j/\partial T)_{P_j,\underline{n}_j}$, $(\partial Z_j/\partial P_j)_{T,\underline{n}_j}$, and $(\partial Z_j/\partial n_{j,i})_{T,P_j,\underline{n}_{j,i}}$ can be calculated from an EOS. These derivatives are easily obtained from software such as Mathematica.

$(\partial n_{j,t}/\partial P_2)_{S,\underline{n}}$ — the expression for $(\partial n_{j,t}/\partial P_2)$ is given by

$$(\partial n_{j,t}/\partial P_2)_{S,\underline{n}} = \sum_{i=1}^{c}(\partial n_{j,i}/\partial P_2)_{S,\underline{n}} \qquad j = 1, 2. \qquad (3.88)$$

Equations (3.85), (3.86), and (3.88) contain $2(c+1)$ unknowns on the right side: (1) one unknown $(\partial P_1/\partial P_2)_{S,\underline{n}}$, (2) $2c$ unknowns $(\partial n_{j,i}/\partial P_2)_{S,\underline{n}}$, and (3) one unknown $(\partial T/\partial P_2)_{S,\underline{n}}$. Once these unknowns are available, the expression for $(\partial V_t/\partial P_2)_{S,\underline{n}}$ can be evaluated. We need, therefore, $2(c+1)$ equations to solve for the same number of unknowns. The material balance, equilibrium criterion, entropy constraint, and the expression that relates P_1 and P_2 provide $2(c+1)$ equations.

Material balance. From Eq. (3.71), the variation in mole number of component i in phases 1 and 2 can be expressed by

$$(\partial n_i/\partial P_2)_{S,\underline{n}} = (\partial n_{1,i}/\partial P_2)_{S,\underline{n}} + (\partial n_{2,i}/\partial P_2)_{S,\underline{n}} = 0 \qquad i = 1, \ldots, c. \qquad (3.89)$$

Therefore, material balance provides c equations.

Equilibrium criterion. The equilibrium criterion $f_{1,i} = f_{2,i}$ can be written in differential form,

$$df_{1,i} = df_{2,i}, \qquad (3.90)$$

where $f_{j,i} = f_{j,i}(T, P_j, n_{j,1}, \ldots, n_{j,c})$. Therefore,

$$df_{j,i} = (\partial f_{j,i}/\partial T)_{P_j,\underline{n}_j}dT + (\partial f_{j,i}/\partial P_j)_{T,\underline{n}_j}dP_j + \sum_{k=1}^{c}(\partial f_{j,i}/\partial n_{j,k})_{T,P_j,n_{j,k}}dn_{j,k}$$

$$j = 1, 2; i = 1, \ldots, c, \qquad (3.91)$$

where $\underline{n}_{j,k} = (n_{j,1}, \ldots, n_{j,k-1}, n_{j,k+1}, \ldots, n_{j,c})$. Substituting Eq. (3.91) into

Eq. (3.90) and then dividing by ∂P_2 while keeping S and $\underline{n}$ constant, one obtains

$$[(\partial f_{1,i}/\partial T)_{P_1,\underline{n}_1} - (\partial f_{2,i}/\partial T)_{P_2,\underline{n}_2}](\partial T/\partial P_2)_{S,\underline{n}}$$
$$+ [(\partial f_{1,i}/\partial P_1)_{T,\underline{n}_1}(\partial P_1/\partial P_2)_{S,\underline{n}} - (\partial f_{2,i}/\partial P_2)_{T,\underline{n}_2}]$$
$$+ \left[\sum_{k=1}^{c}(\partial f_{1,i}/\partial n_{1,k})_{T,P_1,n_{1,k}}(\partial n_{1,k}/\partial P_2)_{S,\underline{n}} \right.$$
$$\left. - \sum_{k=1}^{c}(\partial f_{2,i}/\partial n_{2,k})_{T,P_2,n_{2,k}}(\partial n_{2,k}/\partial P_2)_{S,\underline{n}} \right] = 0 \qquad i = 1,\ldots,c. \quad (3.92)$$

The coefficients $(\partial f_{j,i}/\partial T)_{P_j,\underline{n}_j}$, $(\partial f_{j,i}/\partial P_j)_{T,\underline{n}_j}$, and $(\partial f_{j,i}/\partial n_{j,k})_{T,P_j,n_{j,k}}$ can be obtained from an EOS. We need two more equations to complete the $2(c+1)$ equations.

Entropy constraint. From $S = \sum_{j=1}^{2} S_j$,

$$(\partial S/\partial P_2)_{S,\underline{n}} = \sum_{j=1}^{2}(\partial S_j/\partial P_2)_{S,\underline{n}} = 0. \qquad (3.93)$$

Since $S_j = S_j(T, P_j, n_{j,1}, \ldots, n_{j,c})$, one can write the differential form of S_j, then divide it by ∂P_2 at constant $\underline{n}$ and S, and substitute the results into Eq. (3.93),

$$\sum_{j=1}^{2}\left[(\partial S_j/\partial T)_{P_j,\underline{n}_j}(\partial T/\partial P_2)_{S,\underline{n}} + (\partial S_j/\partial P_j)_{T,\underline{n}_j}(\partial P_j/\partial P_2)_{S,\underline{n}} \right.$$
$$\left. + \sum_{k=1}^{c}(\partial S_j/\partial n_{j,k})_{T,P_j,n_{j,k}}(\partial n_{j,k}/\partial P_2)_{S,\underline{n}} \right] = 0. \qquad (3.94)$$

The coefficients $(\partial S_j/\partial T)_{P_j,\underline{n}_j}$, $(\partial S_j/\partial P_j)_{T,\underline{n}_j}$, and $(\partial S_j/\partial n_{j,k})_{T,P_j,n_{j,k}}$ can be estimated from the expression for entropy of phase j derived in Example 3.2. Note that the entropy constraint provides $(\partial T/\partial P_2)_{S,\underline{n}}$.

Curvature effect. For a curved interface, the gas and liquid phase pressures are related by the well-known Young-Laplace equation derived in Chapter 2,

$$P_c = P_1 - P_2 = 2\sigma/r. \qquad (3.95)$$

In the process sketched in Fig. 3.27, when P_2 changes not only P_1 changes, but also σ and r may change. From these changes, we want to establish the unknown $(\partial P_1/\partial P_2)_{S,\underline{n}}$, which depends on how σ and r change. In the special case that r stays constant (such as in a capillary tube with a fixed contact angle) then $(\partial P_1/\partial P_2)$ is related only to the variations of σ. The variation of σ is due to the variations in temperature, pressure, and composition and curvature. The Weinaug–Katz model (1943) can be used to calculate composition, pressure, and temperature effect on σ:

$$\sigma^{1/4} = \sum_{i=1}^{c} P_i[x_{2,i}(d_2/M_2) - x_{1,i}(d_1/M_1)]. \tag{3.96}$$

In Eq. (3.96), $x_{2,i}$ and $x_{1,i}$ are the mole fractions of component i in the liquid and gas phases, respectively, and d and M are mass density and molecular weight; P_i is the parachor of component i.

The curvature may vary with P_2. In porous media, the relationship between P_1 and P_2 is given by

$$P_c = P_1 - P_2 = \sigma F(S_2) \tag{3.97}$$

where S_2 is the saturation of the liquid phase defined by $S_2 = V_2/(V_1 + V_2)$. Equation (3.97) takes into account the variation of curvature with P_1 and P_2. Taking the derivative of Eq. (3.97) with respect to P_2 at constant S and $\underline{n}$,

$$(\partial P_1/\partial P_2)_{S,\underline{n}} = 1 + F(S_2)(\partial\sigma/\partial P_2)_{S,\underline{n}} + \sigma(\partial F/\partial P_2)_{S,\underline{n}}. \tag{3.98}$$

The terms $(\partial\sigma/\partial P_2)_{S,\underline{n}}$ and $(\partial F/\partial P_2)_{S,\underline{n}}$ are evaluated next. For evaluation of the term $(\partial\sigma/\partial P_2)_{S,\underline{n}}$, a more useful form of Eq. (3.96) is employed,

$$\sigma^{1/4} = \sum_{i=1}^{c} P_i\left[\left(\frac{n_{2,i}}{V_2}\right) - \left(\frac{n_{1,i}}{V_1}\right)\right]. \tag{3.99}$$

From the above equation,

$$(\partial\sigma/\partial P_2)_{S,\underline{n}} = 4\sigma^{3/4} \sum_{i=1}^{c} P_i[-(n_{2,i}/V_2^2)(\partial V_2/\partial P_2)_{S,\underline{n}} + (n_{1,i}/V_1^2)$$

$$\cdot(\partial V_1/\partial P_2)_{S,\underline{n}} + (\partial n_{2,i}/\partial P_2)_{S,\underline{n}}/V_2 - (\partial n_{1,i}/\partial P_2)_{S,\underline{n}}/V_1]. \tag{3.100}$$

The second derivative term in Eq. (3.98) is estimated from

$$(\partial F/\partial P_2)_{S,\underline{n}} = (dF/dS_2)(\partial S_2/\partial P_2)_{S,\underline{n}}, \tag{3.101}$$

where $\quad (\partial S_2/\partial P_2)_{S,\underline{n}} = \dfrac{V_1(\partial V_2/\partial P_2)_{S,\underline{n}} - V_2(\partial V_1/\partial P_2)_{S,\underline{n}}}{(V_1 + V_2)^2}. \tag{3.102}$

We showed earlier how to calculate $(\partial V_1/\partial P_2)_{S,\underline{n}}$ and $(\partial V_2/\partial P_2)_{S,\underline{n}}$. Note that in Eq. (3.100) the effect of interface curvature on σ is neglected;

Eq. (3.99) takes into account only pressure, temperature, and composition variation of σ.

The above system of equations may then be used to iteratively calculate C_S^{tp}. In the first iteration, we may assume that the interface is flat.

Two-phase sonic velocity. The thermodynamic speed of sound, a, is given by the following expression:

$$a^2 = -v^2(\partial P/\partial v)_s, \tag{3.103}$$

which assumes that all irreversibilities, including heat conduction, and diffusion are excluded. We also neglect, in the thermodynamic definition, the frequency dependence of the sonic velocity. Note that in the above equation, v is molar volume. Equation (3.103) applies to both single-phase and two-phase systems. In the two-phase state, one of the phases should be in a dispersed state. For two-phase sonic velocity, Eq. (3.103) becomes

$$\boxed{a^2 = \frac{v_t}{C_S^{tp}}}, \tag{3.104}$$

where v_t is the two-phase molar volume. Since v_t and C_S^{tp} are available once the two-phase isentropic compressibility is calculated, then the two-phase sonic velocity is calculated readily.

Using the equations presented in this section, one can calculate the two-phase isentropic compressibility and the two-phase sonic velocity. In the following, some numerical results are presented from Firoozabadi and Pan (1997), who employed the PR-EOS for the calculation of coefficient derivatives.

Figure 3.28 shows the calculated compressibilities and sonic velocity for a mixture of C_1/C_3 (30 mole% C_1, 70 mole% C_3) at 130°F. In Fig. 3.28a, C_T and C_S are plotted *vs.* pressure. This figure indicates that there is a discontinuity in both isothermal and isentropic compressibilities, when the phase boundaries are crossed. From a pressure of 1200 psia to a bubblepoint pressure of about 977 psia, there is a small increase in C_S of the undersaturated liquid; the C_T increase is, however, more noticeable. At the bubblepoint, there is a sudden increase in both C_T and C_S. Similar behavior is also observed at the dewpoint of about 453 psia. It is interesting to note that the compressibilities in the two-phase region approaching the dewpoint are higher than the corresponding gas-phase compressibilities. Figure 3.28a also reveals that the variation of C_S in the two-phase region is less than the variation of C_T. This figure also provides the experimental isothermal compressibility data of Sage *et al.* (1933). The results in Fig. 3.28a are for a flat interface

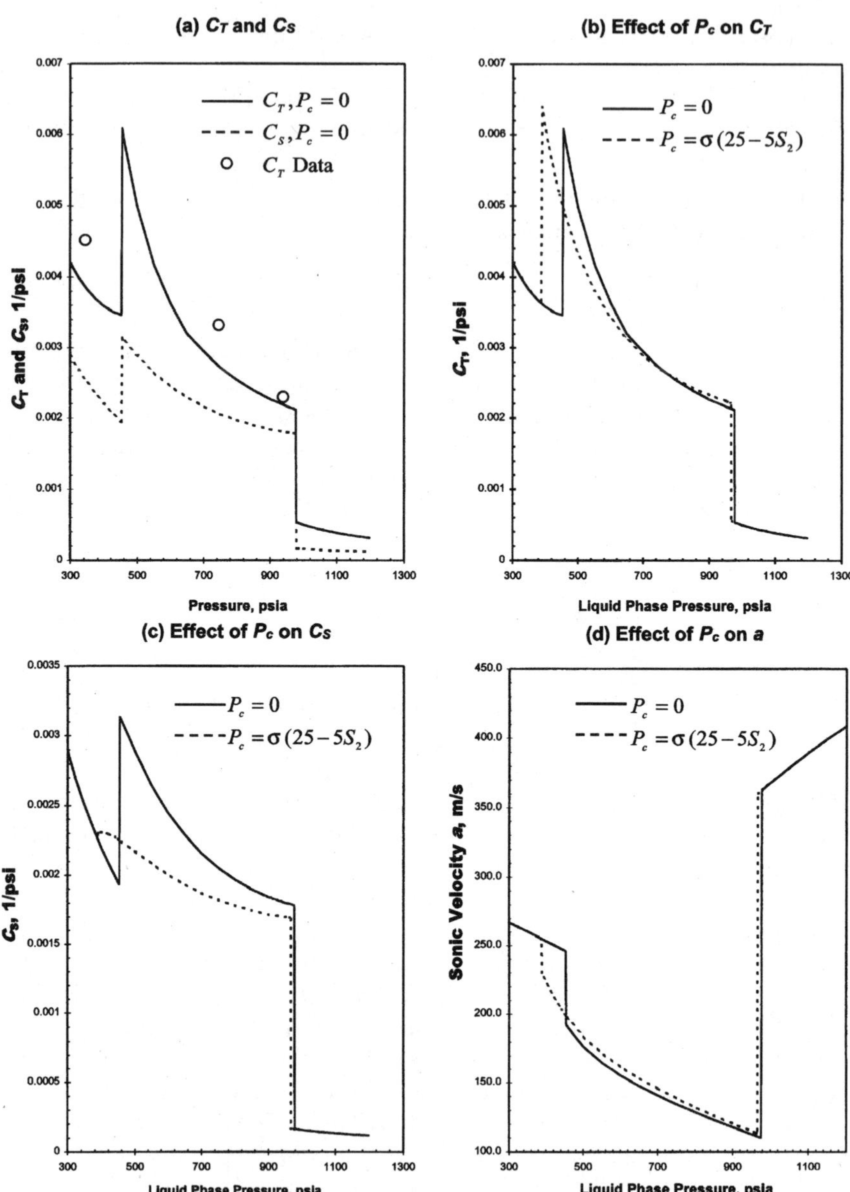

Figure 3.28   Compressibilities and sonic velocity for the C_1/C_3 mixture (30 mole% C_1 and 70 mole% C_3) at 130°F (adapted from Firoozabadi and Pan, 1997).

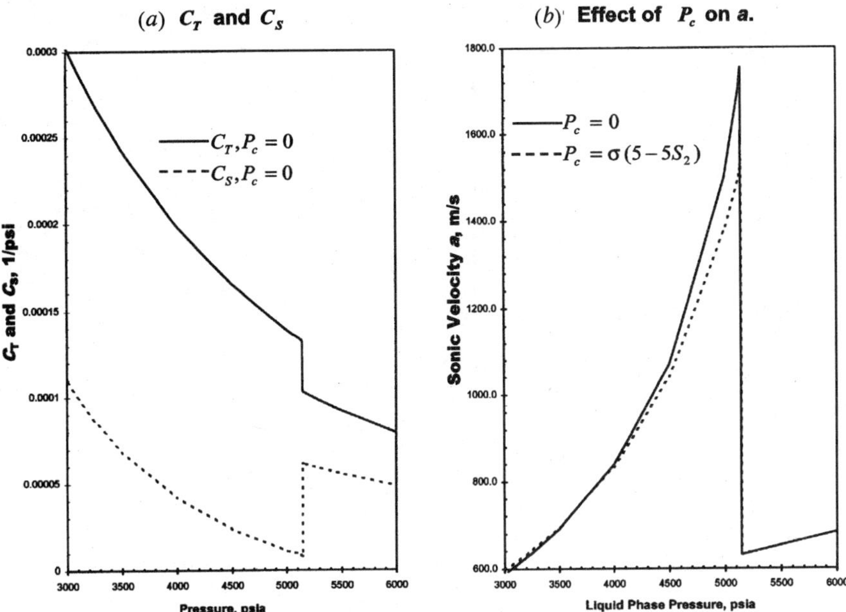

Figure 3.29   Calculated compressibilities and sonic velocity for the C_1/nC_{10} mixture (95 mole% C_1 and 5 mole% nC_{10}) at 160°F (adapted from Firoozabadi and Pan, 1997).

between the phases. Figures 3.28b and 3.28c show the effect of a curved interface on C_T and C_S, respectively. For the preparation of these two figures, $F(S_2) = 25 - 5S_2$ is assumed with the units of (psia · cm)/dyne (see Eq. (3.97)). As we have discussed in Chapter 2, a curved interface may increase or decrease the saturation pressure of mixtures. The calculated bubblepoint pressures of the flat and the curved interfaces are 977.2 and 966.4 psia, respectively. The dewpoint pressures of the flat and curved interfaces are 453.4 and 389.3 psia, respectively. The parachors (see Eq. (3.96)) used in the calculation of interfacial tension are $P_{C_1} = 77$ and $P_{C_3} = 150$ (Katz $et\ al.$, 1959). The results in Figs. 3.28b and 3.28c show that while the interface curvature does not have a significant effect on C_T, its effect on C_S is pronounced. Figure 3.28d plots the sonic velocity both in the single-phase and two-phase regions. In the single-phase liquid, the sonic velocity decreases as the pressure decreases. There is a sharp decrease in the sonic velocity at the bubblepoint, from 370 to about 100 m/s. There are plenty of data on the sonic velocity of two-phase gas–liquid mixtures of water–air and water–steam (Kieffer, 1977). Those data reveal that (1) the presence of one percent by volume air in the form of gas bubbles reduces the sonic velocity from 1500 to 100 m/s and (2) the sonic velocity in the water–steam

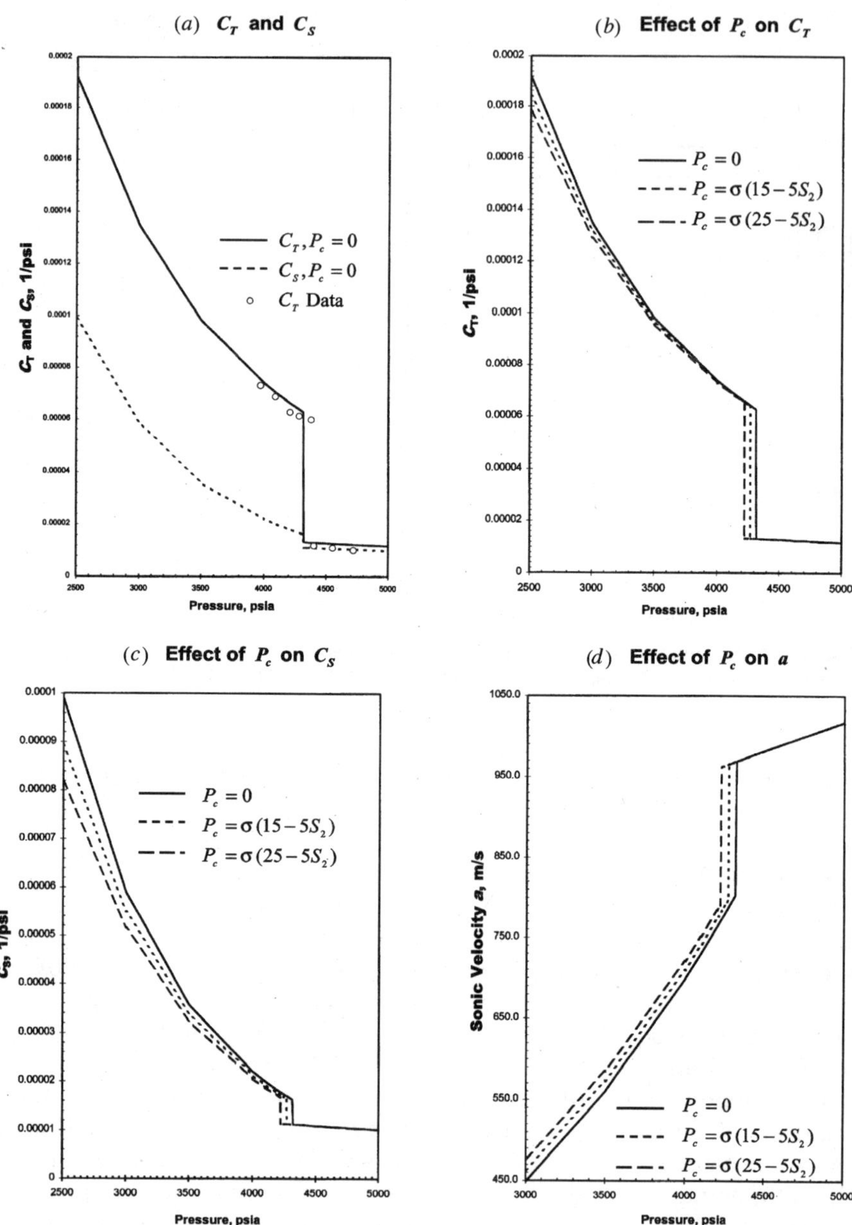

Figure 3.30 Compressibilities and sonic velocity for the North Sea crude at 224.6°F (adapted from Firoozabadi and Pan, 1997).

system can be as low as 10 m/s. However, there is not much data for hydrocarbon mixtures in the two-phase region.

Figure 3.29 shows the compressibilities and sonic velocity for a mixture of 95 mole% C_1 and 5 mole% nC_{10} at 160°F. This fluid system has a retrograde dewpoint pressure of 5146.7 psia when the interface between the phases is flat. Figure 3.29a shows that at the dewpoint, C_T increases while C_S decreases. This behavior is different from the bubblepoint system presented in Fig. 3.28. There is a significant difference between C_T and C_S—a factor of 20. The sonic velocity is plotted in Fig. 3.29b. This figure shows that the sonic velocity increases in the two-phase region, which is in contrast to the results in Fig. 3.28d. Sonic velocity is initially effected by the interface curvature. The dewpoint pressure with the interface curvature between the phases is 5151.3 psia for the capillary pressure shown in Fig. 3.29b. Therefore, there is an increase in dewpoint pressure due to interface curvature.

Figure 3.30 depicts C_T, C_S, and sonic velocity for a North Sea crude oil. The composition and characterization of the plus-fractions are provided by Firoozabadi *et al.* (1988b). In the single-phase liquid, C_T and C_S do not change appreciably with pressure; the difference between C_T and C_S is also not appreciable. However, this difference becomes appreciable in the two-phase region as is indicated in Fig. 3.30a. The measured isothermal compressibility is also plotted in Fig. 3.30a, which shows good agreement with the calculated results. The effects of interface curvature on C_T and C_S are shown in Figs. 30b and 30c, respectively. Because of interface curvature, the bubblepoint pressure decreases, but the effect on C_T and C_S is not significant. The parachors for plus-fractions of the crude were adopted from Firoozabadi *et al.* (1988a). Figure 3.30d shows the calculated sonic velocity. The effect of interface curvature is also shown in the same figure. The results show that (1) there is not a significant drop in the sonic velocity as the crude enters the two-phase region and (2) the interface curvature does not appreciably change the sonic velocity.

The next section presents the sonic velocity and temperature changes due to expansion in the single-phase region.

Single-phase sonic velocity and temperature change due to expansion

The sonic velocity and the thermal properties of fluids are related to their volumetric behavior. The basic equations for the single-phase state are derived in this section.

Speed of sound is useful in the determination of liquid level in gas wells and in the estimation of gas flow rate in critical flow provers, in addition to applications enumerated in the preceding section. Speed of sound correlates with volumetric data and can be used to calibrate equations of state.

Here we will limit our derivation of sonic velocity to a single-phase system (see Eq. (3.103)) where composition is constant. The derivations are very similar to those of Thomas *et al.* (1970).

The molar enthalpy and entropy, h and s, can be assumed to be functions of temperature T, and pressure P. Then

$$dh = (\partial h/\partial T)_P dT + (\partial h/\partial P)_T dP \tag{3.105}$$

$$ds = (\partial s/\partial T)_P dT + (\partial s/\partial P)_T dP. \tag{3.106}$$

Let us define the molar heat capacity at constant pressure as

$$c_P = (\partial h/\partial T)_P. \tag{3.107}$$

For a reversible process in a constant-composition system (see Eq. (1.52) of Chapter 1),

$$dh = Tds + vdP. \tag{3.108}$$

Dividing Eq. (3.108) by dT and holding P constant results in

$$(\partial h/\partial T)_P = T(\partial s/\partial T)_P. \tag{3.109}$$

Combining Eqs. (3.107) and (3.109) provides

$$c_P = T(\partial s/\partial T)_P. \tag{3.110}$$

Now divide Eq. (3.108) by dP, holding the temperature constant:

$$(\partial h/\partial P)_T = T(\partial s/\partial P)_T + v. \tag{3.111}$$

Using the Maxwell relation given by Eq. (3.198) of Chapter 1,

$$(\partial h/\partial P)_T = -T(\partial v/\partial T)_P + v. \tag{3.112}$$

Combining Eqs. (3.105), (3.107), and (3.112),

$$\boxed{dh = c_P dT + [v - T(\partial v/\partial T)_P]dP} . \tag{3.113}$$

An expression for ds can be obtained by combining Eqs. (1.106), (1.110), and Eq. (1.199) of Chapter 1:

$$ds = \frac{c_P}{T}dT - (\partial v/\partial T)_P dP. \tag{3.114}$$

Now express $u = u(T, v)$ and $s = s(T, v)$. Then

$$du = (\partial u/\partial T)_v dT + (\partial u/\partial v)_T dv \tag{3.115}$$

$$ds = (\partial s/\partial T)_v dT + (\partial s/\partial v)_T dv. \tag{3.116}$$

Define the molar heat capacity at constant v as

$$c_V = (\partial u/\partial T)_v. \tag{3.117}$$

For a reversible process,

$$du = Tds - Pdv. \tag{3.118}$$

Dividing Eq. (3.118) by dT while holding v constant and using the definition of c_V from Eq. (3.117),

$$(\partial u/\partial T)_v = T(\partial s/\partial T)_v = c_V. \tag{3.119}$$

Also dividing Eq. (3.118) by dv and holding T constant,

$$(\partial u/\partial v)_T = T(\partial s/\partial v)_T - P. \tag{3.120}$$

Using the Maxwell relation given by Eq. (1.201) of Chapter 1,

$$(\partial u/\partial v)_T = T(\partial P/\partial T)_v - P. \tag{3.121}$$

Combining Eqs. (3.115), (3.117), and (3.121), the following expression for du is obtained:

$$\boxed{du = c_V dT + [T(\partial P/\partial T)_v - P]dv}. \tag{3.122}$$

An expression for ds is also obtained by combining Eqs. (3.116) and (3.119) and the Maxwell relation given by Eq. (1.201) of Chapter 1:

$$ds = \frac{c_V}{T}dT + (\partial P/\partial T)_v dv. \tag{3.123}$$

By combining Eqs. (3.114) and (3.123), an expression for dT can be obtained:

$$dT = \frac{T}{(c_P - c_V)}(\partial v/\partial T)_P dP + \frac{T}{(c_P - c_V)}(\partial P/\partial T)_v dv. \tag{3.124}$$

Now express T as a function of P and v, $T = T(P, v)$; then

$$dT = (\partial T/\partial P)_v dP + (\partial T/\partial v)_P dv. \tag{3.125}$$

From Eqs. (3.124) and (3.125),

$$\boxed{c_P - c_V = T(\partial v/\partial T)_P(\partial P/\partial T)_v}\,.\tag{3.126}$$

Note that in the above expression, the volumetric behavior fully describes the heat capacity difference. Note also that for an ideal gas, $c_P - c_V = R$.

From Eqs. (3.110) and (3.119),

$$(c_P/c_V) = [(\partial s/\partial T)_P]/[(\partial s/\partial T)_v].\tag{3.127}$$

Using Eq. (1.79) of Chapter 1, Eq. (3.127) becomes

$$(\partial P/\partial v)_s = (c_P/c_V)(\partial P/\partial v)_T.\tag{3.128}$$

The term $(\partial P/\partial v)_T$ can be evaluated from $Pv = ZRT$.

$$(\partial P/\partial v)_T = -\frac{ZRT}{v^2} + \frac{RT}{v}\left(\frac{\partial Z}{\partial P}\right)_T\left(\frac{\partial P}{\partial v}\right)_T.\tag{3.129}$$

Rearranging the above equation and using $Pv = ZRT$ provides

$$(\partial P/\partial v)_T = \frac{-P/v}{\left[1 - \dfrac{P}{Z}(\partial Z/\partial P)_T\right]}.\tag{3.130}$$

Combining Eqs. (3.128) and (3.130),

$$(\partial P/\partial v)_S = (c_P/c_V)\left(\frac{-P/v}{\left[1 - \left(\dfrac{P}{Z}\right)(\partial Z/\partial P)_T\right]}\right).\tag{3.131}$$

The isothermal compressibility expression $C_T = -\dfrac{1}{v}(\partial v/\partial P)_T$ and $Pv = ZRT$ can be combined to provide

$$C_T = \frac{1}{P} - \frac{1}{Z}(\partial Z/\partial P)_T.\tag{3.132}$$

From Eqs. (3.103), (3.131), and (3.132), the expression for the sonic velocity becomes

$$a = \left[\frac{(c_P/c_V)v}{MC_T}\right]^{1/2}.\tag{3.133}$$

Eq. (3.133) gives the sonic velocity of a single-phase fluid of constant composition in terms of the volumetric behavior and the c_P/c_V ratio.

The next step is the calculation of the c_P/c_V ratio in terms of volumetric behavior. Let us derive the expression for c_P. At constant temperature, Eq. (3.113) reduces to

$$dh = [v - T(\partial v/\partial T)_P]dP. \tag{3.134}$$

Integrating Eq. (3.134) from pressure 0 to pressure P,

$$h(T, P) - h^*(T, 0) = \int_0^P [v - T(\partial v/\partial T)_P]dP, \tag{3.135}$$

or $\quad h(T, P) - h^*(T, 0) = RT(Z - 1) + \int_\infty^v [T(\partial P/\partial T)_v - P]dv, \tag{3.136}$

where $h^*(T, 0)$ is the ideal gas enthalpy at zero pressure, which is a function of temperature only. Using the PR-EOS, Eq. (3.136) simplifies to

$$\boxed{h(T, P) - h^*(T, 0) = RT(Z - 1) + \frac{Tda/dT - a}{2\sqrt{2}b}\ln\frac{Z + 2.414B}{Z - 2.414B}}. \tag{3.137}$$

From the definition of c_P given by Eq. (3.107),

$$c_P = c_P^* + R(Z - 1) + RT(\partial Z/\partial T)_P + \frac{\partial}{\partial T}\left\{\frac{T\left(\dfrac{da}{dT}\right) - a}{2\sqrt{2}b}\ln\frac{Z + 2.414B}{Z - 2.414B}\right\}, \tag{3.138}$$

where c_P^* is the ideal-gas heat capacity. Passut and Danner (1972) provide correlations to provide ideal gas enthalpy, heat capacity, and entropy of some 90 substances—mostly hydrocarbons.

When c_P is available, then Eq. (3.126) can be used to calculate c_V; the ratio c_P/c_V becomes readily known. Thomas *et al.* (1970) provide the following equation for the low-pressure heat capacity of natural-gas mixtures as a function of temperature and gas gravity G ($G = M$ of gas/M of air):

$$c_P^* = A + BT + CG + DG^2 + E(TG). \tag{3.139}$$

Table 3.5 gives the coefficients of the above equation. Equation (3.139) is valid for gas gravity G in the range of 0.55 to 2.0 and temperature in the 0 to 600°F interval; the units are c_P^* in Btu/(lb·R) and T in °F.

Figures 3.31 to 3.33 show the measured and computed sonic velocities for methane, nitrogen, and *n*-hexane. The computed enthalpy is

TABLE 3.5   Coefficients of Eq. (3.139)

Coefficient	Value
A	3.7771
B	−0.00110
C	7.5281
D	0.65621
E	0.014609

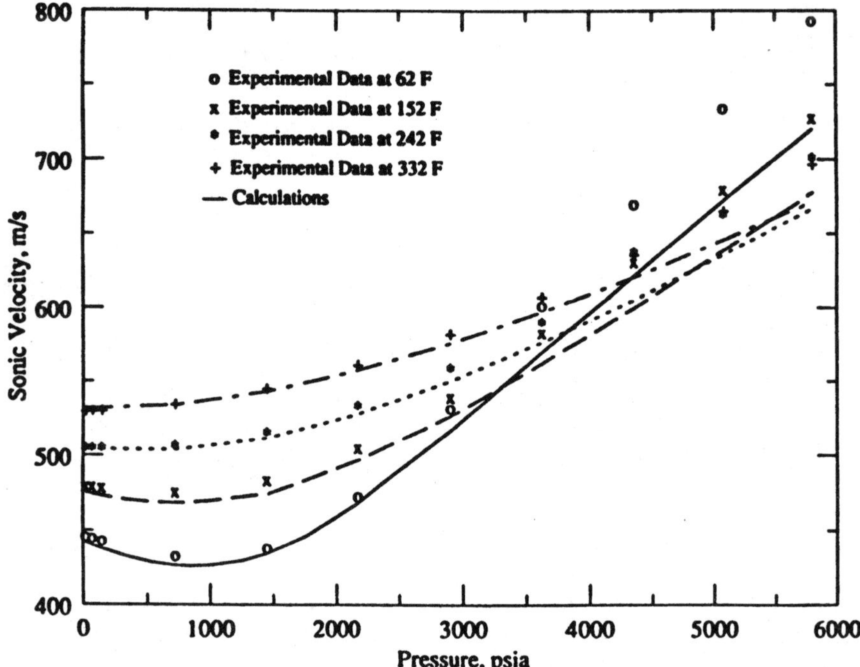

Figure 3.31   Sonic velocity in methane gas (from Nutakki, 1989).

compared with experimental data in Tables 3.6 and 3.7. In these two tables, the computed results are from the Peng-Robinson EOS (1976) and the Schmidt-Wenzel EOS (1980).

Heating and cooling due to expansion

Hydrocarbon fluid systems undergo rapid expansion around the well-bore and in production facilities. As a result of expansion, the temperature may rise or fall. The following establishes the criteria for cooling and heating due to free expansion. The derivations are limited to

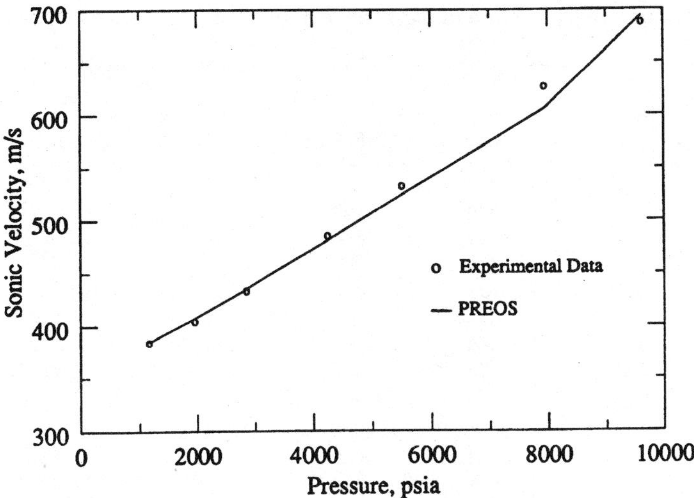

Figure 3.32   Sonic velocity in nitrogen gas at 122°F (from Nutakki, 1989).

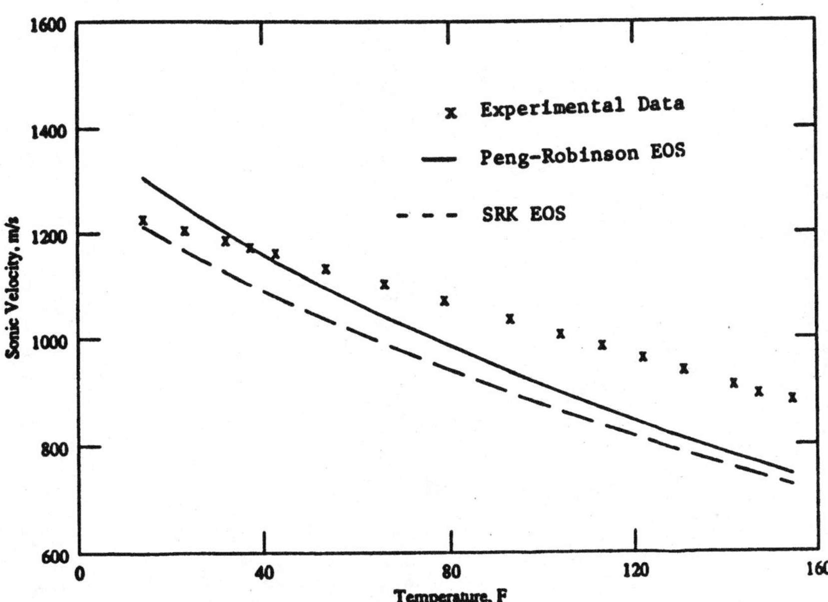

Figure 3.33   Sonic velocity in hexane liquid (from Nutakki, 1989).

TABLE 3.6   Enthalpy of ternary mixture of methane, ethane, and CO_2 (data of Ng and Mather, 1978)

Temperature, °F	Pressure, psia	$(H° - H)$, J/mole			
		Exp.	SRK-EOS	SW-EOS	PR-EOS
194	435	722	725	727	763
194	1015	1715	1739	1734	1807
194	1450	2489	2504	2485	2575
194	1886	3242	3223	3185	3283
140	435	841	865	864	897
140	1015	2113	2156	2138	2203
140	1450	3163	3188	3140	3211
140	1886	4175	4146	4061	4127
104	435	967	983	981	1011
104	1015	2534	2565	2533	2586
104	1450	3924	3924	3836	3878
104	1886	5174	5095	4956	4979

TABLE 3.7   Enthalpy of *n*-pentane (data of Erbar *et al.*, 1964)

Temperature, °F	Pressure, psia	Enthalpy of nC_5, Btu/lbmole		
		Exp.	PRE-OS	SW-EOS
190.6	600	3827	3485	3524
190.6	800	3850	3499	3539
190.6	1000	3875	3516	3557
190.6	1250	3907	3541	3583
190.6	1500	3941	3570	3613
190.6	1750	3975	3601	3646
190.6	2000	4015	3635	3681
280.2	600	8107	7855	7847
280.2	800	8089	7812	7804
280.2	1000	8080	7782	7776
280.2	1250	8082	7760	7754
280.2	1500	8091	7750	7745
280.2	1750	8108	7748	7745
280.2	2000	8129	7754	7752
370.7	600	13070	13267	13163
370.7	800	12844	12925	12815
370.7	1000	12716	12727	12612
370.7	1250	12612	12568	12449
370.7	1500	12547	12462	12340
370.7	1750	12506	12389	12266
370.7	2000	12482	12339	12215

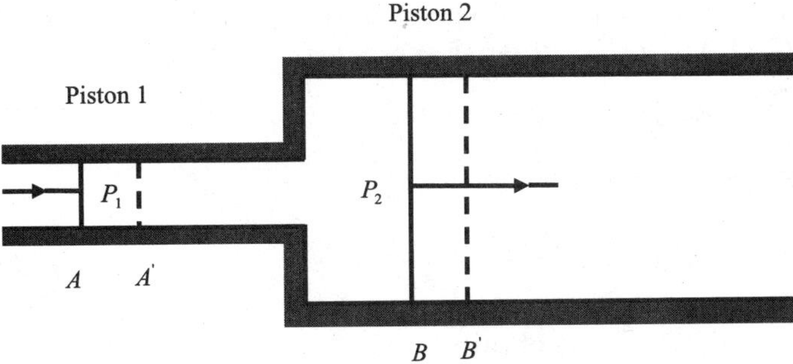

Figure 3.34   Free expansion process.

single-phase systems where the phase composition does not change. In Example 3.4, we provide an expression that can be used in the free expansion of a two-phase system.

Consider the sketch, Fig. 3.34, in which a fluid is expanding from pressure P_1 to pressure P_2 at steady state conditions. The system is rigid and insulated. Assume two imaginary pistons at points A and B at time t. During time interval Δt, piston 1 moves from point A to A' and piston 2 moves from point B to B'. The fluid contained between pistons 1 and 2 is our thermodynamic system. The work done on the system in the time interval Δt is given by

$$W = -P_2 v_2 \Delta m + P_1 v_1 \Delta m, \tag{3.140}$$

where v_1 and v_2 are volume per unit mass at the upstream and downstream of the system, respectively, and Δm is the mass within the volume between the piston in the old and new positions, which is the same in the upstream and downstream sides. If one further assumes that there is no heat flow in the upstream and downstream sides, then

$$\Delta U = W, \tag{3.141}$$

where
$$\Delta U = \Delta m(u_2 - u_1); \tag{3.142}$$

u_2 and u_1 are the internal energies per unit mass in the upstream and downstream sides, respectively. Combining Eqs. (3.140) to (3.142),

$$u_2 - u_1 = -P_2 v_2 + P_1 v_1. \tag{3.143}$$

Since $h = u + Pv$,

$$h_2 = h_1, \tag{3.144}$$

which implies that in a free expansion, the enthalpy remains constant. Now apply Eq. (3.113) to a free-expansion process:

$$c_P dT = -[v - T(\partial v/\partial T)_P]dP. \tag{3.145}$$

Rearranging the above equation,

$$(\partial T/\partial P)_h = \frac{T(\partial v/\partial T)_P - v}{c_P}. \tag{3.146}$$

From the EOS, $v = ZRT/P$, one obtains

$$(\partial v/\partial T)_P = \frac{RZ}{P} + \frac{RT}{P}(\partial Z/\partial T)_P. \tag{3.147}$$

Combining Eqs. (3.146), (3.147), and the EOS, $Pv = ZRT$,

$$(\partial T/\partial P)_h = \frac{(RT^2/P)(\partial Z/\partial T)_P}{c_P}. \tag{3.148}$$

The Joule-Thomson coefficient μ is defined as

$$\mu = (\partial T/\partial P)_h. \tag{3.149}$$

Eq. (3.149) implies that if μ is positive, the free expansion results in cooling. For a negative μ, the free expansion leads to heating.

Combining Eqs. (3.148) and (3.149),

$$\boxed{\mu = \frac{(RT^2/P)(\partial Z/\partial T)_P}{c_P}}. \tag{3.150}$$

Equation (3.150) indicates that if the term $(\partial Z/\partial T)_P$ is positive, the free expansion results in cooling and if it is negative, the expansion leads to an increase in temperature.

For natural gases, the term $(\partial Z/\partial T)_P$ may become negative at pressures above 4000 psia. An EOS conveniently can be used to estimate this term. The method of corresponding states can also be used to compute $(\partial Z/\partial T)_P$.

Jones (1988) reports temperature data during production testing. The data show an increase in bottomhole temperature during flow. The temperature rise from several fields was of the order of 2 to 5°C. All these reservoirs had an undersaturated oil. Figure 3.35 shows a typical downhole pressure and temperature profile during a production test. Jones suggests that the interpretation of temperature fall-off may exhibit the same features as seen on the pressure buildup. Therefore, temperature data may be a useful source of information that can supplement the pressure data.

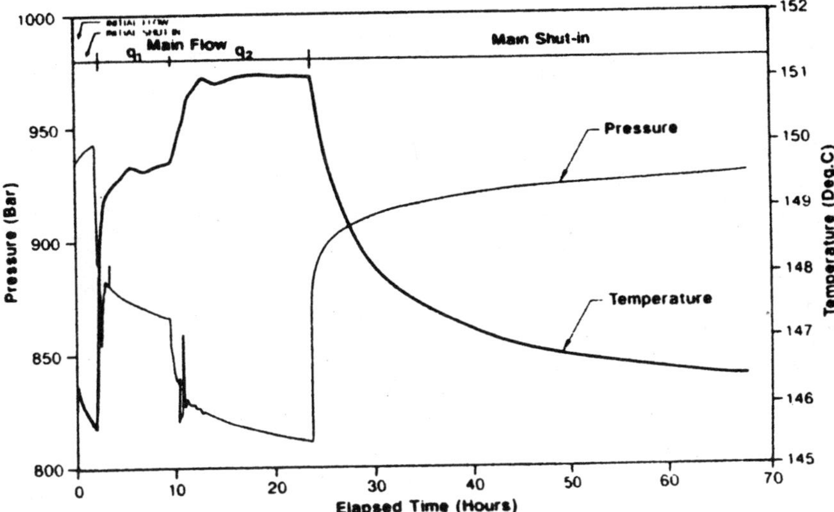

Figure 3.35   Downhole pressure and temperature profiles during a production test (from Jones, 1988).

Examples and theory extension

Example 3.1   *Gas, oil, near-critical oil, and gas and volatile oil*   Sometimes when a hydrocarbon reservoir is discovered, it is not an easy task to describe the state of the fluid in the reservoir. How would one describe a gas, an oil, and a volatile oil? Is the density of the liquid of the fluid flashed to ambient conditions a good measure?

Solution   Perhaps the best way to describe a high-pressure complex hydro-carbon mixture is through a pressure-temperature diagram. Figure 3.36 is a plot of saturation pressure of a given reservoir fluid as a function of tempera-ture. In this figure, a point on the right side of line AB is called a "gas" and on the left side is called an "oil" (line AB passes through the critical point CP). Neither the API gravity nor the color of the stock tank liquid could be used as a yardstick to designate a reservoir fluid as a gas or an oil. As an example, a gas-condensate fluid could produce a liquid with an API gravity of 29° (Kilgrin, 1966).

Figure 3.37 shows a further subdivision of a reservoir fluid. The area in the vicinity of line AB, both on the right and left, is the near-critical region. One simple indication of a fluid being in the near-critical region is the magnitude of methane K-values. Whenever methane K-value is less than 1.1 and the resi-dual (least volatile fraction) has a K-value greater than, say 0.3, near-criticality is assured. On the left side of line AB, the region next to near-critical is the volatile-oil region. The production characteristics of the near-critical and vola-tile-oil fluids can be examined by inspecting Fig. 3.38. This figure represents the constant-volume depletion of a fluid. The dashed curves represent an oil

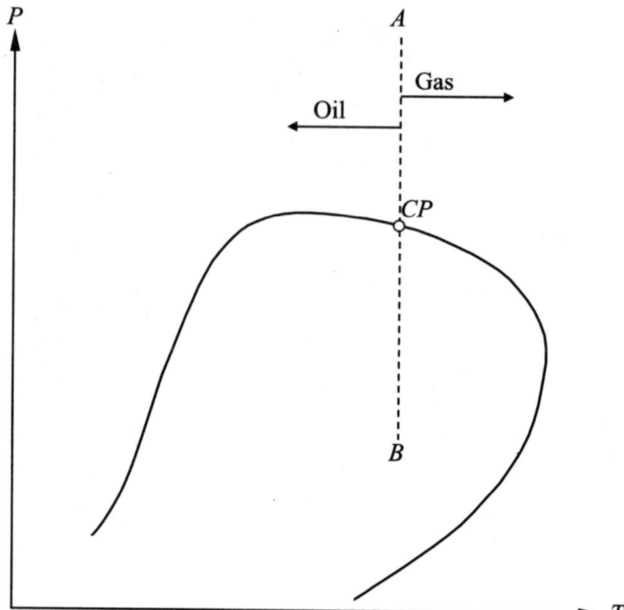

Figure 3.36   Definition of oil and gas systems.

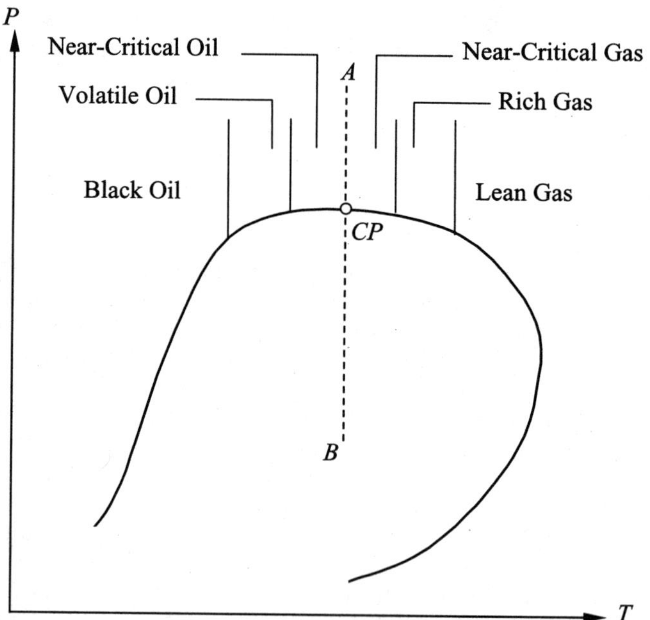

Figure 3.37   Definition of various reservoir-fluid systems.

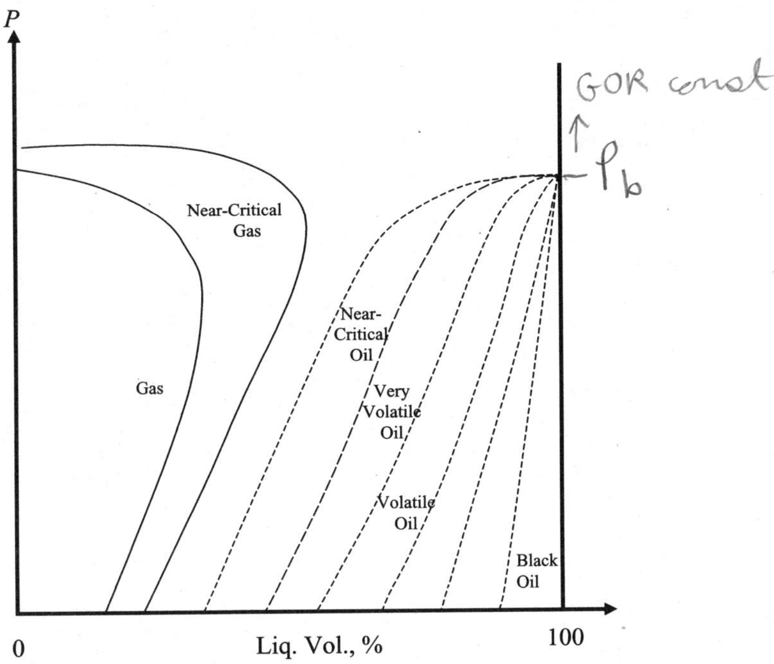

Figure 3.38   Constant-volume depletion or constant-mass expansion of a reservoir-fluid system.

fluid and the solid curves a gas-condensate system. On the "oil" fluid side, this figure implies that as the degree of volatility increases, the producing GOR below the bubblepoint increases. This means that the recovery factor for a very volatile oil, and especially for a near-critical oil, could be low (provided gravity drainage and rock compressibility effects are negligible). Therefore, natural depletion (in the absence of water drive) could be very inefficient. The reason for high GOR production and low recovery is a sudden increase in gas saturation. At high gas saturation, relative permeability to gas is high and therefore, gas is produced at high rates. As a result, oil recovery from pressure depletion is low. Fluid injection, either gas or water, could become viable options.

The diagram in Fig. 3.38 can be generated by starting with a black-oil fluid and then a series of successively increasing temperatures to change the fluid to near-critical oil and then a gas condensate. Figure 3.39, taken from Katz *et al.* (1940), shows the behavior of a hydrocarbon mixture as temperature changes. It should be mentioned that not all reservoir fluids could undergo such a behavior on changing the temperature. An oil with asphaltene materials may not have a measurable dewpoint and, therefore, does not follow the trends of Figs. 3.38 and 3.39.

Figure 3.39 shows the effect of temperature on the volatility of a reservoir oil and the degree of near-criticality for a rich gas condensate. For an oil fluid, the degree of volatility increases as the temperature increases. For a gas-

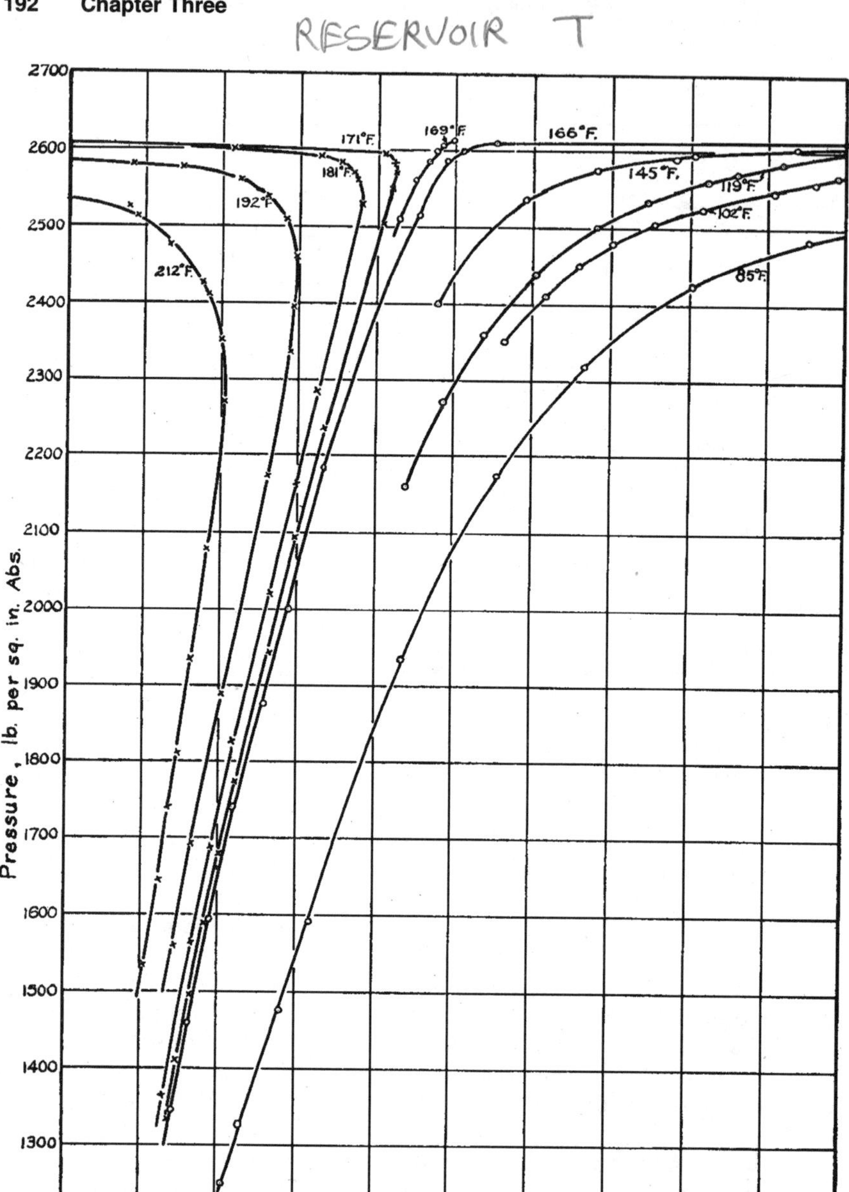

Figure 3.39   Liquid percentage isotherms (adapted from Katz *et al.*, 1940).

condensate fluid, the trend is the reverse. As the temperature increases, the gas moves away from critical behavior. For a gas-condensate reservoir, the liquid recovery also decreases as the gas becomes more near-critical. This is mainly due to high gas mobility compared to condensate liquids, even at high liquid saturations. Recovery of condensate liquids could be around 15 percent or less for some near-critical gas-condensate reservoirs. Recycling is an attractive option, provided liquid saturation is above 15 to 20 percent.

Example 3.2   Derive the following entropy expression for mixtures using the PR-EOS:

$$
s = R\ln(Z-B)/Z + \left\{[da(T)/dT]/2\sqrt{2}b\right\}\ln\left\{\left[Z+(1+\sqrt{2})B\right]\right.
$$
$$
\left./[Z+\left(1-\sqrt{2}\right)B]\right\} + R\sum_{i=1}^{c} x_i \ln(Z/(x_i P)) + \sum_{i=1}^{c} x_i s_i^0(T).
$$

In the above equation s is the molar entropy of the mixture at T, P, and $\boldsymbol{x}$.

Solution   From Chapter 1, Example 1.6,

$$
S = \int_V^\infty [(nR/V) - (\partial P/\partial T)_{V,\underline{n}}]dV + R\sum_{i=1}^{c} n_i \ln V/(n_i RT) + \sum_{i=1}^{c} n_i s_i^0(T).
$$

The above equation in terms of molar entropy is

$$
s = \int_v^\infty [(R/v) - (\partial P/\partial T)_{v,x}]dv + R\sum_{i=1}^{c} x_i \ln v/(x_i RT) + \sum_{i=1}^{c} x_i s_i^0(T).
$$

From the PR-EOS,

$$
(\partial P/\partial T)_{v,\boldsymbol{x}} = R/(v-b) - [da(T)/dT]/[v(v+b)+(v-b)].
$$

Combining the above equations and using

$$
\int_v^\infty [R/v - R/(v-b)]dv = R\ln(v-b)/v
$$

and

$$
\int_v^\infty dv/[v(v+b)+b(v-b)] = (1/(2\sqrt{2}b))\ln\left\{v+(1+\sqrt{2})b/\left[v+(1-\sqrt{2})b\right]\right\},
$$

the expression sought is obtained.

Example 3.3 Show   that the simple volume-translation expression, $v^{true} = v^{EOS} + c$ where $c = \sum_{i=1}^{c} x_i c_i$, does not effect the prediction of phase compositions. It effects only the phase densities.

Solution   Consider the expression for fugacity coefficient given by Eq. (1.100) of Chapter 1. Let us denote the fugacity coefficient using v^{true} as φ_i^{true} and the one from v^{EOS} by φ_i^{EOS}. Then

$$\bar{V}_i^{EOS} = (\partial V^{EOS}/\partial n_i)_{T,P,n_i} = (\partial v^{EOS}/\partial x_i)_{T,P,x_i}$$

$$\bar{V}_i^{true} = (\partial V^{true}/\partial n_i)_{T,P,n_i} = (\partial v^{true}/\partial x_i)_{T,P,x_i}$$

$$\bar{V}_i^{true} = \partial/\partial x_i \left[v^{EOS} + \sum_{j=1}^{c} x_j c_j \right]_{T,P,x_i} = \bar{V}_i^{EOS} + c_i$$

Substituting $\bar{V}_i^{true}$ and $\bar{V}_i^{EOS}$ into Eq. (1.100) of Chapter 1:

$$\varphi_i^{true}/\varphi_i^{EOS} = \exp[-c_i P/RT].$$

The equilibrium ratios which provide the phase compositions are the same whether one uses an EOS with or without volume translation,

$$K_i = \left[\varphi_i^L/\varphi_i^V\right]^{true} = \left[\varphi_i^L/\varphi_i^V\right]^{EOS}.$$

The term $\exp[-c_i P/RT]$ drops out from both the numerator and denominator.

Example 3.4 *Cooling and heating due to expansion* Consider the following three gases at the given temperature and pressure. If these gases undergo a free expansion around the wellbore, would you expect heating or cooling? Compute the final temperature for each case (use the EOS approach to perform all of the computations).

Component	Gas A Mole%	Gas B Mole%	Gas D Mole%
CO_2	2.37	2.44	
N_2	0.31	0.08	
C_1	73.19	82.10	100
C_2	7.8	5.78	
C_3	3.55	2.87	
iC_4	0.71	0.56	
nC_4	1.45	1.23	
C_5	0.64	0.52	
C_5	0.68	0.60	
C_6	1.09	0.72	
C_{7+}	8.21	3.10	
$M_{C_{7+}}$	184.0	132.0	
Init.pres.,psia	8500	6500	7500, 12 000
Fin.pres.,psia	7500	5500	6000, 3000
Init. temp., °F	280	180	180

Solution Equation (3.144) can be used to solve this problem. In a free expansion,

$$h_{initial}(T, P) = h_{final}(T', P').$$

Initial pressure and temperature and final pressure, as well as the composition, are provided. Solution of the above equation requires knowledge of $\sum_{i=1}^{c} x_i h_i^0(T)$ and $\sum_{i=1}^{c} x_i h_i^0(T')$. Passut and Danner (1972) provide the ideal-gas enthalpy data. The results of free expansion for gases A, B, and D are shown in Fig. 3.40. The PR-EOS was used in the calculations. Note that the above equation is not limited to the single-phase state; it also applies to the two-phase state. In the two-phase state at T' and P',

$$h(T', P') = V h^G(T', P') + (1 - V) h^L(T', P'),$$

where h^G and h^L are the molar enthalpies of the gas and liquid phases, respectively, and V is the mole fraction of the gas phase.

Example 3.5 Derive (a) Eq. (3.62) and (b) Eq. (3.61) of the text.

Solution (a) The expression for the fugacity coefficient of component i in a mixture is (see Eq. (1.109) in Chapter 1)

$$RT \ln \varphi_i = \int_V^\infty [(\partial P/\partial n_i)_{T,V,n_i} - RT/V] dV - RT \ln Z.$$

In a mixture with an associating substance,

$$P = P^{ph} + P^{ch} - RT/v.$$

Differentiating with respect to n_i at constant T, V, and $n_i = (n_1, n_2, \ldots, n_{i-1}, n_{i+1}, \ldots, n_c)$,

$$\left(\frac{\partial P}{\partial n_i}\right)_{T,V,n_i} = \left(\frac{\partial P^{ph}}{\partial n_i}\right)_{T,V,n_i} + \left(\frac{\partial P^{ch}}{\partial n_i}\right)_{T,V,n_i} - RT\left(\frac{\partial}{\partial n_i}\right)\left(\frac{1}{v}\right)_{T,V,n_i}$$

$$i = 1, \ldots, c, \text{ except for } H_2O.$$

From $1/v = n/V$

$$\frac{\partial}{\partial n_i}\left(\frac{n}{V}\right)_{T,V,n_i} = \frac{1}{V}.$$

Combining $P^{ch} v = Z^{ch} RT$ and Eq. (3.57),

$$\frac{P^{ch} v}{RT} = \frac{\xi x_a}{\sqrt{\xi} + \left(\dfrac{RTKx_a}{v}\right) + \beta\left(\dfrac{RTKx_a}{v}\right)^2} + 1 - x_a$$

or

$$P^{ch} = \frac{\xi n_a RT}{V\left[\sqrt{\xi} + \left(\dfrac{RTKx_a}{v}\right) + \beta\left(\dfrac{RTKx_a}{v}\right)^2\right]} + RT\left(\frac{n}{V} - \frac{n_a}{V}\right)$$

$$\left(\frac{\partial P^{ch}}{\partial n_i}\right)_{T,V,n_i} = \frac{RT}{V}.$$

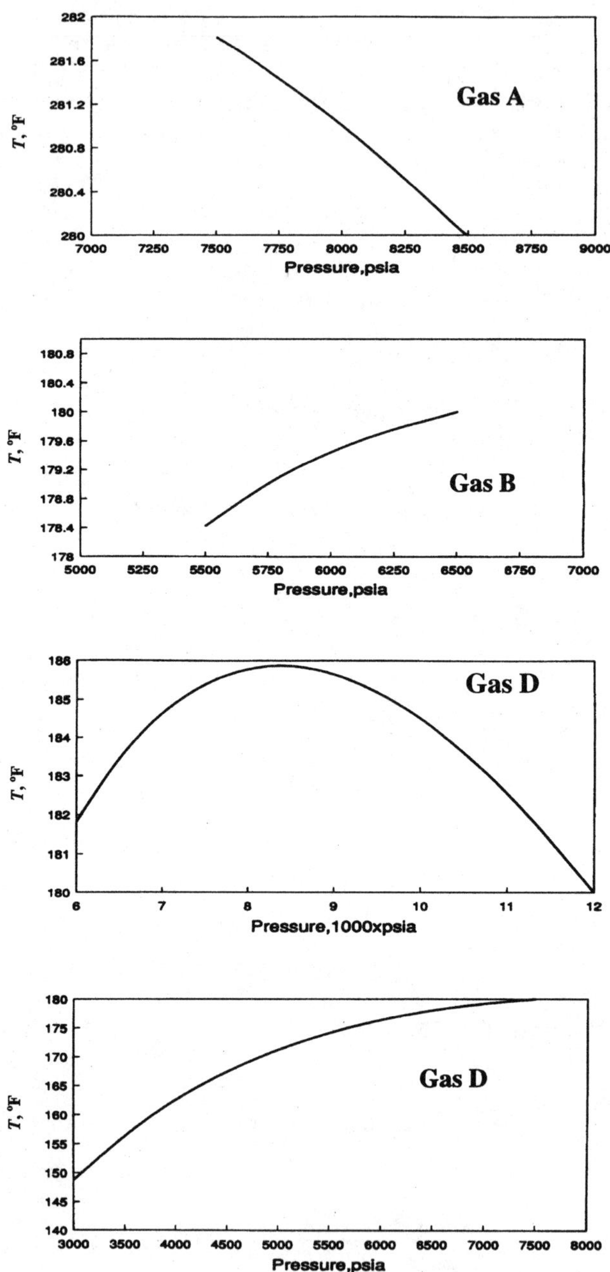

Figure 3.40   Cooling and heating due to expansion.

Combining the above equations,

$$(\partial P/\partial n_i)_{T,V,n_i} = (\partial P^{ph}/\partial n_i)_{T,V,n_i} + RT/V - RT/V = (\partial P^{ph}/\partial n_i)_{T,V,n_i}.$$

Therefore, $RT\ln(\varphi_i Z) = RT\ln(\varphi_i^{ph} Z^{ph})$, or

$$\varphi_i Z = \varphi_i^{ph} Z^{ph}.$$

(b) The expression for $(\partial P/\partial n_i)_{T,V,n_i}$ above can also be written as

$$\left(\frac{\partial P}{\partial n_i}\right)_{T,V,n_i} = \left(\frac{\partial P^{ph}}{\partial n_i}\right)_{T,V,n_i} + \left(\frac{\partial P^{ch}}{\partial n_i}\right)_{T,V,n_i} - \frac{RT}{V}$$

After substitution of the above equation in the expression for fugacity coefficient, one obtains

$$RT\ln\varphi_i Z = \int_V^\infty [(\partial P^{ph}/\partial n_i)_{T,V,n_i} - (RT/V)]dV$$
$$+ \int_V^\infty [(\partial P^{ch}/\partial n_i)_{T,V,n_i} - (RT/V)]dV.$$

The first and the second terms on the right side are $RT\ln(\varphi_i^{ph} Z^{ph})$ and $RT\ln(\varphi_i^{ch} Z^{ch})$, respectively. Therefore, the derivation of Eq. (3.61) of the text is established.

Example 3.6 *Effect of curvature on the interfacial tension of a binary mixture.* Consider a mixture of 95 mole% C_1 and 5 mole% nC_{10} at 327.5 K and 276 bar. Use the PR-EOS and the expression for the interfacial tension given by Eq. (3.96) to calculate the influence of compositional changes due to curvature on the interfacial tension for $r = \infty, 10^{-4}, 10^{-5}$, and 10^{-6} cm. Keep the liquid pressure constant at 276 bar.

Solution The following expressions define the problem:

$$x_1 + x_2 = 1$$

$$y_1 + y_2 = 1$$

$$z_1 = x_1 L + (1 - L)y_1$$

$$f_i^L(T, P^L, x_1) = f_i^V(T, P^V, y_1) \qquad i = 1, 2$$

$$P^V = P^L + \frac{2}{r}\left[\sum_{i=1}^{2} P_i\left(x_i\frac{d^L}{M^L} - y_i\frac{d^V}{M^V}\right)\right]$$

where P_i is the parachor of component i ($P_{C_1} = 70$, $P_{nC_{10}} = 500$) and L is the liquid fraction (see Chapter 4, the section on Two-phase flash calculations). The unknowns in the above set of 6 equations are x_1, x_2, y_1, y_2, L, and P^V. Once these unknowns are calculated, at given r, T, P^L, z_1, and z_2, then σ can be estimated. We used the PR-EOS for the fugacity and phase density calculations. The predicted densities were not adjusted by volume translation. The results are summarized below.

r, cm	σ, dyne/cm	L, liquid mole fraction	$P^V - P^L$, bar
∞	1.06	0.1071	0
10^{-3}	1.06	0.1071	0.002
10^{-4}	1.05	0.1072	0.02
10^{-5}	1.05	0.1078	0.21
10^{-6}	0.97	0.1135	1.95

Note that as r decreases, σ also decreases. In the above calculation we have neglected the effect of curvature itself on σ, which was discussed in Chapter 2.

Example 3.7   Use the PR-EOS to calculate the partial molar volume of C_2 and nC_7 at 80°C and 57 atm for x_{nC_7} in the range of 0 to 0.2. Compare the results with the data of Wu and Ehrlich (1973).

solution   The PR-EOS is pressure-explicit. Therefore,

$$\bar{V}_i = (\partial V/\partial n_i)_{T,P,n_i} = -[(\partial P/\partial n_i)_{T,V,n_i}]/[(\partial P/\partial V)_{T,\underline{n}}]$$

$(\partial P/\partial n_i)_{T,V,n_i}$ and $(\partial P/\partial V)_{T,\underline{n}}$ can be easily calculated from Eq. (3.6). The results are shown in Fig. 3.41. There is a good agreement between data and predicted results.

Example 3.8   Derive the following relationships.

(a) between the isothermal and isentropic compressibilities: $C_T = (c_p/c_V)C_S$, and (b) between c_P and c_V : $c_P = c_V + Tve^2/C_T$.

solution   (a) We can start from Eq. (3.132):

$$C_T = \frac{1}{P} - \frac{1}{Z}(\partial Z/\partial P)_T$$

From Eq. (3.131),

$$(\partial P/\partial v)_s = -\frac{(c_P/c_V)}{v}\left[\frac{1}{1/P - (1/Z)(\partial Z/\partial P)_T}\right].$$

Combining the above two equations results in

$$(\partial P/\partial v)_s = -\frac{(c_P/c_V)}{vC_T}.$$

The definition of isentropic compressibility is

$$C_S = -\frac{1}{V}(\partial V/\partial P)_S.$$

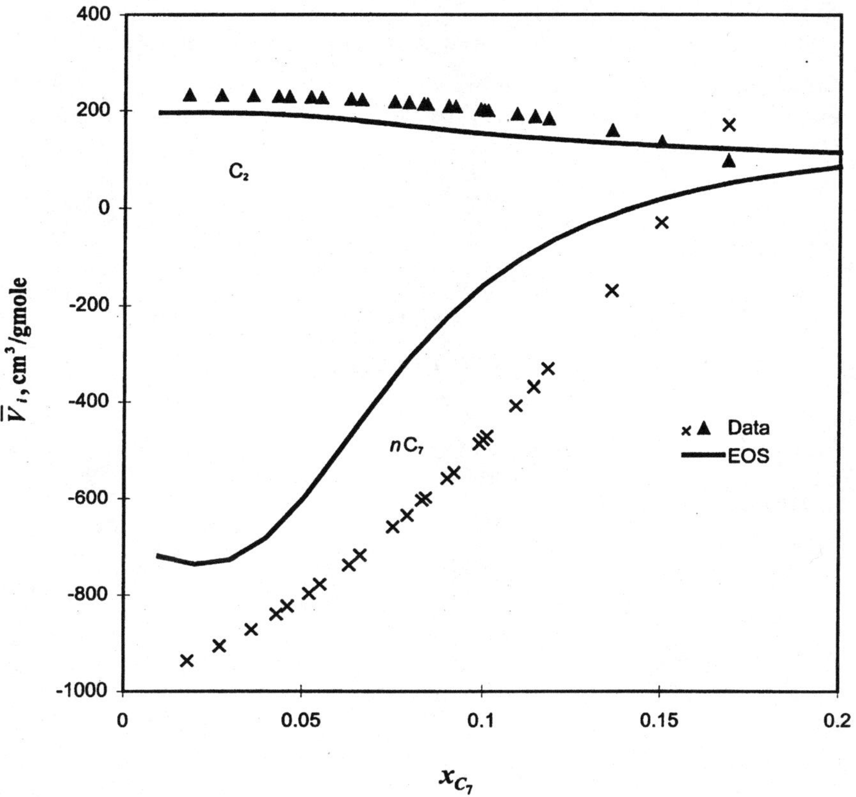

Figure 3.41   Partial molar volumes of C_2 and nzC_7 at 80°C and 57 atm (data from Wu and Ehrlich, 1973).

Therefore,

$$\boxed{C_T = (c_P/c_V)C_S}\,.$$

(b) From Eq. (3.126),

$$c_P - c_V = Tve(\partial P/\partial T)_V.$$

But

$$(\partial P/\partial T)_v = -(\partial v/\partial T)_P/(\partial v/\partial P)_T, \quad (\partial v/\partial T)_P = ve$$

and

$$(\partial v/\partial P)_T = -vC_T.$$

Therefore,

$$\boxed{c_P = c_V + Tve^2/C_T}\,.$$

Example 3.9   Calculate $(\partial n_{j,t}/\partial T)_{P,\underline{n}}$ and $(\partial Z_j/\partial T)_{P,\underline{n}}$ terms in Eq. (3.78). These terms are necessary for the calculation of thermal expansivity.

Solution $(\partial n_{j,t}/\partial T)_{P,\underline{n}}$ can be expressed in terms of $(\partial n_{j,i}/\partial T)_{P,\underline{n}}$ (see Eq. (3.79)), and the latter are obtained from the material balance and equilibrium criterion:

$$\left(\frac{\partial n_{1,i}}{\partial T}\right)_{P,\underline{n}} + \left(\frac{\partial n_{2,i}}{\partial T}\right)_{P,\underline{n}} = 0 \qquad i = 1,\ldots,c,$$

$$\left[\sum_{k=1}^{c}\left(\frac{\partial f_{1,i}}{\partial n_{1,k}}\right)_{n_{1,k}}\left(\frac{\partial n_{1,k}}{\partial T}\right)_{P,\underline{n}} - \sum_{k=1}^{c}\left(\frac{\partial f_{2,i}}{\partial n_{2,k}}\right)_{n_{2,k}}\left(\frac{\partial n_{2,k}}{\partial T}\right)_{P,\underline{n}}\right]$$

$$+ \left[\left(\frac{\partial f_{1,i}}{\partial T}\right)_{P,\underline{n}_1} - \left(\frac{\partial f_{2,i}}{\partial T}\right)_{P,\underline{n}_2}\right] = 0 \qquad i = 1,\ldots,c.$$

Note that we have dropped some of the subscripts on the fugacity derivatives. The above $2c$ equations provide the $2c$ composition variations. $(\partial Z_j/\partial T)_{P,\underline{n}}$ can be calculated from

$$\left(\frac{\partial Z_j}{\partial T}\right)_{P,\underline{n}} = \left(\frac{\partial Z_j}{\partial T}\right)_{P,\underline{n}_j} + \sum_{k=1}^{c}\left(\frac{\partial Z_j}{\partial n_{j,k}}\right)\left(\frac{\partial n_{j,k}}{\partial T}\right) \qquad j = 1,2.$$

An EOS can be used to calculate $(\partial Z_j/\partial T)_{P,\underline{n}_j}$, $(\partial Z_j/\partial n_{j,k})_{P,T,n_{j,k}}$, as well as the fugacity derivatives of phase j. The calculation of $(\partial Z_j/\partial T)_{P,\underline{n}_j}$ can be simplified if one uses

$$\frac{\partial Z_j}{\partial T} = \frac{\partial Z_j}{\partial A}\frac{\partial A}{\partial T} + \frac{\partial Z_j}{\partial B}\frac{\partial B}{\partial T}$$

where the A and B expressions are provided by Eqs. (3.15) and (3.16) for the PR-EOS. In Problem 4.6 we comment how to calculate the derivative of Z with respect to P, which is required for the estimation of isothermal compressibility.

Example 3.10   Consider pure nC_{10} at 5000 psia, and 100° F. What would be the temperature after free expansion to 1000 psia?

solution   A procedure similar to Example 3.4 can be used. The results are plotted in Fig. 3.42. There is heating due to expansion, which is often the case for liquids.

Problems

3.1   From the PR-EOS, derive $Z_c = 0.307$.

3.2   Derive the van der Waals equation of state in the reduced form given by Eq. (3.4).

3.3   Use the van der Waals equation of state to derive the following expression

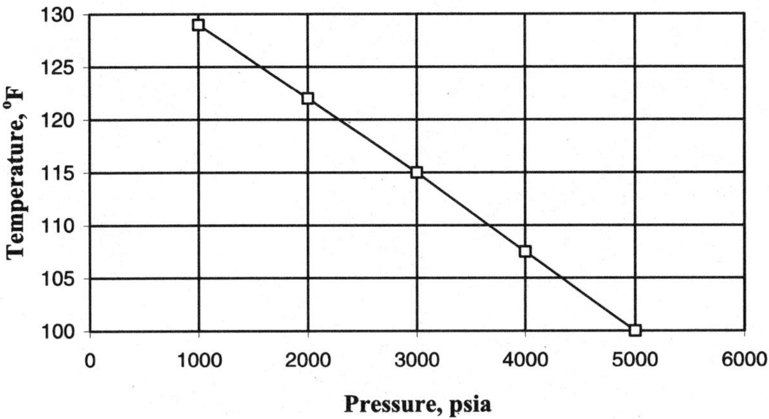

Figure 3.42   Temperature rise of nC_{10} due to expansion.

for the compressibility factor in terms of T_r and P_r:

$$Z^3 - [1 + (P_r)/(8T_r)]Z^2 + [27P_r/(64T_r)]Z - (27P_r)/(512T_r) = 0.$$

Compare the results from the above equation and the generalized compressibility factor chart for natural gases shown in Fig. 3.43. Note that away from the critical point, the agreement is very good.

3.4   Use the PR-EOS to prove that at the critical point of a pure substance $(\partial^3 P/\partial v^3) < 0$. *Hint*: You may use software such as Mathematica for this purpose.

3.5   Derive the expression for the partial molar volume of component i in a mixture using the PR-EOS. *Hint*: You may use software such as Mathematica for this purpose.

3.6   Derive the following expression for the derivative of Z with respect to pressure at constant T and composition from the PR-EOS

$$\frac{\partial Z}{\partial P} = \frac{a}{R^2 T^2}\left[\frac{B - Z}{3Z^2 - 2(1 - B) + A - 3B^2 - 2B}\right]$$
$$+ \frac{b}{RT}\left[\frac{-Z^2 + 2(3B + 1)Z + (A - 2B - 3B^2)}{3Z^2 - 2(1 - B)Z + (A - 3B^2 - 2B)}\right].$$

Note: to facilitate the above derivation, use

$$\frac{\partial Z}{\partial P} = \frac{\partial Z}{\partial A}\frac{\partial A}{\partial P} + \frac{\partial Z}{\partial B}\frac{\partial B}{\partial P}.$$

3.7   Derive Eq. (3.32) of the text.

3.8   Prove that at the critical point of a pure component, $(\partial Z/\partial v) = P_c/RT_c$ and $(\partial^2 Z/\partial v^2) = 0$. Use the criterion of the second derivative to calculate the critical

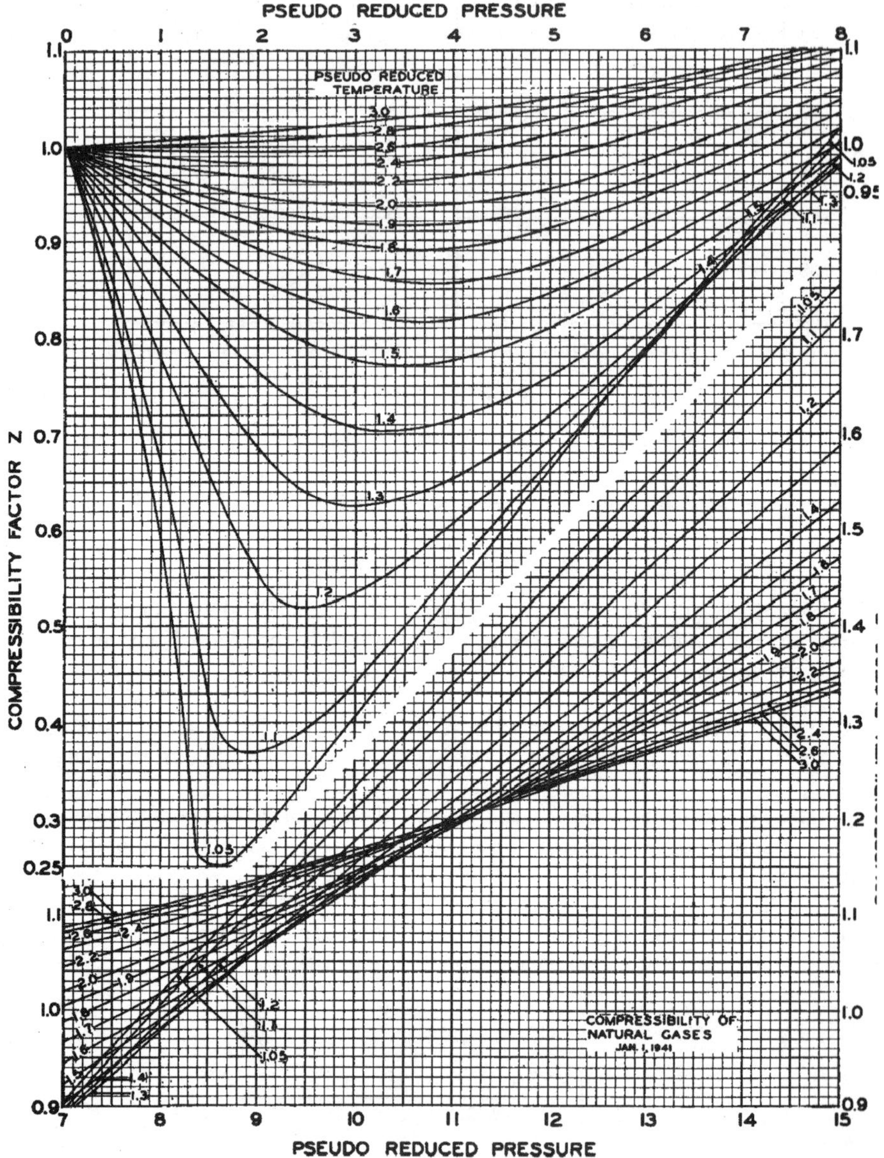

Figure 3.43   Compressibility factor chart for gases from (from Standing and Katz, 1942).

compressibility factor for the SRK equation of state from $Z^3 - Z^2 + Z(A - B^2 - B) - AB = 0$, where $A = aP/(RT)^2$ and $B = (bP/RT)$.

3.9 *Relationship between infinite dilution activity coefficient and interaction coefficient in binary mixtures* At infinite dilution, the activity coefficient is solely a measure of solvent–solute interactions. The activity coefficient of component i at infinite dilution is shown by γ_i^∞, that is, $x_i \to 0$. Derive the following expression for γ_1^∞ from the PR-EOS for a binary mixture and show that $\ln \gamma_1^\infty$ is linear in δ_{12}.

$$\ln \gamma_1^\infty = \frac{b_1}{b_2}(Z_2 - 1) - (Z_1 - 1) + \ln\left(\frac{Z_1 - B_1}{Z_2 - B_2}\right) + \frac{A_1}{2\sqrt{2}B_1}\ln\left[\frac{Z_1 + 2.414B_1}{Z_1 - 0.414B_1}\right]$$
$$- \left(\frac{a_{12}P}{R^2T^2}\right)\frac{1}{\sqrt{2}B_2}\ln\left[\frac{Z_2 + 2.41B_2}{Z_2 - 0.414B_2}\right] + \frac{b_1 A_2}{2\sqrt{2}B_2 b_2}\ln\left[\frac{Z_2 + 2.414B_2}{Z_2 - 0.414B_2}\right].$$

In the above equation subscripts 1 and 2 denote component indices; $Z_1, Z_2, B_1, B_2, A_1,$ and A_2 are pure liquid parameters (at T and P), and $a_{12} = a_1^{1/2} a_2^{1/2}(1 - \delta_{12})$.

Hint: The activity coefficient of component i can be written as

$$\ln \gamma_i = \ln \varphi_i(T, P, \boldsymbol{x}) - \ln \varphi_i^\circ(T, P).$$

3.10 Use the definition of the solubility parameter from Eqs. (1.155) and (1.156) of Chapter 1, and the PR-EOS to derive the following expression for the solubility parameter,

$$\delta = \left[\frac{1}{2\sqrt{2}bv^L}\left(a - T\frac{da}{dT}\right)\ln\frac{v^L + (\sqrt{2}+1)b}{v^L - (\sqrt{2}-1)b}\right]^{1/2}.$$

3.11 *Effect of temperature on the pressure in gas and liquid systems* Two students were arguing in a laboratory on the effect of temperature increase on two cylinders, one containing methane gas and the other cylinder containing n-decane liquid. The pressures of the two cylinders at the laboratory temperature of $25°$C were the same — both were 100 atm.

They were planning to move the two cylinders outside the laboratory on a hot summer day to a room at $40°$C temperature and leave them there for the whole day. Student A was concerned that the pressure of liquid decane may increase to 195 atm while he thought the methane cylinder pressure would increase to only 110 atm.

Use the PR-EOS to calculate the pressure of methane and decane cylinders at $40°$C and to test the skill of the student A in EOS calculations. If student A is correct, then there is a basis that airlines would accept only fluids in the two-phase state for the transportation of liquids in a container.

3.12 Use the PR-EOS with and without association to calculate the maximum value for the superheating of water at atmospheric pressure. The measured value is $270°$C (Kenrick *et al.*, 1924).

3.13 Consider the P–T diagram of a hydrocarbon mixture when the interface between the phases is flat (see the sketch below). Now suppose the interface between the phases is curved with the gas on the convex side. The dewpoint curve is sketched below for the curved interface system. Sketch the bubblepoint curve of the same system with a curved interface.

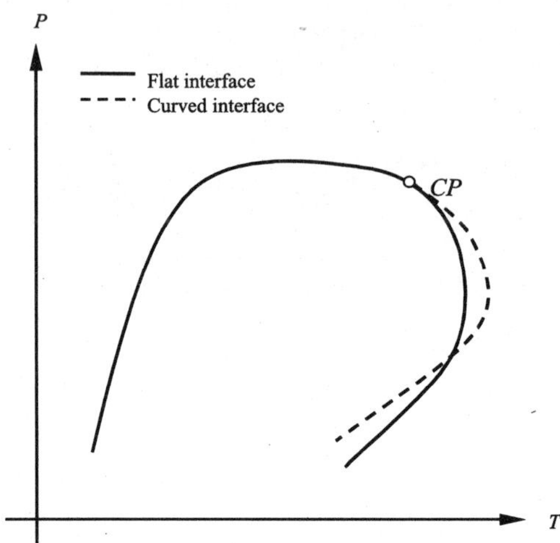

3.14 The following figure shows a plot of volume *vs.* pressure for a reservoir crude at constant temperature both in the single-phase and two-phase states. The change in compressibility provides the bubblepoint pressure, P_b.

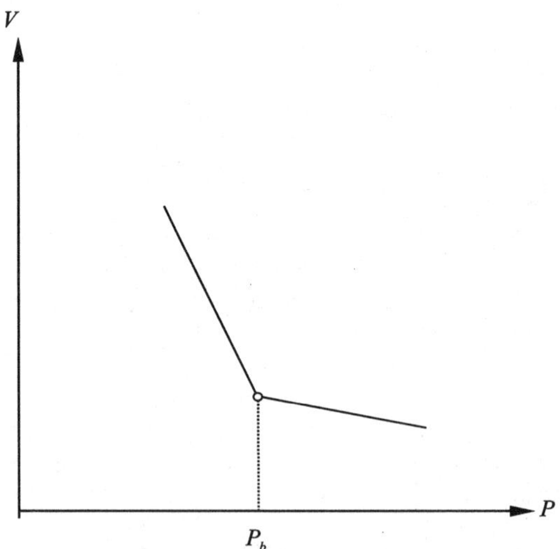

(a) When a fluid is in the near-critical region, then the V *vs.* P plot at constant temperature is sketched as the following:

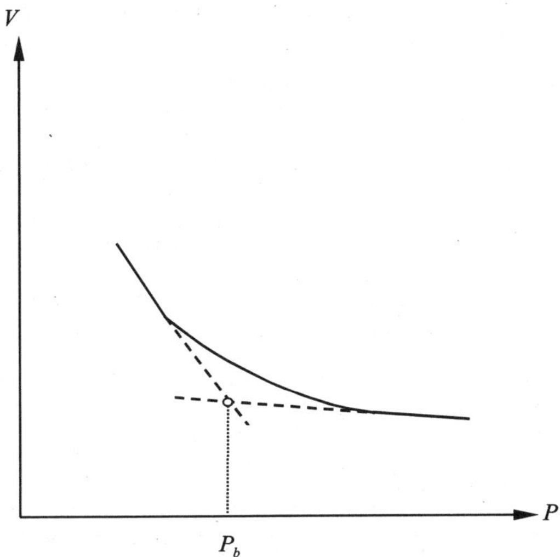

Explain the reason for the gradual increase in the compressibility from single-phase to two-phase states.

(b) Why cannot the V *vs.* P plot be used to estimate the retrograde dewpoint pressure?

References

1. Acs, G., S. Dolescholl, and E. Farkas: "General Purpose Compositional Model," *Soc. Pet. Eng. J.*, p. 543, Aug. 1985.
2. Anderko, A.: "Phase Equilibria in Aqueous Systems for an Equation of State Based on the Chemical Approach," *Fluid Phase Equilibria*, vol. 65, p. 89, 1991.
3. Anselme, M.J.: "The Critical Properties of Thermally Stable and Unstable Fluids and Dilute Fluid Mixtures," *Ph.D. Dissertation,* Georgia Institute of Technology, Atlanta, GA, 1988.
4. Arbabi, S., and A. Firoozabadi: "Near-Critical Phase Behavior of Reservoir Fluids Using Equations of State," *SPE Advanced Techology*, p. 139, March 1995.
5. Benedict, M., G.B. Webb, and L.C. Rubin: "An Empirical Equation for Thermodynamic Properties of Light Hydrocarbons and Their Mixtures II: Mixtures of Methane, Ethane, Propane, and n-Butane," *J. Chem. Physics*, vol. 10, p. 747, 1942.
6. Bjorlykke, O., and A. Firoozabadi: "Measurements and Computation of Retrograde Condensation and Near-Critical Phase Behavior," *SPE Res. Eng.*, p. 271, May 1992.
7. Brent, R.P.: *Algorithms for Minimization Without Derivatives*, Prentice-Hall Inc., N.J., 1973.
8. Briggs, L.J.: "Limiting Negative Pressure of Water," *J. of Applied Physics*, vol. 21, p. 721, July 1950.
9. Cavett, R.H.: "Physical Data for Distillation Calculations—Vapor-Liquid Equilibria," *Proc. API 27th Mid-Year Meeting*, vol. 52, p. 351, 1962.
10. Chu, G.F., and J.M. Prausnitz: "A Phenomenological Correction to an Equation of State," *AIChE J.*, vol. 35, p. 1487, 1989.
11. Debenedetti, Pablo, G.: *Metastable Liquids, Concepts and Principles,* Princeton University Press, Princeton, NJ, 1996.

12. Erbar, J.H., C.L. Persyn, and W.C. Edmister: "Enthalpies and K-ratios for Hydrocarbon Mixtures by New Improved Computer Program," Proceedings of the 1964 GPA Annual Convention, p. 26, 1964.
13. Firoozabadi, A. and K. Aziz: "Analysis and Correlation of Nitrogen and Lean Gas Misicibility," *SPE Res. Eng.*, p. 575, Nov.
14. Firoozabadi, A., and H. Pan: "Two-Phase Isentropic Compressibility and Two-Phase Sonic Velocity for Multicomponent-Hydrocarbon Mixtures," SPE 38844, *SPE Form. Eval. and Eng.* (in Review, 1997).
15. Firoozabadi, A., D.L. Katz, H. Soroosh, and V.A. Sajjadian: "Surface Tension of Reservoir Crude Oil Systems Recognizing the Asphalt in the Heavy Fraction," *SPE Res. Eng.*, p. 265, Feb. 1998a.
16. Firoozabadi, A., R. Nutakki, T. W. Wong, and K. Aziz: "EOS Predictions of Compressibility and Phase Behavior in Systems Containing Water, Hydrocarbons, and CO_2," *SPE Res. Eng.*, p. 673, May 1988b.
17. Firoozabadi, A.: "Reservoir-Fluid Phase Behavior and Volumetric Prediction with Equations of State," *J. Pet. Tech.*, p. 397, April 1988.
18. Gas Processors Association, "Engineering Data Book," 9th ed., Tulsa, OK, 1972.
19. Heidemann, R.A., and J.M. Prausnitz: "A van der Waals Type Equation of State for Fluids with Associating Molecules," *Proc. Nat. Acad. Sci*, vol. 73, p. 1773, 1976a.
20. Heidemann, R.A., and J.M. Prausnitz: "Phase Equilibria from an Equation of State for Fluids with Associating Molecules," paper presented at the *AIChE Annual Meeting,* Chicago, 1976b.
21. Hoffman, A.E., J.S. Crump, and C.R. Hocott: "Equilibrium Constants for a Gas-Condensate System," *Trans, AIME*, vol. 198, p. 1, 1953.
22. Jhaveri, B.S., and G.K. Youngren: "Three-Parameter Modification of the Peng-Robinson Equation of State to Improve Volumetric Predictions," *SPE Res. Eng.*, p. 1033, 1988.
23. Johnson, N.L., and S. Kotz: *Continuous Univariate Distributions*, Houghton Mifflin Co., Boston, 1970.
24. Jones, C.: "The Use of Bottomhole Temperature Variations in Production Testing," SPE 18381, paper presented at the SPE European Petroleum Conference, London, Oct. 16–19, 1988.
25. Katz, D.L., D.J. Vink, and R.A. David: Phase Diagram of a Mixture of Natural Gas and Natural Gasoline Near the Critical Conditions, *Trans. AIME*, vol. 136, p. 106, 1940.
26. Katz, D.L. and A. Firoozabadi,: "Predicting Phase Behavior of Condensate/Crude-Oil Systems Using Methane Interaction Coefficients," *J. Pet. Tech.*, p. 1649, Nov. 1978.
27. Katz, D., L. Cornell, R. Kobayashi, F. Poettman, J. Vary, J. Elenbaas, and C. F. Weinaug "Handbook of Natural Gas Engineering," McGraw-Hill Book Co., New York, 1959.
28. Kenrick, F. B., C. S. Gilbert, and K. L. Wismer: "The Superheating of Liquids," *J. Phys. Chem.*, vol. 28, p. 1297, 1924.
29. Kieffer, S.W.: "Sound Speed in Liquid-Gas Mixtures: Water-Air and Water-Steam," *J. Geophysical Research*, vol. 82, p. 2895, July 10, 1977.
30. Kilgren, K.H.: "Phase Behavior of a High-Pressure Condensate Reservoir Fluid," *J. Pet. Tech.*, p. 1001, August 1966.
31. Lee, B.I., and M.G. Kesler: "A Generalized Thermodynamic Correlation on Three-Parameter Corresponding States," *AIChE J.*, vol. 21, p. 510, May 1975.
32. Macias, L.: "Multiphase, Multicomponent Compressibility in Petroleum Engineering," Ph.D. dissertation, Stanford U., Stanford, CA, March 1985.
33. Mathias, P.M., and T.W. Copeman: "Extension of the Peng-Robinson Equation of State to Complex Mixtures: Evaluation of the Various Forms of the Local Composition Concept," *Fluid Phase Equilibria*, vol. 13, p. 91, 1983.
34. Mathias, P.M., T. Naheiri, and E.M. Oh.: "A Density Correction for the Peng-Robinson Equation of State," *Fluid Phase Behavior*, vol. 47, p. 77, 1989.
35. McKetta, J.J., Jr., and D.L. Katz: "Methane-n-Butane-Water-System in Two- and Three-Phase Regions," *Ind. Eng. Chem.*, vol. 40, p. 835, 1948.
36. Michel, S., H.H. Hooper, and J.M. Prausnitz: "Mutual Solubilities of Water and Hydrocarbons from an Equation of State. Need for an Unconventional Mixing Rule," *Fluid Phase Equilibria*, vol. 45, p. 173, 1989.

37. Ng, H.J., C.J. Chen, and D.B. Robinson: "Vapor Liquid Equilibrium and Condensing Curves in the Vicinty of the Critical Point for a Typical Gas Condensate," Research Report PR-96, Gas Processors Assn., Tulsa, OK, 1986.

38. Ng, H.J., and A.E. Mather: "Enthalpy of Gaseous Mixtures of Methane+Ethane+Carbon Dioxide Under Pressure," *J. Chem. Eng. Data*, vol. 23, p. 224, 1978.

39. Nikitin, E.D., P.A. Pavlov, and A.P. Popov: "Vapor-Liquid Critical Temperatures and Pressures of Normal Alkanes from 19 to 36 Carbon Atoms, Naphthalene and m-terphenyl Determined by the Pulse-Heating Technique," *Fluid Phase Equilibria*, vol. 141, p. 155, 1997.

40. Nikitin, E.D., P.A. Pavlov, and P.V. Skripov: "Measurement of the Critical Properties of Thermally Unstable Substances and Mixtures by the Pulse-Heating Method,", *J. Chem. Thermodyn.*, vol. 25, p. 869, 1993.

41. Nutakki, R.: "Phase Behavior Calculations for Systems with Hydrocarbons, Water, and CO_2," Ph.D. dissertation, Stanford University, Stanford, CA, Nov. 1989.

42. Passut, C.A., and R.P. Danner: "Correlation of Ideal Gas Enthalpy, Heat Capacity, and Entropy," *Ind. Eng. Chem. Proc. Res. Dev.*, vol. 11, p. 543, 1972.

43. Peneloux, A., E. Rauzy, and R. Freze: "A Consistent Correction for Redlich-Kwong-Soave Volumes," *Fluid Phase Equilibria*, vol. 8, p. 7, 1982.

44. Peng, D.Y., and D.B. Robinson: "A New Two-Constant Equation of State," *Ind. Eng. Chem. Fund.*, vol. 15, p. 59, 1976.

45. Pitzer, K.S.: "Corresponding States for Perfect Liquids," *Journal of Chemical Physics*, vol. 7, p. 583, 1939.

46. Pitzer, K.S., D.Z. Lippmann, R.F. Curl, C.M. Higgins, and D.E. Petersen: "The Volumetric and Thermodynamic Properties of Fluids. II. Compressibility Factor, Vapor Pressure, and Entropy of Vaporization," *J. Amer. Chem. Soc.*, vol. 77, p. 3433, 1955.

47. Prausnitz, J.M., R.N. Lichtenthaler, and E.G. de Azevedo: *Molecular Thermodynamics of Fluid-Phase Equilibria,* 2nd ed., Prentice-Hall Inc., Englewood Cliff, N.J., 1986.

48. Riazi, M.R., and T.E. Daubert: "Simplify Property Predictions," *Hydrocarbon Processing*, p. 115, March 1980.

49. Richardson, W.C., M.F. Fontaine, and S. Haynes: "Compositional Changes in Distilled, Steam-Distilled, and Steamflooded Crude Oils," paper SPE 24033 presented at the 1992 Western Regional Meeting, Bakersfield, CA, March 30-April 1, 1992.

50. Robinson, D.B., D.Y. Peng, and S.Y.K. Chung: "Development of the Peng-Robinson Equation and its Application to Phase Equilibrium in a System Containing Methanol," *Fluid Phase Equilibria*, vol. 24, p. 25, 1985.

51. Roland, C.H.: "Vapor-Liquid Equilibria for Natural Gas Crude Oil Mixtures," *Ind. Eng. Chem.*, vol. 37, p. 930, Oct. 1945.

52. Rowlinson, J.S., and F.L. Swinton: *Liquid and Liquid Mixtures*, 3rd ed., Butterworth Scientific, London, 1982.

53. Sage, B.H., W.N.B. Lacey, and J.G. Schaafsma: "Report on API Research Project 37—Covers Behavior of Hydrocarbon Mixtures," *Oil and Gas J.*, p. 12, Nov. 1933.

54. Schmidt, G., and H. Wenzel: "A Modified Van der Waals Type Equation of State," *Chem. Eng. Sci.*, vol. 35, p. 1503, 1980.

55. Shinta, A.A., and A. Firoozabadi: "Equation of State Representation of Aqueous Mixtures Using an Association Model," *Can. J. Chem. Eng.*, p. 367, June 1995.

56. Shinta, A.A., and A. Firoozabadi: "Predicting Phase Behavior of Water/Reservoir-Crude Systems with the Association Concept," *SPE Res. Eng.*, p. 131, May 1997.

57. Skuse, B., A. Firoozabadi, and H. Ramey Jr.: "Computation and Interpretation of Capillary Pressure from a Centrifuge," *SPE Form. Eva.*, p. 17, March 1992.

58. Soave, G.: "Equilibrium Constant From a Modified Redlich-Kwong Equation of State," *Chem. Eng. Sci.*, vol. 27, p. 1197, 1972.

59. Starling, K.E.: *Fluid Thermodynamic Properties for Light Petroleum Systems*, Gulf Publishing Company, 1973.

60. Teja, A.S., R.J. Lee, D. Rosenthal, and M. Anselme: "Correlation of the Critical Properties of Alkanes and Alkanols," *Fluid Phase Equilibria*, vol. 56, p. 153, 1990.

61. Thomas, L.K., R.W. Hankinson, and K.A. Phillips: "Determination of Acoustic Velocities for Natural Gas," *J. Pet. Tech.*, p. 889, July 1970.

62. Twu, C.H.: "An Internally Consistent Correlation for Predicting the Critical Properties and Molecular Weight of Petroleum and Coal-Tar Liquids," *Fluid Phase Equilibria*, vol. 16, p. 137, 1984.

63. Voronel, A.V.: "In Phase Transitions and Critical Phenomena," C. Domb, and M.S. Green, (eds.), Academic Press, London, vol. 5, 1976.

64. Watts, J.W.: "A Compositional Formulation of the Pressure and Saturation Equations," *SPE Res. Eng.*, p. 243, May 1986.

65. Weinaug, C. F., and D. L. Katz: "Surface Tension of methane–propane mixtures," *Ind. Eng. Chem.*, vol. 36, p. 239, Feb. 1943.

66. Whitson, C.H., "Characterizing Hydrocarbon Plus Fractions," *Soc. Pet. Eng. J.*, p. 683, Aug. 1983.

67. Wu, P. C., and Ehrlich, P.: "Volumetric Properties of Supercritical Ethane-n-Heptane Mixtures: Molar Volumes and Partial Molar Volumes," *AIChe. J.*, p. 553, May 1973.

68. Yarborough, L.: "Application of a Generalized Equation of State to Petroleum Reservoir Fluids," *Equations of State in Engineering*, Advances in Chemistry Series, K.C. Chao and R.L. Robinson (eds.), American Chem. Soc., Washington, DC, vol. 182, p. 385, 1979.

69. Zudkevitch, D., and J. Jaffe: "Correlation and Prediction of Vapor-Liquid Equilibrium with the Redlich-Kwong Equation of State," *AIChE J.*, p. 496, May 1970.

4

Equilibrium, stability, and criticality

Equilibrium, stability, and criticality are important concepts that are closely related. In this chapter, after the formulation of simple methods for phase-equilibria calculations, the concepts of stability and criticality are introduced, and the application of the Gibbs free energy surface analysis to phase-equilibrium calculations is demonstrated. Then the stability and criticality concepts are presented in detail. These concepts are useful for a broad range of problems in engineering and physics.

Suppose we are given a mixture of different species at a given temperature and pressure. If we know that the given mixture will split into two phases, say a gas phase and an oil phase, then we want to calculate the amount as well as the composition of each phase. This type of calculation is called the two-phase flash. In a more complicated case, we may not know into how many phases the mixture will split. Use of phase stability analysis provides a sound basis for such a situation.

The stability concept is also useful in establishing the maximum supersaturation that can occur for both pure components and multicomponent mixtures. Suppose we are given a mixture of C_1 and nC_5 that has a fixed bubblepoint pressure at a given temperature. If the pressure of this mixture at the given temperature is lowered from a pressure above to a pressure below the bubblepoint, the gas phase may not appear if the pressure reduction is carried out rapidly. However, there is a theoretical limit for the pressure at which the gas phase will appear no matter how fast is the pressure reduction. This maximum supersaturation can be calculated from the stability limit to be discussed in this chapter. In the last part of this chapter, we will present methods that can be used in calculating the critical point of complex mixtures. Methods that were introduced around 1980 will be emphasized.

In this chapter, there is a strong emphasis on step by step derivations of the criteria of stability and criticality, mainly relying on the Second Law and its equivalent forms. We would like to comment that, to the best of our knowledge, the stability criteria and concepts have not yet been developed for multicomponent systems with curved interfaces.

Two-phase flash calculations

A two-phase flash problem can be stated as follows. Given the number of moles of feed, F, the mole fraction of components in the feed $z_i, i = 1, \ldots, c$ (i.e., $\sum_{i=1}^{c} z_i = 1$), and pressure and temperature, find the number of the moles in the gas and liquid phases, V, and L, respectively, and the mole fractions of the gas and liquid phases, y_i and x_i, respectively. We assume that the interface between the gas and liquid phases is flat, which implies that the gas and liquid phase pressures are the same. Figure 4.1 shows a sketch for the process. The application of this type of calculation is shown in Fig. 4.2, where a gas phase from the gas cap of a hydrocarbon reservoir at pressure P_{res} and temperature T_{res} is flashed at a lower pressure P and a lower temperature T. Similarly, the oil phase from the oil column at temperature T_{res} and pressure P_{res} can be flashed at a lower pressure P and a lower temperature T (see Fig. 4.2).

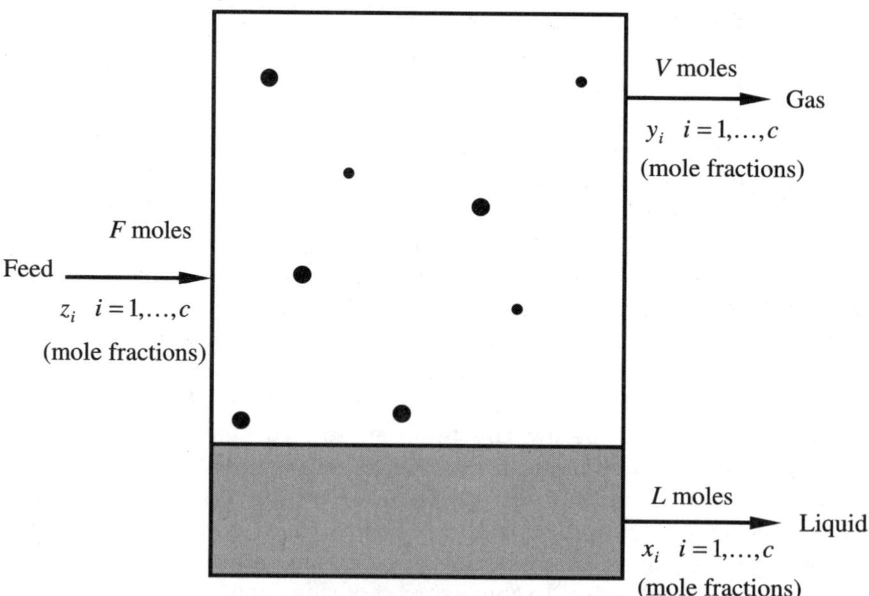

Figure 4.1   Schematic of two-phase flash at constant T and P.

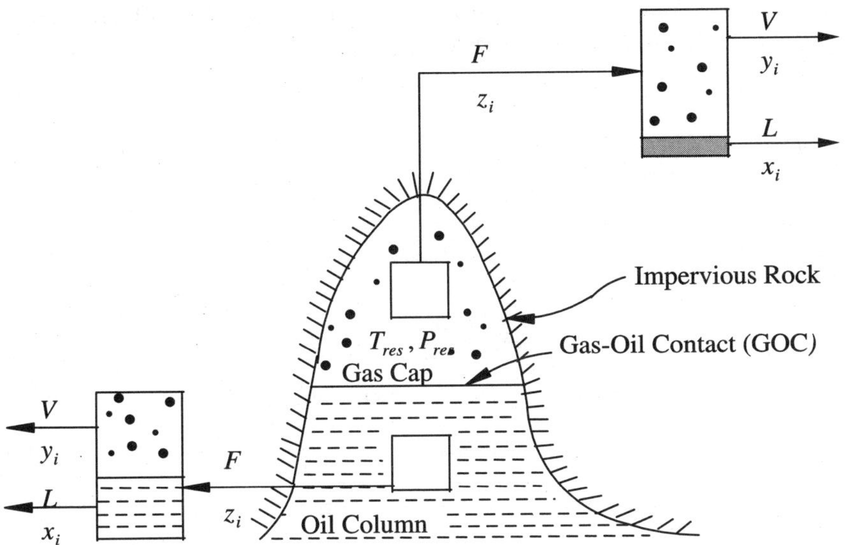

Figure 4.2   Schematic of the flash of the gas-cap gas and oil from the oil column.

Let us write the equations that define the two-phase flash.

(1) Equality of chemical potentials or fugacities provides the equilibrium condition.

$$f_i^L(T, P, \boldsymbol{x}) = f_i^V(T, P, \boldsymbol{y}), \qquad i = 1, \ldots, c \tag{4.1}$$

(2) Material balance for components provides the mass conservation.

$$F z_i = x_i L + y_i V \qquad i = 1, \ldots, c \tag{4.2}$$

(3) Mole fraction constraints assure that the sum of the mole fractions in a given phase are equal to unity.

$$\sum_{i=1}^{c} x_i = 1 \tag{4.3}$$

$$\sum_{i=1}^{c} y_i = 1 \tag{4.4}$$

In the above equations, there are $2c$ unknowns in x_i and y_i and two unknowns in V and L. Various solution methods for the above $(2c + 2)$ equations and unknowns are available. Two solution techniques will be presented in the following. A third technique will be briefly described.

Successive substitution technique. In this technique, through an iterative procedure, only one unknown, the fraction of liquid or vapor, is searched. Let us define the equilibrium ratio, K_i, as

$$K_i = y_i/x_i \qquad i = 1, \ldots, c. \tag{4.5}$$

Then

$$y_i = K_i x_i \qquad i = 1, \ldots, c. \tag{4.6}$$

Combining Eqs. (4.2) and (4.6),

$$Lx_i + VK_i x_i = Fz_i \qquad i = 1, \ldots, c. \tag{4.7}$$

Taking the summation over $i = 1, \ldots, c$ and using Eqs. (4.3) and (4.4),

$$L = F - V. \tag{4.8}$$

Combining Eqs. (4.7) and (4.8) and solving for x_i results in

$$\boxed{x_i = \frac{z_i}{1 + (K_i - 1)(V/F)}} \qquad i = 1, \ldots, c. \tag{4.9}$$

Similarly, one also obtains

$$y_i = \frac{K_i z_i}{1 + (K_i - 1)(V/F)} \qquad i = 1, \ldots, c. \tag{4.10}$$

Combining the above two equations and the two constraint equations, Eqs. (4.3) and (4.4),

$$\sum_{i=1}^{c} \frac{(K_i - 1)z_i}{1 + (K_i - 1)(V/F)} = 0. \tag{4.11}$$

Define $\alpha = V/F$; then Eq. (4.11) becomes the Rachford-Rice (1952) expression,

$$h(\alpha) = \sum_{i=1}^{c} \frac{(K_i - 1)z_i}{1 + \alpha(K_i - 1)} = 0. \tag{4.12}$$

The equilibrium ratios, K_i, in Eq. (4.12) defined previously in Eq. (4.5) are functions of T, P, and composition of one of the phases, say $x = (x_1, x_2, \ldots, x_{c-1})$. Let us examine the dependency. The fugacity of component i in a phase is given by

$$f_i^V = y_i \varphi_i^V P \qquad i = 1, \ldots, c \tag{4.13}$$

$$f_i^L = x_i \varphi_i^L P \qquad i = 1, \ldots, c, \tag{4.14}$$

where φ_i^V and φ_i^L are gas- and liquid-phase fugacity coefficients defined in Chapter 1. At equilibrium $f_i^L = f_i^V$, and therefore

$$\boxed{K_i = y_i/x_i = \varphi_i^L/\varphi_i^V}, \qquad i = 1, \ldots, c \tag{4.15}$$

where φ_i^L is a function of $(T, P, \underline{x})$, and φ_i^V is a function of $(T, P, \underline{y})$. Since composition of the phases, $\underline{x}$ and $\underline{y}$, are related by $f_i^L = f_i^V$, $i = 1, \ldots, c$, then $\underline{x}$ and $\underline{y}$ in Eq. (4.15) are not independent of each other; $K_i = K_i(T, P, \underline{x})$ or $K_i = K_i(T, P, \underline{y})$. Now we can proceed with the iterative solution procedure.

Step 1.   Guess the initial values of K_i at the fixed temperature and pressure. The Wilson (1969) correlation can be used for this purpose. In this correlation, which is based on the direct application of Raoult's law (see Example 1.7, Chapter 1), the ideal K-value of component i is given by

$$\ln K_i = 5.37(1 + \omega_i)\left[1 - T_{ci}/T\right] + \ln(P/P_{ci}), \tag{4.16}$$

where T_{ci} and P_{ci} are the critical temperature and critical pressure of component i, respectively.

Step 2.   Solve Eq. (4.12) for α. This equation is readily solved by Newton's method, to be discussed shortly.

Step 3.   Calculate x_i and y_i from Eqs. (4.9) and (4.10), then the compressibility factors of the liquid and gas phases from an EOS.

Step 4.   Calculate φ_i^L and φ_i^V.

Step 5.   Update K_i. The update is given by $K_i^{new} = K_i^{old} \exp[-\ln(f_i^V/f_i^L)]$.

Step 6.   Test the convergence. Different convergence criteria can be used. One may use the criterion $(1/c) \sum_{i=1}^{c}[\ln(f_i^V/f_i^L)]^2 <$ a given tolerance, say 10^{-12}.

If the convergence criterion is not satisfied, steps 2 through 6 are repeated. When the ratio of component fugacities converge to a value other than one, and α is outside the interval [0,1], a single-phase region may exist. Because of a poor initial guess of K_i, α may be calculated to be outside the interval [0,1], in early iterations, and the mixture may not be single-phase. If $h(\alpha = 0) < 0$, then set $x_i = z_i$, and the vapor phase composition is estimated from $y_i = K_i z_i / \sum_{j=1}^{c} K_j z_j$ (see Problem 4.1). Then in the first iteration, we start from Step 4 and the process is repeated.

The number of iterations depends on closeness to the critical point. If away from the critical point, depending on the initial guess and the tightness of the tolerance, the number of iterations is often less than 10 to 20. When close to the critical point, the number of iterations may exceed several thousand and other methods become necessary.

One may use Newton's method to solve Eq. (4.12). The first step is to evaluate the derivative of h with respect to α, $h'(\alpha)$:

$$h'(\alpha) = -\sum_{i=1}^{c} \frac{(K_i - 1)^2 z_i}{[1 + (K_i - 1)\alpha]^2} \tag{4.17}$$

Note that $h'(\alpha)$ is negative and therefore h is a monotonically decreasing function of α. The expression for updating α is given by

$$\alpha^{new} = \alpha^{old} - (h/h'). \tag{4.18}$$

The solution in the interval 0 to 1 for $\alpha = V/F$ exists provided

$$h(0) = \sum_{i=1}^{c} K_i z_i - 1 > 0 \tag{4.19}$$

and

$$h(1) = 1 - \sum_{i=1}^{c} (z_i/K_i) < 0. \tag{4.20}$$

When $h(0) \leq 0$, the fluid may be a subcooled liquid and for $h(1) \geq 0$, the fluid may be a superheated vapor.

Now we examine the shape of h vs. α and demonstrate that in a given mixture in the two-phase region, always at least one of the K_i should be greater than one and at least one K_i should be less than one. In other words, in a binary mixture, the K_i-value of the more volatile component, say component "1", is greater than one and the K_i-value of the less volatile component, say component "2", is less than one.

The asymptotes of h from Eq. (4.12) are given by

$$1 + (K_i - 1)\alpha_{i\infty} = 0 \qquad i = 1, \ldots, c, \tag{4.21}$$

where $\alpha_{i\infty}$ represents the asymptote of component i. From Eq. (4.21), if $K_i < 1, \alpha_{i\infty} > 1$, and if $K_i > 1, \alpha_{i\infty} < 0$. Figure 4.3 depicts $h(\alpha)$ vs. α for different values of K_i; the order is $K_1 > K_2 > K_3 > \cdots > K_c > 0$ (component "1" is the most volatile component). Note that in Fig. 4.3, the solution lies between the two asymptotes corresponding to $K_1 > 1$ and $K_2 < 1$. Also note that because of the monotonically decreasing trend, there is only one solution, $0 < \alpha^* < 1$. If $\alpha^* = 1$, then $L = F$, i.e., liquid phase, and if $\alpha^* = 0$, $V = F$, i.e., gas phase.

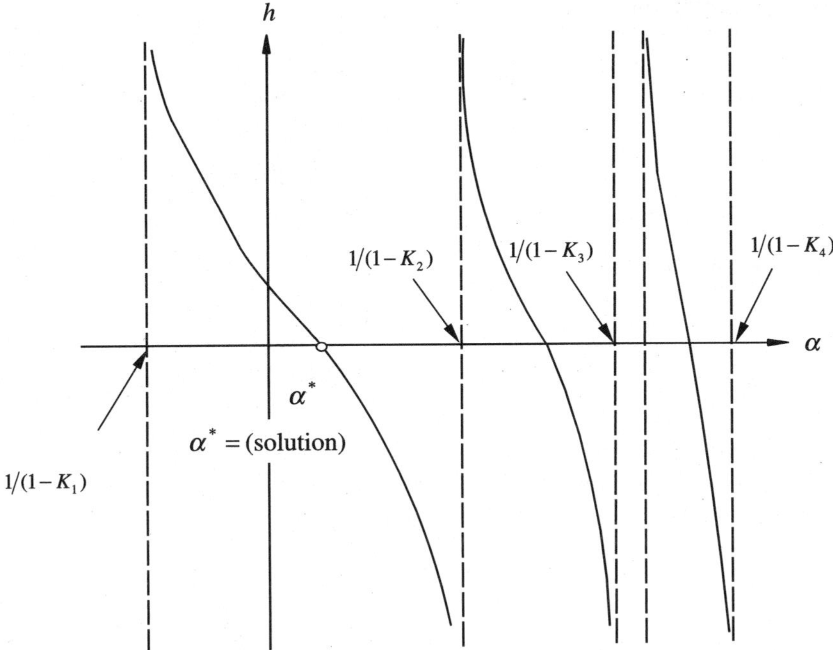

Figure 4.3   The asymptotes of h vs. α (see Eqs. (4.12) and (4.21)).

Newton's method   The nonlinear system of Eqs. (4.1) and the linear Eqs. (4.2) to (4.4) that define the two-phase flash can be solved by Newton's method to obtain $2(c+1)$ unknowns, V, F, x_i, and y_i from the $2(c+1)$ equations. In the following, we will use Newton's method to solve the problem of bubblepoint pressure of a c-component system, a special case of two-phase flash, when $\alpha = 0$.

Given T and the overall composition of the fluid z_i, find the bubblepoint pressure P and the composition of the gas bubble, y_i. At the bubblepoint, the composition of the liquid phase, x_i, is the same as the overall composition, z_i; therefore $\sum_{i=1}^{c} z_i = \sum_{i=1}^{c} x_i = 1$. The expressions defining the bubblepoint pressure and bubble composition are provided by Eqs. (4.1) and (4.4).

There are $(c+1)$ equations and $(c+1)$ unknowns, P and $y_i (i = 1, \ldots, c)$. From $f_i^L = \varphi_i^L x_i P$ and $f_i^V = \varphi_i^V y_i P$, Eq. (4.1) can be expressed as

$$y_i = x_i \varphi_i^L / \varphi_i^V \qquad i = 1, \ldots, c. \tag{4.22}$$

Equations (4.22) and (4.4) can be written as

$$\mathscr{F}_i = y_i - x_i \varphi_i^L / \varphi_i^V \qquad i = 1, \ldots, c. \tag{4.23}$$

$$\mathscr{F}_{c+1} = 1 - \sum_{i=1}^{c} y_i. \tag{4.24}$$

Next we write the Taylor's series expansion of the above two equations,

$$\mathscr{F}_i^{n+1} \approx \mathscr{F}_i^n + (\partial \mathscr{F}_i / \partial P) \Delta P + \sum_{k=1}^{c} (\partial \mathscr{F}_i / \partial y_k) \Delta y_k \qquad i = 1, \ldots, c \tag{4.25}$$

$$\mathscr{F}_{c+1}^{n+1} \approx \mathscr{F}_{c+1}^n + \sum_{k=1}^{c} (\partial \mathscr{F}_{c+1} / \partial y_k) \Delta y_k \tag{4.26}$$

where n is the iteration level; the derivatives are evaluated at the nth iteration level (the iteration level n will be dropped in the expression for derivatives in the following). From Eq. (4.24),

$$(\partial \mathscr{F}_{c+1} / \partial y_k) = -1 \qquad k = 1, \ldots, c. \tag{4.27}$$

Equations (4.25) and (4.26) can be cast into the forms

$$\sum_{k=1}^{c} \left(\frac{\partial \mathscr{F}_i}{\partial y_k} \right) \Delta y_k + \frac{\partial \mathscr{F}_i}{\partial P} \Delta P = \mathscr{F}_i^{n+1} - \mathscr{F}_i^n = R_i \qquad i = 1, \ldots, c \tag{4.28}$$

$$- \sum_{k=1}^{c} \Delta y_k + (0)\Delta P = \mathscr{F}_{c+1}^{n+1} - \mathscr{F}_c^n = R_{c+1}. \tag{4.29}$$

In the matrix form,

$$\begin{pmatrix} \dfrac{\partial \mathscr{F}_1}{\partial y_1} & \dfrac{\partial \mathscr{F}_1}{\partial y_2} & \cdots & \dfrac{\partial \mathscr{F}_1}{\partial y_c} & \dfrac{\partial \mathscr{F}_1}{\partial P} \\[2ex] \dfrac{\partial \mathscr{F}_1}{\partial y_1} & \dfrac{\partial \mathscr{F}_2}{\partial y_2} & \cdots & \dfrac{\partial \mathscr{F}_2}{\partial y_c} & \dfrac{\partial \mathscr{F}_2}{\partial P} \\[2ex] \vdots & \vdots & & \vdots & \vdots \\[1ex] \dfrac{\partial \mathscr{F}_c}{\partial y_1} & \dfrac{\partial \mathscr{F}_c}{\partial y_2} & \cdots & \dfrac{\partial \mathscr{F}_c}{\partial y_c} & \dfrac{\partial \mathscr{F}_c}{\partial P} \\[2ex] -1 & -1 & \cdots & -1 & 0 \end{pmatrix} \begin{pmatrix} \Delta y_1 \\ \Delta y_2 \\ \vdots \\ \Delta y_c \\ \Delta P \end{pmatrix} \begin{pmatrix} R_1 \\ R_2 \\ \vdots \\ R_c \\ R_{c+1} \end{pmatrix} \tag{4.30}$$

We can write Eq. (4.30) in a condensed form,

$$\vec{J} \Delta \underline{y} = \vec{R}, \tag{4.31}$$

where $\vec{J}$ is the Jacobian matrix. Next we write the expression for the elements of the Jacobian matrix $\vec{J}$. Using Eq. (4.23), the elements of the Jacobian can be determined:

$$\left(\frac{\partial \mathscr{F}_i}{\partial y_j}\right)_{T,P,y_j} = -x_i \frac{\partial}{\partial y_j}(\varphi_i^L/\varphi_i^V)_{T,P,y_j} \qquad i = 1,\ldots,c \text{ and } j = 1,\ldots,c, i \neq j$$

$$(4.32)$$

$$\left(\frac{\partial \mathscr{F}_i}{\partial y_i}\right)_{T,P,y_i} = 1 - x_i \frac{\partial}{\partial y_i}(\varphi_i^L/\varphi_i^V)_{T,P,y_i} \qquad i = 1,\ldots,c \qquad (4.33)$$

$$\left(\frac{\partial \mathscr{F}_i}{\partial P}\right)_{T,y} = -x_i \frac{\partial}{\partial P}\left(\frac{\varphi_i^L}{\varphi_i^V}\right)_{T,y} \qquad i = 1,\ldots,c. \qquad (4.34)$$

The above equations can be further simplified to

$$\left(\frac{\partial \mathscr{F}_i}{\partial y_j}\right)_{T,P,y_j} = +x_i \frac{\varphi_i^L}{(\varphi_i^V)^2}\left(\frac{\partial \varphi_i^V}{\partial y_j}\right)_{T,P,y_j} \qquad i = 1,\ldots,c \text{ and } j = 1,\ldots,c, i \neq j$$

$$(4.35)$$

$$\left(\frac{\partial \mathscr{F}_i}{\partial y_i}\right)_{T,P,y_i} = 1 + x_i \frac{\varphi_i^L}{(\varphi_i^V)^2}\left(\frac{\partial \varphi_i^V}{\partial y_i}\right)_{T,P,y_i} \qquad i = 1,\ldots,c \qquad (4.36)$$

$$\left(\frac{\partial \mathscr{F}_i}{\partial P}\right)_{T,y} = -x_i \left[\varphi_i^V\left(\frac{\partial \varphi_i^L}{\partial P}\right)_{T,y} - \phi_i^L\left(\frac{\partial \varphi_i^V}{\partial P}\right)_{T,y}\right]\bigg/ (\varphi_i^V)^2 \qquad i = 1,\ldots,c.$$

$$(4.37)$$

The derivatives $(\partial \varphi_i^V/\partial y_j)$, $(\partial \varphi_i^L/\partial P)$, and $(\partial \varphi_i^V/\partial P)$ can be calculated either analytically or numerically using an EOS. One could start with the Wilson equation to obtain the initial guess for y and Raoult's law for the initial guess for P, then use Eqs. (4.32) to (4.37) to calculate the elements of the Jacobian and follow an iterative process to calculate y and P. Note that the mixture at temperature T might have two bubblepoint pressures—an upper bubblepoint and a lower bubblepoint. A very low guess for pressure may give the lower bubblepoint and a high value may give the upper bubblepoint. Upper and lower bubblepoints are not common for reservoir fluids. However, two dewpoints exist for reservoir gases. The upper dewpoint is called the retrograde dewpoint.

Other methods The method of successive substitution (SS), and the full Newton method for solving the system of nonlinear algebraic equations of flash have certain limitations. They have also certain desirable features. The successive substitution method, in addition to its simplicity, is also a stable method that results in the reduction of Gibbs free energy in every iteration (see Michelsen, 1982b). The SS method, unlike Newton's method, does not require good initial estimates of K_i-values. As we have seen, the SS method relies on the assumption that in every iteration the K_i-values are constant, which is not true for nonideal phases. As the nonideality of phases increases, the rate of convergence decreases; for the near-critical region, the rate of convergence becomes extremely slow.

The classical Newton method has quadratic convergence properties whereas the successive substitution method has linear rate of convergence. However, because of the overshoot, the Newton iteration may fail to converge when the initial estimate is not a good estimate of the solution of the system of nonlinear equations.

A very large array of methods have been proposed to accelerate the convergence rate of successive substitution. These include the use of the Dominant Eigenvalue Method of Crowe and Nishio (1975). In spite of significant increase in the rate of convergence, such methods may cause the loss of basic stability of successive substitution.

In Newton's method, various approaches have been suggested to adjust the step length to alleviate the overshoot problem. These adjustments do not guarantee convergence of the Newton's method in early iterations.

The combination of successive substitution and Newton's method is a good choice and has the desirable features of both. In this approach, the successive substitution comprises the first few iterations and later, when a switching criterion is met, Newton's method is used. To our knowledge, some commercial reservoir simulation models have adopted the combined successive substitution-Newton approach after the experience with various methods of solving nonlinear flash calculation including Powell's method (1970). The application of a reduction method to phase equilibrium calculations has also been proposed (Michelsen, 1986; Hendriks, and Van Bergen, 1992). In this approach, the dimensionality of phase equilibrium problems for multicomponent mixtures can be drastically reduced. The application of reduction methods and its implementation in reservoir compositional models is under evaluation.

Even when the nonlinear-flash equations are properly solved and convergence is achieved, there is no guarantee that the solution obtained is a true solution. The equilibrium condition given by the equality of chemical potentials or fugacities is a necessary but not a sufficient condition. However, for gas-liquid equilibria, the true solution is nearly always obtained from the equality of chemical potentials. For liquid–liquid and vapor–liquid–liquid and higher equilibria calculations, the equality of chemical potentials alone may lead to a false solution. One needs to use an additional relationship to examine the validity of the calculated results. The additional relationship is derived from the Gibbs free energy surface analysis. Phase stability analysis is also useful to check if a given composition is in a single- or two-phase state. In the following, we will give a simple geometrical interpretation of the phase-stability analysis, mainly drawing from the work of Baker, Pierce, and Luks (1982) using the Gibbs free energy.

Gibbs free energy surface analysis

As stated before, Eq. (4.1) is not equivalent to the global minimum of the Gibbs free energy. The system of predicted phases from Eqs. (4.1) to (4.4) must have the lowest possible Gibbs free energy at system temperature T and pressure P. The global minimum of Gibbs free energy is the statement of the Second Law, which is equivalent to the statement that the entropy of an isolated system must be a maximum.

In three- and higher-phase calculations, there are many instances where the equality of chemical potentials or fugacities does not guarantee a global Gibbs free energy minimum. As a result, the solution may be false. Examples of three and higher phases are CO_2–hydrocarbon systems and water–hydrocarbon systems, as well as mixtures of rich gases and crude oils. When there is solid precipitation, the occurrence of several phases is a norm rather than an exception, as we will discuss in the next chapter. All these multiple phases occur in the temperature range of 50 to 300°F, commonly observed in the production facilities and in the reservoir.

Let us study a two-component mixture comprised of a less volatile component 1 and a more volatile component 2 at temperature T. For this binary mixture, the pressure–composition diagram is sketched in Fig. 4.4; P_A and P_B are the vapor pressures at T of pure components 1 and 2, respectively. Several different regions exist in this figure. Toward the right of curve AC, the state of the system is gas. Along the line FDC, the mixture exists as three phases, gas and two liquids; the pressure for this state is called the three-phase pressure, $P_{3\phi}$. In the region $ACDF$ gas and a liquid phase rich in the less volatile component 1 coexist. To the left of AFH, only a liquid phase rich in component 1 is present. Two liquid phases, L_A and L_B, exist in the region bounded by $HFDE$. Liquid L_B and gas exist in the $CBDC$ region, and only liquid phase L_B exists in the region bounded by the BD and DE curves.

Now consider the mixing of components 1 and 2 at constant temperature and pressure. The molar Gibbs free energy of mixing, Δg_{mix}, can be expressed as

$$\Delta g_{mix} = g - \sum_{i=1}^{2} x_i \mu_i^0(T, P), \tag{4.38}$$

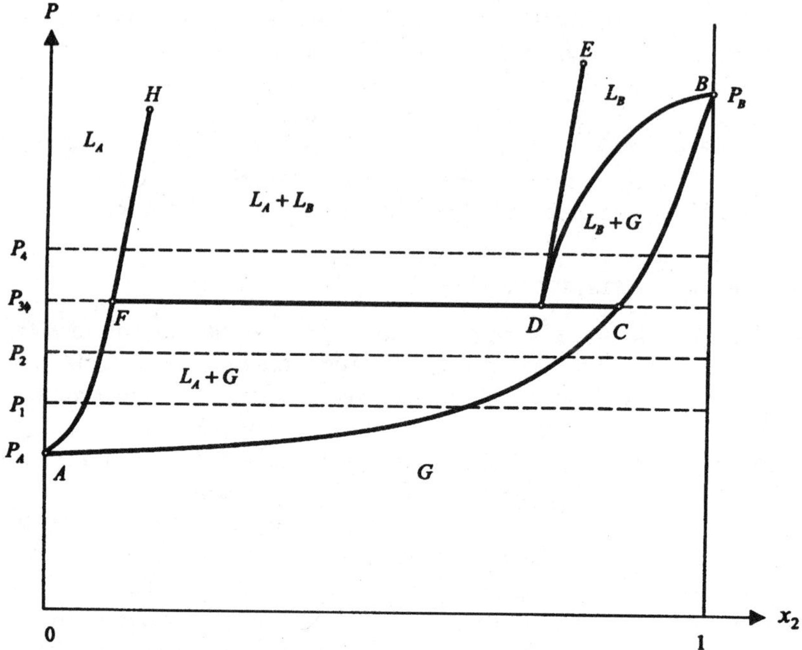

Figure 4.4   Pressure-composition diagram of the binary mixture at temperature T (adapted from Baker *et al.*, 1982).

where $g = \sum_{i=1}^{2} x_i \mu_i(T, P, x_1)$ is the molar Gibbs free energy of the mixture and the term $\sum_{i=1}^{2} x_i \mu_i^0(T, P)$ represents the Gibbs free energy of one mole of components 1 and 2 before mixing. Equation (4.38) can be expressed as

$$\Delta g_{mix} = x_1(\mu_1 - \mu_1^0) + x_2(\mu_2 - \mu_2^0). \tag{4.39}$$

In Example 4.11, we will illustrate how to calculate Δg_{mix} over the whole range of composition (that is, $0 \leq x_2 \leq 1$), making the assumption that over the whole range, the mixture stays in the hypothetical homogeneous single-phase state.

From Eq. (4.39), one may obtain the following expression:

$$(\partial \Delta g_{mix}/\partial x_2) = -(\mu_1 - \mu_1^0) + (\mu_2 - \mu_2^0). \tag{4.40}$$

The above relationship can be derived by taking the derivative of Eq. (4.39) with respect to x_2 and using the Gibbs-Duhem equation $\sum_{i=1}^{2} x_i(\partial \mu_i/\partial x_2) = 0$ at constant temperature and pressure. Finally, by combining Eqs. (4.39) and (4.40),

$$\boxed{\Delta g_{mix} = (\mu_1 - \mu_1^0) + x_2(\partial \Delta g_{mix}/\partial x_2)}. \tag{4.41}$$

Figure 4.5a shows a plot of Δg_{mix} vs. x_2 at pressure P_1 and temperature T for the binary mixture (see Fig. 4.4). At a given overall composition z_2, the composition and amount of the gas and liquid phases can be obtained from this figure. The compositions are obtained by drawing a tangent line to the Δg_{mix} curve. At the points of tangency, the chemical potentials of components 1 and 2 are equal; $\mu_1^{L_A} = \mu_1^V$ and $\mu_2^{L_A} = \mu_2^V$. The proof of the equality of chemical potentials is simple. From Fig. 4.5a,

$$\Delta g_{mix}^V = \Delta g_{mix}^{L_A} + (\partial \Delta g_{mix}/\partial x_2)(x_2^V - x_2^{L_A}), \qquad (4.42)$$

where $(\partial \Delta g_{mix}/\partial x_2)$ is the slope of the common tangent, and superscripts L_A and V indicate the points of tangency. Writing Eq. (4.41) for the points of tangency and substituting the results in Eq. (4.42) gives $\mu_1^{L_A} = \mu_1^V$ which is the sought relation. The molar Gibbs free energy of the hypothetical single-phase system is Δg_{mix1} (see Fig. 4.5a). The molar Gibbs free energy of the two-phase mixture is Δg_{mix2}. From $z_2 = Lx_2^{L_A} +$

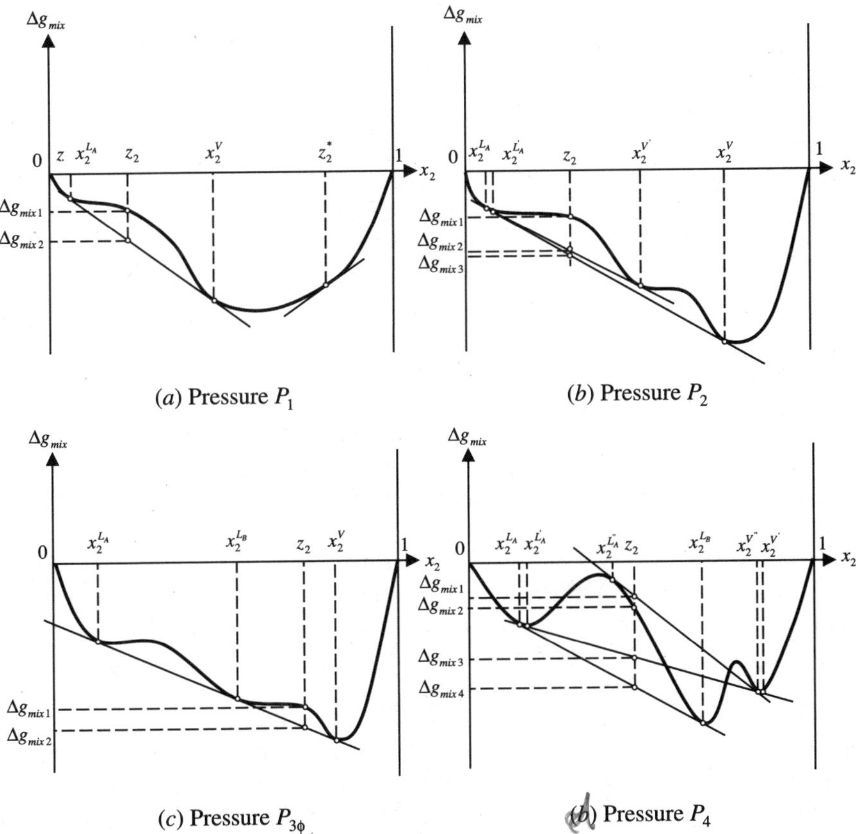

(a) Pressure P_1

(b) Pressure P_2

(c) Pressure P_3

(d) Pressure P_4

Figure 4.5 Δg_{mix} vs. x_2 of the binary mixture at temperature T (adapted from Baker, *et al.*, 1982).

$(1 - L)x_2^V$ ($F = 1$ in Eq. (4.2)), L and $V = 1 - L$ can be computed. Note that $x_2^{L_A}$ and x_2^V are the compositions of the liquid and vapor phases. Since $\Delta g_{mix2} < \Delta g_{mix1}$, then the two-phase state for the given overall mole fraction, z_2, is more stable than the single phase. For $0 \le z_2 < x_2^{L_A}$ and $x_2^V < z_2 \le 1$, there will be only one single phase. Outside the two-phase region, the tangent to Δg_{mix} curve lies below this curve, and only one phase exists. As an example at z_2^*, $x_2^V < z_2^* \le 1$, only the gas phase exists.

At pressure P_2, the Δg_{mix} plot is sketched in Fig. 4.5b. At overall composition z_2, two different tangent lines can be drawn that satisfy the material balance and the equality of chemical potentials. One tangent line results in the vapor composition $x_2^{V''}$ and liquid composition $x_2^{L'_A}$. The other tangent line provides vapor and liquid compositions x_2^V and $x_2^{L_A}$ but does not intersect the Δg_{mix} curve except at the tangency points. Note that $\Delta g_{mix1} > \Delta g_{mix2} > \Delta g_{mix3}$, and therefore Δg_{mix3} is a minimum. The single phase with the corresponding Δg_{mix1} has the highest Δg_{mix}. Outside the range $x_2^{L_A} \le z_2 \le x_2^V$, the tangent has one point of tangency and the system does not split into two phases. We will later show that for this range $(\partial \mu_2 / \partial x_2) < 0$, which ensures the stability of the single phase.

At $P = P_{3\phi}$, the plot of Δg_{mix} is shown in Fig. 4.5c. Any z_2 that is in the $x_2^{L_A}$ and x_2^V range $x_2^{L_A} < z_2 < x_2^V$ may split into three phases: two liquid phases and one gas phase. The common tangent line implies that the chemical potentials are the same for all three phases for component 2, as well as for component 1. Outside the interval $x_2^{LA} \le z_2 \le x_2^V$, the tangent intersects the Δg_{mix} curve at the point of tangency and therefore one phase exists. Note that at z_2, $\Delta g_{mix1} > \Delta g_{mix2}$ and therefore Δg_{mix2} corresponds to the more stable three phases.

Figure 4.5d shows the Δg_{mix} plot at P_4. For the overall composition z_2, there are several tangent lines. Three tangent lines are shown. Since Δg_{mix4} provides the lowest Δg_{mix} and does not intersect the Δg_{mix} curve, the corresponding tangent is the most stable solution giving liquid–liquid equilibria with composition $x_2^{L_A}$ for phase L_A and $x_2^{L_B}$ for phase L_B.

The above simple geometrical description for a binary system can be readily extended to a multicomponent mixture. For a multicomponent mixture ($c \ge 3$), the corresponding Δg_{mix} becomes a hypersurface instead of a curve, and the tangent line becomes a hyperplane. The criterion, however, remains the same. The tangent hyperplane corresponding to a stable equilibrium state cannot lie above the Gibbs free energy hypersurface for any composition. One approach to avoid false solutions is to search for all possible values of x_2 that provide the global minimum of Δg_{mix}. (In Fig. 4.5d, Δg_{mix4} is the global minimum for a fixed z_2 and all possible values of x_2.) This straightforward approach can be attacked mathematically using a minimization algorithm (as we will discuss later in this chapter), but it may be computationally expensive. A more practical approach is the method suggested by Michelsen (1982a) to determine certain minima of a distance function.

Tangent plane distance (TPD) analysis

Consider N_i moles of a homogeneous mixture at temperature T and pressure P. The overall mole fractions are $z_i = N_i / \sum_{i=1}^{c} N_i = N_i/N$. The Gibbs free energy of the homogeneous single-phase system (see Fig. 4.6a) is

$$G^I = \sum_{i=1}^{c} N_i \mu_i(\pmb{z}, T, P), \qquad (4.43)$$

where $\pmb{z} = (z_1, z_2, \ldots, z_{c-1})$. If a very small amount of a second phase is formed at the same pressure and temperature with mole numbers n_i, and $n_i \ll N_i$, then the Gibbs free energy of the new system II (see Fig. 4.6b) is

$$G^{II} = G(\underline{\pmb{n}}, T, P) + G(\underline{\pmb{N} - \pmb{n}}, T, P), \qquad (4.44)$$

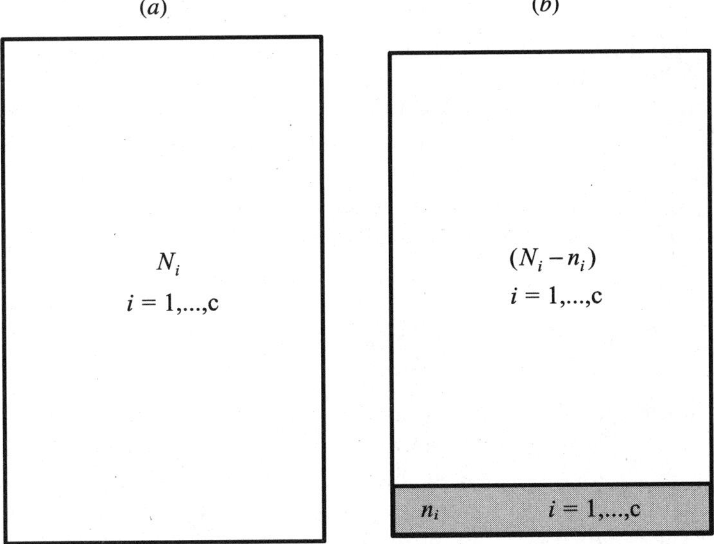

Figure 4.6   Systems with and without a second phase at temperature T and pressure P.

where $G(\underline{n}, T, P)$ and $G(\underline{N - n}, T, P)$ are the Gibbs free energies of the two phases of system II.

$$G(\underline{n}, T, P) = \sum_{i=1}^{c} n_i \mu_i(\mathbf{x}, T, P) \tag{4.45}$$

$$x_i = n_i / \sum_{i=1}^{c} n_i = n_i/n \tag{4.46}$$

and

$$G(\underline{N - n}, T, P) \approx G^I(\underline{N}, T, P) - \sum_{i=1}^{c}(\partial G/\partial N_i)\Big|_{N_i} (n_i)$$

$$= G^I(\underline{N}, T, P) - \sum_{i=1}^{c} n_i \mu_i(\mathbf{z}, T, P). \tag{4.47}$$

Equation (4.47) is a Taylor's series expansion of $G(\underline{N - n}, T, P)$ around $G(\underline{N}, T, P)$. The change in Gibbs free energy from I to II is

$$\Delta G = G^{II} - G^I. \tag{4.48}$$

Combining Eqs. (4.43) to (4.45), and (4.47) and (4.48),

$$\Delta G = \sum_{i=1}^{c} n_i[\mu_i(T, P, \mathbf{x}) - \mu_i(T, P, \mathbf{z})]. \tag{4.49}$$

If $\Delta G > 0$ for all feasible values of $\mathbf{x}$, then the original state at I is stable and G^I cannot be further reduced (that is, G is a global minimum at constant T and P). However, if $\Delta G < 0$ for any feasible value of $\mathbf{x}$, then the original state is unstable and the single phase will split into more than one phase; it may split into two phases, three phases, or more. When $\Delta G = 0$, the system is said to be neutral.

If we divide Eq. (4.49) by n (the total number of moles in the new phase),

$$\Delta g(\mathbf{x}) = \Delta G/n = \sum_{i=1}^{c} x_i[\mu_i(\mathbf{x}) - \mu_i(\mathbf{z})], \tag{4.50}$$

where $\Delta g(\mathbf{x})$ is a molar quantity. Note that we have dropped T and P dependencies since T and P are held constant. Similarly to Eq. (4.49), $\Delta g(\mathbf{x}) > 0$ implies that system I is stable, and for $\Delta g(x) < 0$, system I is unstable. Equation (4.50) has a simple geometrical interpretation. For simplicity, let us select a two-component system in which $x_2 = 1 - x_1$. Then Eq. (4.50) reduces to

$$\Delta g(x_1) = \sum_{i=1}^{2} x_i[\mu_i(x_1) - \mu_i(z_1)]. \tag{4.51}$$

Figure 4.7 depicts the plot of $g(x_1)$,

$$g(x_1) = \sum_{i=1}^{2} x_i \mu_i(x_1) \qquad (4.52)$$

vs. x_1; x_1 is the mole fraction of component 1, which can vary from 0 to 1. An EOS can be used to calculate $g(x_1)$. In Fig. 4.7, z_1 is the mole fraction of component 1, a fixed overall composition. Let us draw a tangent line at point z_1 to the curve $g(x_1)$. The equation for this tangent line is

$$T(x_1) = g(z_1) + \partial g(x_1)/\partial x_1 \big|_{@z_1} (x_1 - z_1). \qquad (4.53)$$

From Eq. (4.52) and the expression $x_2 = 1 - x_1$, we can evaluate the slope $\partial g(x_1)/\partial x_1$:

$$\partial g(x_1)/\partial x_1 = \mu_1(x_1) - \mu_2(x_1) + x_1 \partial \mu_1(x_1)/\partial x_1 + x_2 \partial \mu_2(x_1)/\partial x_1 \qquad (4.54)$$

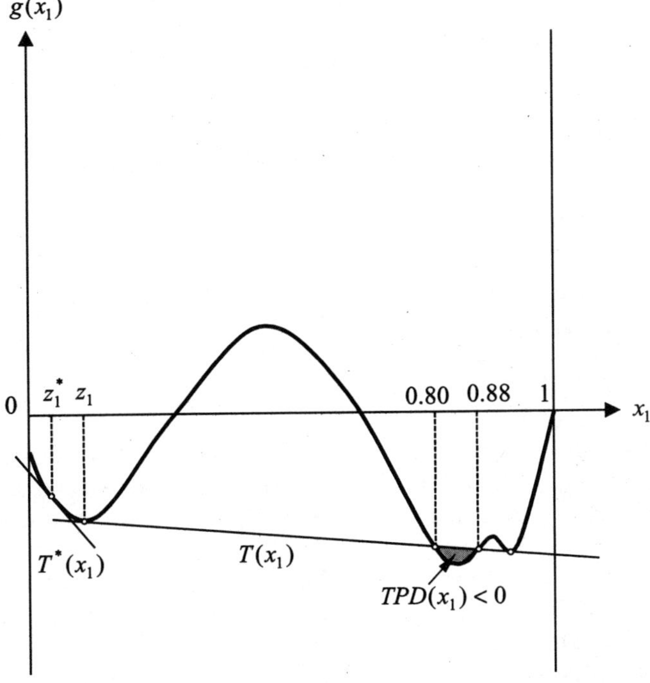

Figure 4.7 Plot of $g(x_1)$ vs. x_1.

From the Gibbs-Duhem equation at constant T and P, (see Eq. (1.40), of Chapter 1) $\sum_{i=1}^{2} x_i d\mu_i = 0$, which after dividing by ∂x_1 becomes

$$\sum_{i=1}^{2} x_i(\partial\mu_i/\partial x_1) = 0. \tag{4.55}$$

Combining Eqs. (4.54) and (4.55),

$$\partial g(x_1)/\partial x_1 = \mu_1(x_1) - \mu_2(x_1) \tag{4.56}$$

At $x_1 = z_1$,

$$\partial g(x_1)/\partial x_1\big|_{@z_1} = \mu_1(z_1) - \mu_2(z_1). \tag{4.57}$$

Combining Eqs. (4.53), (4.57) and $g(z_1) = \sum_{i=1}^{2} z_i\mu_i(z_1)$,

$$T(x_1) = \sum_{i=1}^{2} x_i\mu_i(z_1). \tag{4.58}$$

Now we define the tangent plane distance (TPD) as the difference from $T(x_1)$ to $g(x_1)$ at any point x_1:

$$TPD(x_1) = g(x_1) - T(x_1). \tag{4.59}$$

From Eqs. (4.52), (4.58), and (4.59), one obtains

$$TPD(x_1) = \sum_{i=1}^{2} x_i[\mu_i(x_1) - \mu_i(z_1)]. \tag{4.60}$$

Equation (4.60) is the same as Eq. (4.50) with $c = 2$. For a c-component system,

$$\boxed{TPD(\boldsymbol{x}) = \sum_{i=1}^{c} x_i[\mu_i(\boldsymbol{x}) - \mu_i(\boldsymbol{z})]}, \tag{4.61}$$

which is the same as Eq. (4.50).

The criteria of the stability of system I in Fig. 4.6 now can be stated in terms of the tangent plane distance. $TPD(\boldsymbol{x})$ should be positive over the whole range of $\boldsymbol{x}$. Note that in Fig. 4.7 the $TPD(\boldsymbol{x})$ from z_1 becomes negative for $0.80 < x_1 < 0.88$ and, therefore, the system of overall composition z_1 is not stable. One the other hand, the overall composition z_1^* has a tangent $T^*(x_1)$ that is below the curve $g(x_1)$ and does not have a $TPD(x_1)$ of less than zero; therefore, it is stable and the system cannot split into two phases.

Different methods have been proposed to search for the value(s) of x to test the stability of a system with a fixed composition z. Michelsen (1982a) suggested locating the stationary points (maxima, minima, or saddle points) of $TPD(x)$ rather than conducting an exhaustive search in x-space for values of x where $TPD(x) \geq 0$. Let us examine the $g(x_1)$, the tangent $T(x_1)$ at point z_1, and the $TPD(x_1)$ shown in Fig. 4.8. The stationary points of $TPD(x)$ for a c-component system occur at points where

$$\frac{\partial}{\partial x_i}[TPD(x)] = 0 \qquad i = 1, \ldots, c-1. \tag{4.62}$$

In Eq. (4.62), the derivatives are with respect to x_1 to x_{c-1}. Substituting the expression for $TPD(x)$ from Eq. (4.61) and using the Gibbs-Duhem expression, $\sum_{i=1}^{c} x_i(\partial\mu_i/\partial x_j) = 0$, we obtain

$$\boxed{\mu_i(x) - \mu_i(z) = \mu_c(x) - \mu_c(z) = K} \qquad i = 1, \ldots, c-1. \tag{4.63}$$

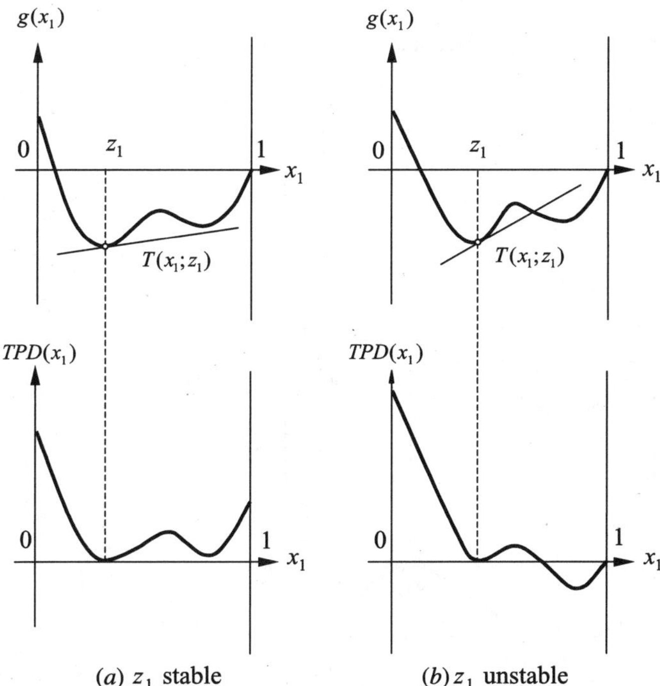

(a) z_1 stable   (b) z_1 unstable

Figure 4.8   Plots of $g(x_1)$, and $TPD(x_1)$ vs. x_1.

The $TPD(\boldsymbol{x})$ at the stationary points are obtained by substituting Eq. (4.63) into Eq. (4.61):

$$TPD(\boldsymbol{x}) = \sum_{i=1}^{c} x_i K = K. \tag{4.64}$$

The distance K is shown in one of the stationary points of Fig. 4.9, at $x_{1,sp}^0$. From the criteria of stability set forth before for a stable system, $K \geq 0$, and for an unstable system, $K < 0$.

Equation (4.63) is a key equation for stability analysis. The solution of Eq. (4.63) provides the $\boldsymbol{x}$ of the stationary points of $TPD(\boldsymbol{x})$, which is not a trivial task. The fugacity form of that equation can be obtained as follows.

From

$$\mu_i(\boldsymbol{x}, T, P) = \mu_i^0(T, P) + RT \ln f_i(\boldsymbol{x}, T, P)/f_i^0(T, P)$$

$$\mu_i(\boldsymbol{z}, T, P) = \mu_i^0(T, P) + RT \ln f_i(\boldsymbol{z}, T, P)/f_i^0(T, P),$$

Eq. (4.63), and the defintion of fugacity coefficients $\varphi_i(\boldsymbol{x}, T, P) = f_i(\boldsymbol{x}, T, P)/(x_i f_i^0(T, P))$ and $\varphi_i(\boldsymbol{z}, T, P) = f_i(\boldsymbol{z}, T, P)/(z_i f_i^0(T, P))$,

$$\ln \varphi_i(\boldsymbol{x}) + \ln x_i - \ln z_i - \ln \varphi_i(\boldsymbol{z}) = k \qquad i = 1, \ldots, c-1, \tag{4.65}$$

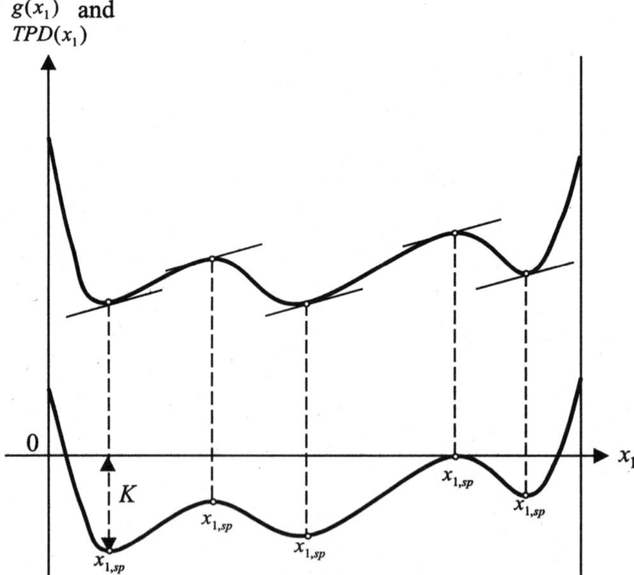

Figure 4.9   Plots of $g(x_1)$, $TPD(x_1)$, and the stationary points.

where $k = K/RT$. Equation (4.65) provides the $\boldsymbol{x}$ at which $TPD(\boldsymbol{x})$ is at a stationary state. This equation is nonlinear and may be transformed to the following form by introducing a new variable X_i,

$$\ln X_i = \ln x_i - k. \tag{4.66}$$

from which

$$x_i = X_i e^k \tag{4.67}$$

follows. Since $\sum_{j=1}^{c} x_j = 1$, then

$$e^k = 1 / \sum_{j=1}^{c} X_j. \tag{4.68}$$

and, therefore,

$$x_i = X_i / \sum_{j=1}^{c} X_j \tag{4.69}$$

Equation (4.69) implies that the new variable X_i is similar to mole numbers for the new phase.

Combining Eqs. (4.65) and (4.66),

$$\ln X_i + \ln \varphi_i(\boldsymbol{x}) - \ln z_i - \ln \varphi_i(\boldsymbol{z}) = 0 \qquad i = 1, \ldots, c - 1, \tag{4.70}$$

where $\boldsymbol{x}$ is given by Eq. (4.67). Solution of the above nonlinear system of equations provides $\underline{X}$ that can be used to examine the stability analysis. From Eq. (4.68), $\sum_{j=1}^{c} X_j = 1/e^k$. Since for a stable system of fixed composition, z_i, when $k \geq 0$, the system is stable; therefore, in terms of $\underline{X}$, when $\sum_{j=1}^{c} X_j < 1$, the same system is stable. If $\sum_{j=1}^{c} X_j > 1$, the system is unstable.

The stability analysis of a given phase with a fixed overall composition $\underline{z}$ is a search for a trial phase of small amount that is taken out of the original single phase (that is, the homogeneous phase). The trial phase when combined with the remainder of the original mixture gives a Gibbs free energy that should be higher than that of the original single-phase mixture for the single phase to be stable. If the feed composition $\underline{z}$ can be identified as a liquid, then we can search for a vapor-like trial phase with composition $\underline{X}$ estimated from

$$X_i = z_i K_i \qquad i = 1, \ldots, c. \tag{4.71}$$

The K_i-values are estimated from the Wilson correlation (Eq. (4.16)) or other suitable correlations. However, if the feed composition $\underline{z}$ is identified as a vapor, then we need to search for a liquid-like trial phase with composition $\underline{X}$ estimated from

$$X_i = z_i/K_i \qquad i = 1, \ldots, c. \tag{4.72}$$

For reservoir and some other fluids at high pressures, it may not be possible to identify the feed as liquid or vapor, and therefore both estimates of $\underline{X}$ from Eqs. (4.71) and (4.72) should be used as the first initial guess. The strategy for the solution of nonlinear Eq. (4.70) can be to first calculate the term $-(\ln z_i + \ln \varphi_i(\mathbf{z}))$. Then with the initial guess from either Eq. (4.71) or Eq. (4.72), use either successive substitution or Newton's method, or a combination of these two methods as we discussed earlier, to solve for $\underline{X}$. In the successive substitution, the updating of $\underline{X}$ is simply

$$X_i^{new} = \exp[(\ln z_i + \ln \varphi_i(\mathbf{z})) - \ln \varphi_i(\mathbf{x})] \qquad i = 1, \ldots, c - 1, \tag{4.73}$$

where $x_i = X_i^{old}/\sum_{j=1}^{c} X_j^{old}$. Then is the testing for convergence.

In case the solution is nontrivial (that is, $\underline{X}$ is not equal to $\mathbf{z}$ within a given tolerance), the original phase is considered unstable if $\sum_{j=1}^{c} X_j > 1$ (of course within a given tolerance). If convergence on X_i is achieved but $\sum_{j=1}^{c} X_j < 1$, another estimate of $\underline{X}$ is made by using Eq. (4.72) if Eq. (4.71) was used first. If all calculations converge to a trivial solution after using Eqs. (4.72) and (4.73) for the first initial guess or if no convergence is achieved, the original phase is assumed to be stable.

When testing the stability of a two-phase mixture, it is necessary to select only one of the phases to test its stability. A complication arises in the initial guess for $\underline{X}$. Four different sets of initial estimates of $\underline{X}$ are recommended for stability analysis calculations (Michelsen, 1982a) to avoid trivial solutions of $\underline{X}$ for either phase. These estimates are calculated from the following equations (Michelsen, 1982a)

$$X_i = 1/2(y_i + x_i) \qquad i = 1, \ldots, c, \tag{4.74}$$

$$X_s = 0.999 \tag{4.75}$$

$$X_i = (1 - X_s)/(c - 1) \qquad i = 1, \ldots, s - 1, s + 1, \ldots, c \tag{4.76}$$

$$X_i = \exp[\ln z_i + \ln \varphi_i(\mathbf{z})] \qquad i = 1, \ldots, c. \tag{4.77}$$

In Eq. (4.74), the trial phase composition is estimated to fall between the gas and liquid phase compositions. If the estimate from Eqs. (4.75) and (4.76) yields convergence to a trivial solution, Eq. (4.77) is used. In Eqs. (4.75) and (4.76), one time the selected component "s" is the most volatile component. The next time X_s is the mole fraction of the heaviest component or the pseudocomponent. The usefulness of Eqs. (4.75) and (4.76) is that the chances of locating a phase instability are improved by using distinct values of composition that are widely separated. The fourth estimate from Eq. (4.77) assumes ideal behavior. Similar to the stability of a single-phase system, if from any of the initial guesses the phase becomes unstable, then further estimates will not be pursued.

One then may observe a certain drawback in the search of a trial phase composition and the trivial solutions. The situation gets quite complicated when testing the stability of three-phase and four-phase systems. A large variety of initial estimates of the trial phase composition are required. An alternative to the stationary points of *TPD* of Michelsen (1982a) has been suggested by Trangenstein (1987). He suggests minimizing ΔG given in Eq. (4.49) for the variables $\underline{n}$. Details of the procedure can be found in Trangenstein (1987) for two-phase systems, and Perschke (1988) and Chang (1990) for three-phase systems.

Example of multiphase flash and stability analysis. We will, in detail, discuss the stability analysis of a three-component system of $C_1/CO_2/nC_{16}$ at $T = 294.0\,\text{K}$ and $P = 67\,\text{bar}$ with $z_{C_1} = 0.05$, $z_{CO_2} = 0.90$, and $z_{nC_{16}} = 0.05$. At fixed temperature and pressure, from the phase rule $F = c + 2 - p$, there can be a maximum of three phases when the interface between the phases is flat. The first question is what types of phases may exist—gas, liquid, or solid. As we will see in Chapter 5, a solid phase does not exist for the above system. Therefore one might expect (1) a single gas phase or a single liquid phase, (2) gas and liquid phases, (3) liquid and liquid phases, or (4) gas–liquid–liquid phase separation. The difficulty in liquid–liquid (L–L) and vapor–liquid–liquid (V–L–L) and higher-phase equilibria (for more than three components) is how many phases should be considered for flash calculations. One approach is to determine whether one, two, or more phases are to be considered without prior knowledge of the true number of phases. In certain cases, as we will see in the next chapter, it is possible from thermodynamic stability analysis to determine the true number of phases a priori without performing a flash. However, in general, we do not know the true number of phases. One may, therefore, follow a sequential approach as outlined next for the $C_1/CO_2/nC_{16}$ example.

1. Perform the stability analysis on the overall mixture to examine the stability of the single-phase mixture. In case the original mixture is unstable, estimates of equilibrium ratios (K_i-values) are then on hand for the two-phase flash.

2. Perform the two-phase flash using the K_i-values from the stability test as the initial guess.

3. Perform stability analysis on only one of the phases of the two-phase flash. If the selected phase is unstable, estimates of the equilibrium ratios for part of the three-phase flash are available.

4. Perform the three-phase flash using the estimates of equilibrium ratios from steps 2 and 3.

5. Perform the stability analysis on only one of the phases of the three-phase flash to check stability of the three phases. The three-phase flash should be stable since single-phase and two-phase were unstable (this is true only for a three-component system).

In Step 1, we can use K_i-values from the Wilson correlation (see Eq. (4.16)) to proceed with the stability of the original phase and to calculate the composition of the trial second phase. The Wilson correlation may not be appropriate if the trial phase and the original phase are in liquid states. From Eq. (4.16), $K_{C_1} = 4.564$, $K_{CO_2} = 0.8709$, and $K_{nC_{16}} = 0.000001$. We can use $x_i = K_i z_i$ and $x_i = z_i/K_i$ as estimates of the trial phase composition. If we minimize the fugacity form of Eq. (4.50), with respect to x_i with the initial estimates above, at the minimum $k = \Delta g/RT = -0.03105$ is obtained. We used software simulated annealing (to be discussed next) to perform the minimization. At this minimum, $x_{C_1} = 0.0408$, $x_{CO_2} = 0.6819$, and $x_{nC_{16}} = 0.2773$. Since $k < 0$, then the single phase is unstable. From the trial phase composition, initial estimates of the K_i-values for the two-phase flash are, therefore, available: $K_{C_1} = 1.2255$, $K_{CO_2} = 1.3198$, $K_{nC_{16}} = 0.1803$.

In Step 2, a two-phase flash is performed with initial estimates of equilibrium ratios from the first step. The results of the two-phase flash are $y_{C_1} = 0.0523$, $y_{CO_2} = 0.9382$, $y_{nC_{16}} = 0.0095$; $x_{C_1} = 0.04007$, $x_{CO_2} = 0.7335$, $x_{nC_{16}} = 0.2265$; $K_{C_1} = 1.3075$, $K_{CO_2} = 1.2791$, $K_{nC_{16}} = 0.0420$; and $V = 0.8137$ (that is, the mole fraction of the vapor phase).

In Step 3, the stability of one of the phases is analyzed to check the stability of the two-phase system. Let us select the liquid phase with $z_{C_1} = 0.0440$, $z_{CO_2} = 0.7335$, and $z_{nC_{16}} = 0.2265$. Note that instead of x, we have used the symbols z to emphasize that the original liquid phase composition is fixed. We need an initial estimate of K_i-values to proceed with the minimization of Eq. (4.50). The simulated annealing minimization is not in general sensitive to the initial guess; one does not need to be concerned with good initial estimates of K_i. Let us use initial K_i-values: $K_{C_1} = 2.0$, $K_{CO_2} = 1.05$, $K_{nC_{16}} = 0.9$, with the mole fraction ratio being that of the trial phase composition to the original phase composition. The initial estimates of K_i are used to estimate x_i (see Eqs. (4.71) and (4.72)), and then minimization of Δg is performed. The minimum is $k = \Delta g/RT = -0.00461$, which implies that the liquid phase from the two-phase flash is unstable. The composition of the trial phase is $y_{C_1} = 0.10852$, $y_{CO_2} = 0.89144$, and $y_{nC_{16}} = 0.00004$. The K_i-values can then calculated from the trial phase composition and the composition of the original liquid phase.

In Step 4, the K_i-values from steps 2 and 3 provide the initial estimates of the three-phase flash. From the three-phase flash, the compositions and the amounts of the three phase are calculated: $y_{C_1} = 0.093768$, $y_{CO_2} = 0.906177$, $y_{nC_{16}} = 0.000055$; $x'_{C_1} = 0.04396$, $x'_{CO_2} = 0.94428$, $x'_{nC_{16}} = 0.01176$; $x''_{C_1} = 0.03532$, $x''_{CO_2} = 0.74527$, $x''_{nC_{16}} = 0.21941$. The mole fractions of the phases are $V = 0.1546$, $L' = 0.6525$, and $L'' = 0.1929$. The primes and double-primes represent the two liquid phases.

Although there is no need to perform stability analysis of the three-phase system, since single-phase and two-phase system were unstable, let us study the stability of each of the three phases of the three-phase flash. Take the vapor phase first: $z_{C_1} = 0.093768$, $z_{CO_2} = 0.906177$ and $z_{nC_{16}} = 0.000055$. Note again the change of symbols from y to z. We can use K_i-values from step 1. The results of the minimization of $\Delta g(\boldsymbol{x})$ are $k = \Delta g(\boldsymbol{x}) = -1 \times 10^{-6}$ and $x_{C_1} = 0.0353$, $x_{CO_2} = 0.7453$, and $x_{nC_{16}} = 0.2194$. Since $k \approx 0$ (within a certain tolerance), the vapor phase is stable. The trial phase composition is the same as the liquid phase composition in step 4. We can also study the stability of the liquid phase with composition $z_{C_1} = 0.04396$, $z_{CO_2} = 0.94428$, and $z_{nC_{16}} = 0.01176$. At the minimum of $\Delta g(\boldsymbol{x})$, $k = \Delta g(\boldsymbol{x})/RT = -1 \times 10^{-7}$. The composition of the trial phase at the minimum is $x''_{C_1} = 0.03532$, $x''_{CO_2} = 0.74527$, and $x''_{nC_{16}} = 0.21941$. The trial phase composition is the same as that of the vapor phase. Since $k \approx 0$ (within a given tolerance), this liquid phase is also stable. Finally, we can study the stability of the third phase, i.e., the heavy liquid phase with $z_{C_1} = 0.03523$, $z_{CO_2} = 0.74527$, and $z_{nC_{16}} = 0.21940$. At the minimum of Eq. (4.50), $k = \Delta g/RT = -0.2 \times 10^{-8}$ and the composition of the trial phase is the same as the heavy liquid phase composition. The fact that the compositions of the trial phase and the heavy liquid phase are the same implies that there is no other phase composition for the trial phase to provide the minimum of Δg.

Direct minimization of Gibbs free energy in multiphase flash

The general problem of multicomponent, multiphase flash calculation at constant temperature and pressure has been attacked by direct minimization of the Gibbs free energy (Pan and Firoozabadi, 1998). Using an algorithm that can provide the global minimization of Gibbs free energy, one can readily perform a multiphase flash to obtain the composition and the number of phases.

Consider the multiphase, multicomponent flash at constant temperature and pressure sketched in Fig. 4.10. The stable-equilibrium state consists of p phases; each phase j consists of $N_{j,1}, N_{j,2}, N_{j,3}, \ldots, N_{j,c}$

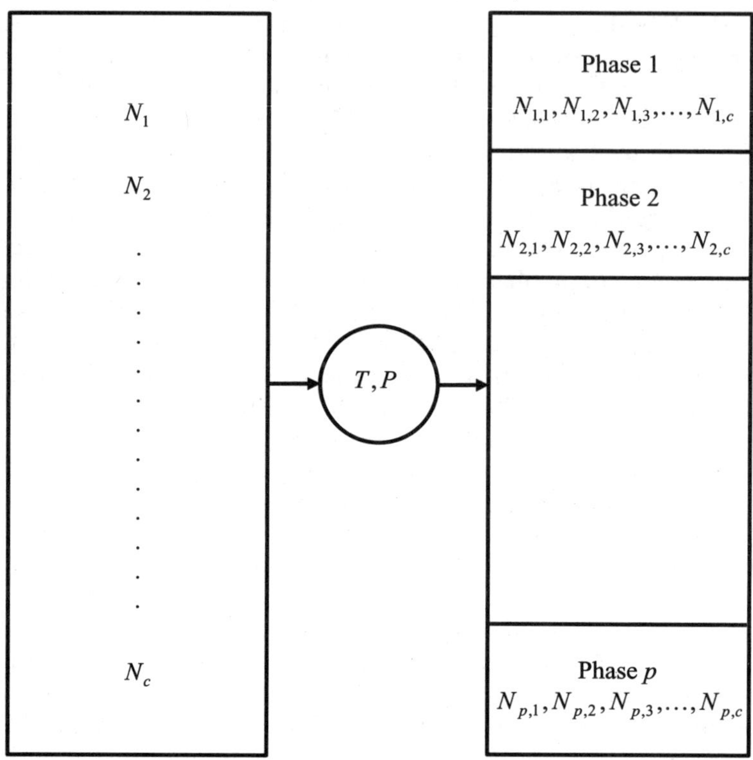

Figure 4.10 Schematic representation of multiphase, multicomponent flash at constant temperature and pressure (from Pan and Firoozabadi, 1998).

moles. The stable-equilibrium state at which the Gibbs free energy of the system is a minimum is a necessary and sufficient condition. At constant temperature and pressure, the Gibbs free energy of the system in the right side of Fig. 4.10 can be written as

$$G = \sum_{j=1}^{p} G_j(N_{j,1}, N_{j,2}, \ldots, N_{j,c}), \qquad (4.78)$$

where G_j is the Gibbs free energy of phase j and G is the total Gibbs free energy of the system. The stable-equilibrium state corresponds to the global minimum of G with respect to $N_{j,i}$ moles $(i = 1, \ldots, c, \ j = 1, \ldots, p)$, subject to the following constraints:

   (a) the material balance for component i

$$N_i = \sum_{j=1}^{p} N_{j,i} \qquad (4.79)$$

(b) non-negative mole numbers of component i in phase j,

$$0 \leq N_{j,i} \leq N_i \qquad i = 1, \ldots, c; \; j = 1, \ldots, p. \tag{4.80}$$

The Gibbs free energy of phase j, G_j, in Eq. (4.78) is given by $G_j = \sum_{i=1}^{c} N_{j,i}\mu_{j,i}(\underline{N}_j)$. Therefore, the Gibbs free energy of the system sketched in Fig. 4.10 can be written as

$$G = \sum_{j=1}^{p} \sum_{i=1}^{c} N_{j,i}\mu_{j,i}(\underline{N}_j), \tag{4.81}$$

where $\mu_{j,i}(\underline{N}_j) = \mu_i^0(T) + RT \ln f_{j,i}(\underline{N}_j)/f_i^0(T)$. Therefore,

$$G = \left[\sum_{i=1}^{c} N_i\mu_i^0(T) - RT \sum_{i=1}^{c} N_i f_i^0(T) \right] + RT \sum_{i=1}^{c} \sum_{j=1}^{p} \ln f_{j,i}(\underline{N}_j). \tag{4.82}$$

In Eq. (4.82), the first term on the right side is a constant, since T, P, and N_i are fixed. The second term on the right is a function of $N_{j,i}$, since $f_{j,i}(\underline{N}_j)$ is a function of $N_{j,i}$. We can eliminate the mole numbers of one of the phases, say phase 1, from the material balance expression, Eq. (4.79):

$$N_{1,i} = N_i - \sum_{j=2}^{p} N_{j,i} \qquad i = 1, \ldots, c \tag{4.83}$$

The above equations reduce the number of unknowns from pc to $c(p-1)$ in the minimization search.

The search for the global minimum of a function of many variables with close minima is a very difficult problem. Disciplines ranging from economics to engineering need to use minimization (or optimization) algorithms. The simulated annealing (SA) algorithm is a powerful tool for this purpose. This algorithm is apparently very effective when the global extremum is hidden among many local extrema. The root of simulated annealing is in thermodynamics. When a molten metal is cooled slowly (annealing process), the system is able to reach a highly ordered crystalline state of the lowest energy. Rapid cooling (quenching process) leads to a polycrystalline or amorphous state having somewhat higher energy. The simulated annealing algorithm is analogous to the annealing process of the molten metal. It decreases the objective function slowly to reach its global minimum. A detailed description of the method can be found in Press *et al.* (1992), Corana *et al.* (1987), and Goffe *et al.* (1994). Mathematically, the SA algorithm globally optimizes an objective function with the constraints of bounds. The sole drawback of the method is the computational cost because of the "slow cooling process." The computational time increases roughly linearly with the number of independent variables. The SA randomly searches for the optimized point. Therefore, the computational time is nearly the same for the near-critical region as for away from the critical region.

The use of the above algorithm for phase-equilibrium calculations is straightforward. For vapor–liquid equilibria, initial estimates of $K_i = y_i/x_i$ can be obtained from the Wilson correlation. Close to the critical point, $K_i = 1$ is provided for light components and $K_i = 0.5$ for heavy components of a multicomponent mixture. Figure 4.11 depicts the example for the multicomponent mixture: $z_{C_1} = 0.35$, $z_{C_2} = 0.03$, $z_{C_3} = 0.04$, $z_{nC_4} = 0.06$, $z_{nC_5} = 0.04$, $z_{nC_6} = 0.03$, $z_{nC_7} = 0.05$, $z_{nC_{10}} = 0.30$, $z_{nC_{14}} = 0.05$ and varying amounts of CO_2. (The introduction of CO_2 in the mixture reduces the mole fraction of the other components proportionally.) Note that the calculation proceeds to the near-critical region where all K-values are close to one.

For V–L–L equilibrium calculations, one can select the light liquid phase as the reference phase and then estimate the initial equilibrium ratios according to the procedure of Example 4.2. Figure 4.12 plots the calculated results for a CO_2–reservoir oil system at various pressures at fixed composition and 307.6 K. The compositions of oil B and CO_2 mixtures investigated by Shelton and Yarborough (1977) are used. Note that up to a pressure of 78.8 bar, two phases, a vapor phase and a hydrocarbon-rich liquid phase, coexist. From $P = 78.8$ to 81.2 bar, three phases exist and above $P = 81.2$ bar, the vapor phase disappears and two liquid phases are present. Details of various parameters can

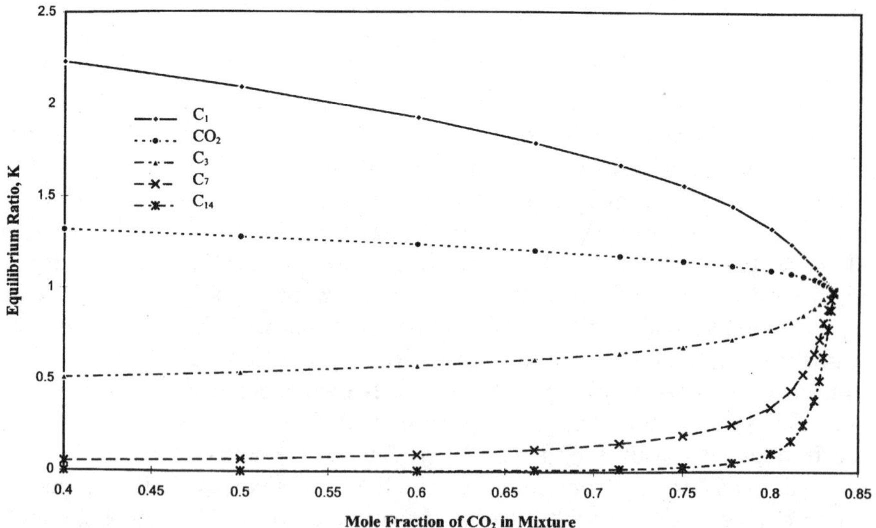

Figure 4.11 Equilibrium ratios vs. CO_2 concentration for the CO_2–synthetic oil mixture at 322.0 K and 105.35 bar (adapted from Pan and Firoozabadi, 1998).

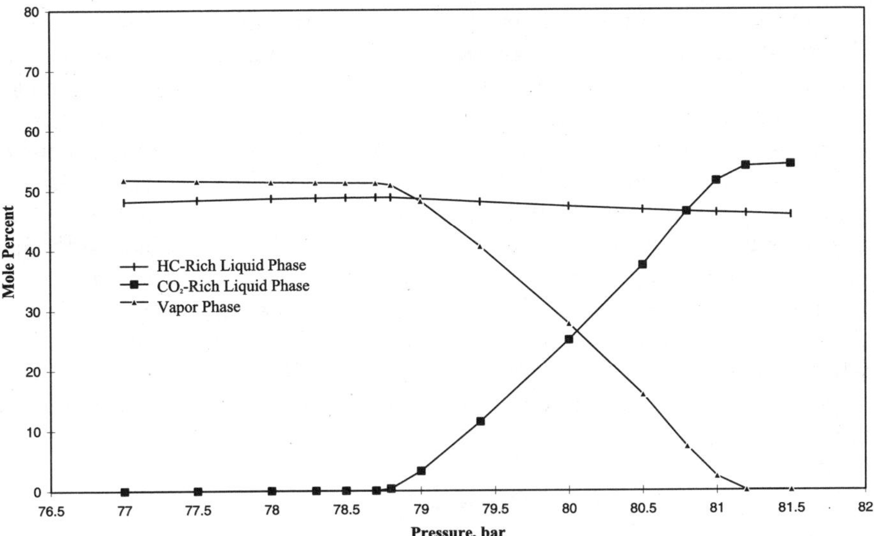

Figure 4.12   Mole percent of different phases vs. pressure for the CO_2–reservoir oil system at 307.6 K (adapted from Pan and Firoozabadi, 1998).

be found in Pan and Firoozabadi (1998). We will later (Chapter 5) use direct minimization of Gibbs free energy to perform wax and asphaltene precipitation calculations.

Stability analysis and stability limit

Considerable effort will be made in this chapter to derive the criteria of stability and criticality from basic principles. In order to appreciate these concepts, we first derive the criteria of stability for a pure component separately and then proceed briefly to binary mixtures. The stability criteria for multicomponent systems will be derived in detail. The derivation for multicomponent systems is very general and relies on the Legendre and Jacobian transformations discussed in Chapter 1 and the mathematical principles for definiteness that will be discussed in this chapter.

Stability analysis for a single component

The basic derivations for the stability analysis of a single-component system are similar to Haase's presentation (1956) with certain changes.

Stability criteria based on entropy   Let us consider a single-component single-phase system with molar internal energy u and molar volume v

at temperature T and pressure P. At this temperature and pressure, if the single-phase system is not stable, it may split into two phases; N' moles of the primed phase and N'' moles of the double-primed phase. The molar internal energy and the molar volume of the primed phase are represented by $u + \Delta u$ and $v + \Delta v$, and those of the double-primed phase are represented by $u + \Delta^0 u$ and $v + \Delta^0 v$. Let us represent the molar entropy of the original single phase by s, the primed phase by $s + \Delta s$, and the double-primed phase by $s + \Delta^0 s$. The criteria of stability of the original single phase in comparison to the two-phase state can be developed in terms of the entropy difference between the single-phase and two-phase states. The problem is sketched in Fig. 4.13: which one of the two states is more stable, a or b?

The expression for the entropy change, ΔS, between the two states can be written as

$$\Delta S = S_{two\ phase} - S_{single\ phase} \tag{4.84}$$

and the entropy of each system is given by

$$S_{two\ phase} = N'(s + \Delta s) + N''(s + \Delta^0 s) \tag{4.85}$$

$$S_{single\ phase} = Ns = (N' + N'')s. \tag{4.86}$$

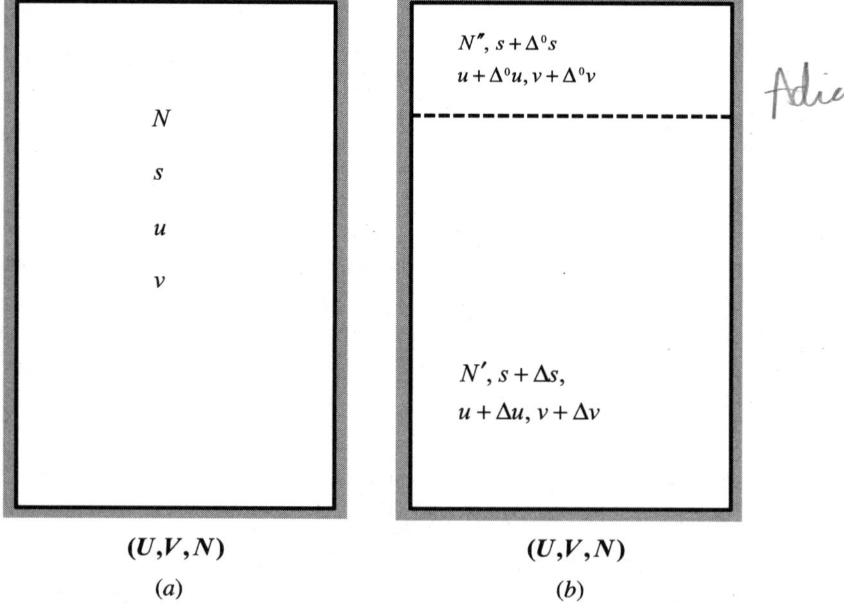

Figure 4.13 Stability of a single component at constant U, V, and $N\,(N = N' + N'')$

Therefore,

$$\Delta S = N'\Delta s + N''\Delta^0 s. \tag{4.87}$$

Based on the Second Law, the single-phase state is stable if $\Delta S < 0$; the entropy of the original single-phase state is greater than the entropy of the two-phase state for the isolated path. The constraints of the isolated path are

$$\Delta U = U_{two\ phase} - U_{single\ phase} = N'\Delta u + N''\Delta^0 u = 0 \tag{4.88}$$

$$\Delta V = V_{two\ phase} - V_{single\ phase} = N'\Delta v + N''\Delta^0 v = 0 \tag{4.89}$$

$$N = N' + N'' = constant. \tag{4.90}$$

The ΔU and ΔV expressions in Eqs. (4.88) and (4.89) are derived in the same way as Eq. (4.87) was derived. Next we perform the Taylor's series expansion of $\Delta s(u, v)$, and $\Delta^0 s(u, v)$ in terms of independent variables u and v.

$$\Delta s = (\partial s/\partial u)\Delta u + (\partial s/\partial v)\Delta v + (1/2)[(\partial^2 s/\partial u^2)(\Delta u)^2$$
$$+ 2(\partial^2 s/\partial u \partial v)(\Delta u\Delta v) + (\partial^2 s/\partial v^2)(\Delta v)^2] + \cdots \tag{4.91}$$

and

$$\Delta^0 s = (\partial s/\partial u)\Delta^0 u + (\partial s/\partial v)\Delta^0 v + (1/2)[(\partial^2 s/\partial u^2)(\Delta^0 u)^2$$
$$+ 2(\partial^2 s/\partial u \partial v)(\Delta^0 u\Delta^0 v) + (\partial^2 s/\partial v^2)(\Delta^0 v)^2] + \cdots . \tag{4.92}$$

The Taylor's series expansions are based on the assumptions that $\Delta u, \Delta v, \Delta^0 u$, and $\Delta^0 v$ are small. Note that in Eqs. (4.91) and (4.92), the partial derivatives are evaluated at the original single-phase state.

From Eqs. (4.88) and (4.89),

$$\Delta^0 u = -(N'/N'')\Delta u \tag{4.93}$$

and

$$\Delta^0 v = -(N'/N'')\Delta v. \tag{4.94}$$

The substitution of $\Delta^0 u$ and $\Delta^0 v$ from Eqs. (4.93) and (4.94) into Eq. (4.92) results in

$$\Delta^0 s = -(N'/N'')[(\partial s/\partial u)\Delta u + (\partial s/\partial v)\Delta v] + (1/2)(N'/N'')^2[(\partial^2 s/\partial u^2)(\Delta u)^2$$
$$+ 2(\partial^2 s/\partial u \partial v)(\Delta u\Delta v) + (\partial^2 s/\partial v^2)(\Delta v)^2] + \cdots . \tag{4.95}$$

Combining Eqs. (4.87), (4.90), (4.91), and (4.95),

$$\Delta S = \frac{1}{2}\frac{N'N}{N''}\left[(\partial^2 s/\partial u^2)(\Delta u)^2 + 2(\partial^2 s/\partial u\partial v)(\Delta u\Delta v) + (\partial^2 s/\partial v^2)(\Delta v)^2\right] + \cdots$$

$$(4.96)$$

Since $\frac{1}{2}N'N/N''$ is positive, then the condition for the stability of the original single phase is

$$(\partial^2 s/\partial u^2)(\Delta u)^2 + 2(\partial^2 s/\partial u\partial v)(\Delta u\Delta v) + (\partial^2 s/\partial v^2)(\Delta v)^2 < 0. \qquad (4.97)$$

Therefore, the stability of the original phase is defined with respect to the second-order variations of s. If the second-order variations are zero, third-order variations must also be zero, and then fourth-order variations should be negative.

Let us review some simple concepts on the negativeness and positiveness of a function $f(u, v)$ of two independent variables u and v to make use of the above expression in terms of simple criteria. The function $f(u, v)$ is said to be positive definite if it is strictly positive at all points other than $u = 0, v = 0$, (i.e., the origin). The same function f, if it is strictly negative at all points except at the origin $u = 0, v = 0$ is called negative definite. If it can take either sign, both $f > 0$ and $f < 0$, it is called indefinite. Semidefiniteness will be defined shortly.

Consider the quadratic form for f:

$$f(u, v) = au^2 + 2buv + cv^2. \qquad (4.98)$$

Equation (4.98) can be written as

$$f(u, v) = a\left(u + \frac{b}{a}v\right)^2 + \left(c - \frac{b^2}{a}\right)v^2. \qquad (4.99)$$

The first term on the right side of Eq. (4.99) is never negative provided $a > 0$. When $a < 0$, this term is never positive. For the second term to be always positive, $(c - b^2/a) > 0$. From the above simple discussion, the criteria for positive definiteness and negative definiteness of the quadratic $f = au^2 + 2buv + cv^2$ are then established: (1) f is positive definite if and only if $a > 0$ and $(ac - b^2) > 0$, and (2) f is negative definite if and only if $a < 0$ and $(ac - b^2) > 0$. For positive definiteness, the above conditions demand $c > 0$ and for negative definiteness $c < 0$. However, the sign of c is necessary but not sufficient for either positiveness or negativeness. When $ac = b^2$, the second term on the right side of Eq. (4.99) disappears and only the first term remains. Then f is said to be positive semidefinite when $a > 0$ and negative semidefinite when $a < 0$. The quadratic f is indefinite if $ac - b^2 < 0$; f can take either sign, both

$f > 0$ and $f < 0$. Later we will discuss the definiteness of functions of more than two variables. Strang (1986, 1988) presents a detailed analysis of definiteness in a simple manner.

With the above background on definiteness, we now examine the quadratic given by Eq. (4.97) for negative definiteness:

$$(\partial^2 s/\partial u^2) < 0 \tag{4.100}$$

$$(\partial^2 s/\partial u^2)(\partial^2 s/\partial v^2) - (\partial^2 s/\partial u \partial v)^2 > 0. \tag{4.101}$$

Note that $(\partial^2 s/\partial v^2) < 0$ is a necessary but not sufficient condition of negative definiteness. To derive a simple criterion for stability from Eqs. (4.100) and (4.101), it is necessary to derive the expressions for $(\partial^2 s/\partial u^2)$, $(\partial^2 s/\partial v^2)$, and $(\partial^2 s/\partial u \partial v)$ in terms of pressure, volume, temperature, and heat capacity.

$(\partial^2 s/\partial u^2)_v$ *expression* The first derivative of s with respect to u at constant volume is given by $(\partial s/\partial u)_v = 1/T$ (from $du = Tds - Pdv$). The second derivative of s with respect to u at constant v is simply

$$(\partial^2 s/\partial u^2)_v = -1/T^2(\partial T/\partial u)_v = -1/(T^2 c_V). \tag{4.102}$$

In the derivation of the above equation, the definition of c_V from Eq. (3.117) of Chapter 3 is used.

$(\partial^2 s/\partial v^2)_u$ *expression* From Eq. (1.22) of Chapter 1, $(\partial s/\partial v)_u = P/T$. The second derivative is

$$(\partial^2 s/\partial v^2)_u = (1/T)(\partial P/\partial v)_u - (P/T^2)(\partial T/\partial v)_u, \tag{4.103}$$

but

$$(\partial T/\partial v)_u = -(\partial u/\partial v)_T/(\partial u/\partial T)_v = -(1/c_V)(\partial u/\partial v)_T. \tag{4.104}$$

From Eq. (1.21) of Chapter 1,

$$(\partial u/\partial v)_T = T(\partial s/\partial v)_T - P. \tag{4.105}$$

Using the Maxwell relation given by Eq. (1.201) of Chapter 1, the above equation simplifies to

$$(\partial u/\partial v)_T = T(\partial P/\partial T)_v - P. \tag{4.106}$$

Combining Eqs. (4.104) and (4.106),

$$(\partial T/\partial v)_u = (1/c_V)\big[P - T(\partial P/\partial T)_v\big]. \tag{4.107}$$

We also need to evaluate $(\partial P/\partial v)_u$ in Eq. (4.103) using $P = P(v, T)$.

$$dP = (\partial P/\partial v)_T dv + (\partial P/\partial T)_v dT \qquad (4.108)$$

and, dividing by ∂v at constant u,

$$(\partial P/\partial v)_u = (\partial P/\partial v)_T + (\partial P/\partial T)_v (\partial T/\partial v)_u. \qquad (4.109)$$

Substituting $(\partial T/\partial v)_u$ from Eq. (4.107) into the above equation,

$$(\partial P/\partial v)_u = (\partial P/\partial v)_T + (\partial P/\partial T)_v (1/c_V)\left[P - T(\partial P/\partial T)_v\right]. \qquad (4.110)$$

Combining Eqs. (4.103), (4.107), and (4.110) gives the sought expression,

$$(\partial^2 s/\partial v^2)_u = (1/T)(\partial P/\partial v)_T - (1/T^2)(1/c_V)\left[P - T(\partial P/\partial T)_v\right]^2. \qquad (4.111)$$

$(\partial^2 s/\partial v \partial u)$ *expression*   From $(\partial s/\partial u)_v = 1/T$, one obtains

$$(\partial^2 s/\partial u \partial v) = -(1/T^2)(\partial T/\partial v)_u. \qquad (4.112)$$

Substituting Eq. (4.107) into Eq. (4.112),

$$(\partial^2 s/\partial u \partial v) = \left[1/(T^2 c_V)\right]\left[T(\partial P/\partial T)_v - P\right] \qquad (4.113)$$

Now one can substitute Eqs. (4.102), (4.111), and (4.113) into Eqs. (4.100) and (4.101). From Eqs. (4.100) and (4.102), $-[1/(T^2 c_V)] < 0$ and since T is positive, then

$$\boxed{c_V > 0} \qquad (4.114)$$

The above relationship is referred to as thermal stability. The substitution of Eqs. (4.102), (4.111), and (4.113) into Eq. (4.101) gives $-[1/c_V T^3][\partial P/\partial v] > 0$ and, since $c_V > 0$, then

$$\boxed{(\partial P/\partial v)_T < 0}, \qquad (4.115)$$

which is referred to as mechanical stability. From the definition of isothermal compressibility $C_T = -(1/v)(\partial P/\partial v)_T$, an alternative statement of mechanical stability is

$$\boxed{C_T > 0}. \qquad (4.116)$$

Therefore $c_V > 0$ and $C_T > 0$ guarantee that the original single phase shown in Fig. 4.13 is stable. Note that for a single-component fluid to be at stable-equilibrium state, both of the relations given by Eqs. (4.114) and (4.115) should be satisfied simultaneously. In Eq. (4.114), the stable equilibrium is based on a thermal quantity and in Eq. (4.115), stable equilibrium is based on mechanical quantities, therefore they are referred to as thermal and mechanical stabilities, respectively.

Figure 4.14 shows a sketch of the P–v plot for a pure substance at three different temperatures; T_c is the critical temperature, which will be discussed in the presentation on criticality. For $T_2 < T_c$, the ABC branch has a negative slope, i.e., $-(\partial P/\partial v) > 0$, and according to Eq. (4.115) is stable. Similarly the EFG branch also has a negative slope and is stable. At C and E, $(\partial P/\partial V)_T = 0$, which defines the limit of stability. At $T > T_1$, the whole curve has a negative slope and therefore is stable. Note that the segment of the P–v curve between C and E has a positive slope, $(\partial P/\partial v) > 0$, and according to Eq. (4.115) is unstable. The criteria can also be derived from stability analysis of the internal energy.

Stability criteria based on internal energy   The stability of the original single phase in comparison to the two phases demands that the internal energy change be positive (see Fig. 4.15):

$$\Delta U = N'(u + \Delta u) + N''(u + \Delta^0 u) - (N' + N'')u = N'\Delta u + N''\Delta^0 u > 0, \tag{4.117}$$

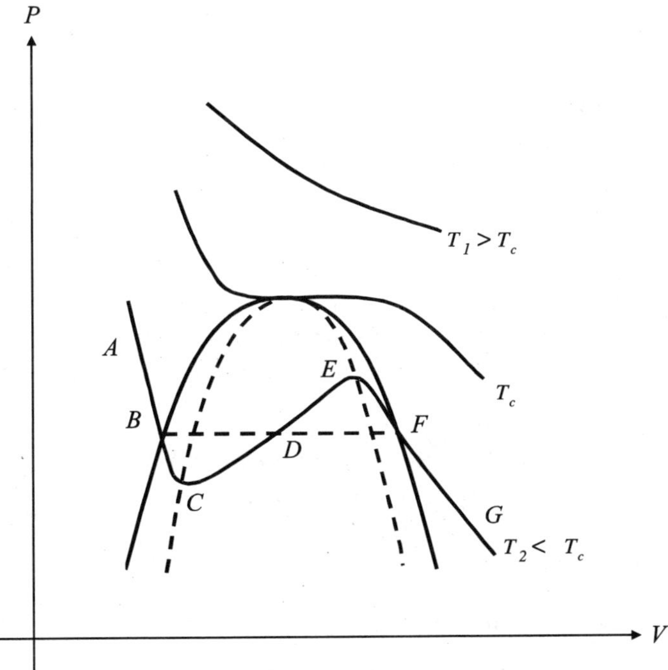

Figure 4.14 P–V plot showing stable and unstable regions.

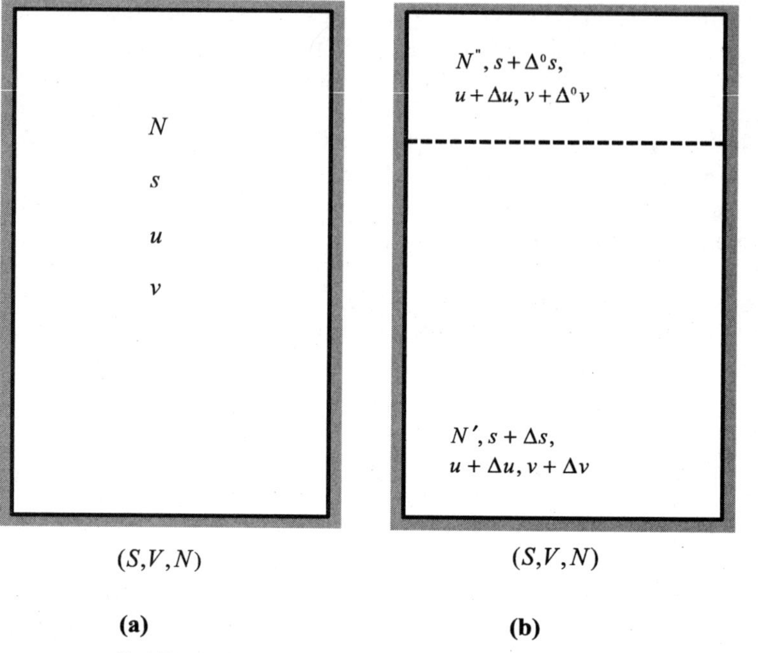

Figure 4.15 Stability of a single component at constant S, V, and N ($N = N' + N''$)

subject to the constraints

$$\Delta S = N'\Delta s + N''\Delta^0 s = 0 \tag{4.118}$$

$$\Delta V = N'\Delta v + N''\Delta^0 v = 0 \tag{4.119}$$

$$N = N' + N'' = constant, \tag{4.120}$$

where ΔS is the entropy change from the single-phase to the two-phase state. Since $u = u(s, v)$, then the Taylor's series expansion of Δu and $\Delta^0 u$ can be expressed by

$$\Delta u = (\partial u/\partial s)\Delta s + (\partial u/\partial v)\Delta v + (1/2)[(\partial^2 u/\partial s^2)(\Delta s)^2$$
$$+ 2(\partial^2 u/\partial s\partial v)(\Delta s\Delta v) + (\partial^2 u/\partial v^2)(\Delta v)^2] + \cdots. \tag{4.121}$$

$$\Delta^0 u = (\partial u/\partial s)\Delta^0 s + (\partial u/\partial v)\Delta^0 v + (1/2)[(\partial^2 u/\partial s^2)(\Delta^0 s)^2$$
$$+ 2(\partial^2 u/\partial s\partial v)(\Delta^0 s\Delta^0 v) + (\partial^2 u/\partial v^2)(\Delta^0 v)^2] + \cdots. \tag{4.122}$$

Similarly to the entropy derivation, Δs, Δv, $\Delta^0 s$, and $\Delta^0 v$ should be small. From the constraints given by Eqs. (4.118) and (4.119),

$\Delta^0 s = -(N'/N'')\Delta s$ and $\Delta^0 v = -(N'/N'')\Delta v$; substitution of these equations into Eq. (4.122) and combining the results with Eqs. (4.117) and (4.121) gives

$$\Delta U = \left(\frac{1}{2}\frac{N'N}{N''}\right)\left[\left(\frac{\partial^2 u}{\partial s^2}\right)(\Delta s)^2 + \left(\frac{\partial^2 u}{\partial v^2}\right)(\Delta v)^2 + 2\left(\frac{\partial^2 u}{\partial v\partial s}\right)(\Delta s\Delta v)\right] + \cdots > 0.$$

$$(4.123)$$

For ΔU to be positive, since $\frac{1}{2}N'N/N''$ is positive, the second-order variations with respect to s and v have to be positive. In case the second-order variations are zero, higher-order even variations should be positive. In other words, the first nonvanishing terms of order two or higher-even order should be positive. Therefore, stability of the single-phase state demands that

$$\boxed{(\partial^2 u/\partial s^2)(\Delta s)^2 + (\partial^2 u/\partial v^2)(\Delta v)^2 + 2(\partial^2 u/\partial v\partial s)(\Delta v\Delta s) > 0}.$$ (4.124)

The positive definiteness of Eq. (4.124) is based on

$$(\partial^2 u/\partial s^2)_v > 0 \tag{4.125}$$

and
$$\left(\frac{\partial^2 u}{\partial s^2}\right)\left(\frac{\partial^2 u}{\partial v^2}\right) - \left(\frac{\partial^2 u}{\partial s\partial v}\right)^2 > 0, \tag{4.126}$$

as was shown before for the positive definiteness condition. The terms in Eq. (4.125) and (4.126) are evaluated readily in terms of T, P, and c_V.

From $du = Tds - Pdv$, $(\partial u/\partial s)_v = T$ and therefore

$$(\partial^2 u/\partial s^2)_v = (\partial T/\partial s)_v. \tag{4.127}$$

From Eq. (3.123) of Chapter 3, $(\partial s/\partial T)_v = c_V/T$, and therefore

$$\boxed{(\partial^2 u/\partial s^2)_v = T/c_V}. \tag{4.128}$$

From $du = Tds - Pdv$, $(\partial u/\partial v)_s = -P$, and $(\partial^2 u/\partial v^2)_s = -(\partial P/\partial v)_s$ which when combined with Eq. (3.128) of Chapter 3 gives

$$(\partial^2 u/\partial v^2)_s = -(c_P/c_V)(\partial P/\partial v)_T. \tag{4.129}$$

From the expression for du, $(\partial u/\partial s)_v = T$, and $(\partial^2 u/\partial s\partial v)_s = (\partial T/\partial v)_s$. The expression for $(\partial T/\partial v)_s$ is given by

$$(\partial T/\partial v)_s = -\frac{(\partial s/\partial v)_T}{(\partial s/\partial T)_v} = -\frac{(\partial s/\partial v)_T}{c_V/T}. \tag{4.130}$$

Combining Eqs. (1.201) of Chapter 1 and (4.130),

$$(\partial^2 u/\partial s\partial v) = T\frac{(\partial P/\partial T)_v}{c_V}. \tag{4.131}$$

We also can write

$$(\partial P/\partial T)_v = -\frac{(\partial v/\partial T)_P}{(\partial v/\partial P)_T} = e/C_T, \tag{4.132}$$

where e is the thermal expansivity and C_T is the isothermal compressibility. Combining Eqs. (4.131) and (4.132),

$$(\partial^2 u/\partial s\partial v) = -Te/c_V C_T \tag{4.133}$$

From Eqs. (4.125) and 1(4.26), $T/c_V > 0$ which gives $c_V > 0$; therefore, the thermal stability condition is established. This is the same relationship as that developed earlier from the entropy approach. Substituting Eqs. (4.128), (4.129), and (4.133) into Eq. (4.126) and using $(\partial P/\partial v)_T = -vC_T$ and $c_P = c_V + Tve^2/C_T$ (see Example 3.8 and Eq. (3.126) of Chapter 3),

$$T/(c_V C_T v) > 0. \tag{4.134}$$

Therefore $C_T > 0$. Note that since c_V and C_T are positive for a stable system, the implication is that $c_P > 0$ when a system is stable.

Stability analysis in a two-component system

Consider N moles of a binary mixture of composition x_1 and x_2 (mole fractions) at temperature T and pressure P. Let us compare the stability of the single-phase system with that of the two-phase system at the same temperature and pressure. Similarly to the single-component system in the preceding section, we assume N' moles of primed phase and N'' moles of double-primed phase, in the two-phase state. The composition of primed and double-primed phases are $x_1 + \Delta x_1$ and $x_1 + \Delta^0 x_1$, respectively, for component 1 (see Fig. 4.16).

This time we will work with the Gibbs free energy function. The Gibbs free energy difference between the two states shown in Fig. 4.16 is

$$\Delta G = G_{two\ phase} - G_{single\ phase}, \tag{4.135}$$

where

$$G_{two\ phase} = N'(g + \Delta g) + N''(g + \Delta^0 g)$$

and

$$G_{single\ phase} = (N' + N'')g.$$

Therefore,

$$\Delta G = N'\Delta g + N''\Delta^0 g. \tag{4.136}$$

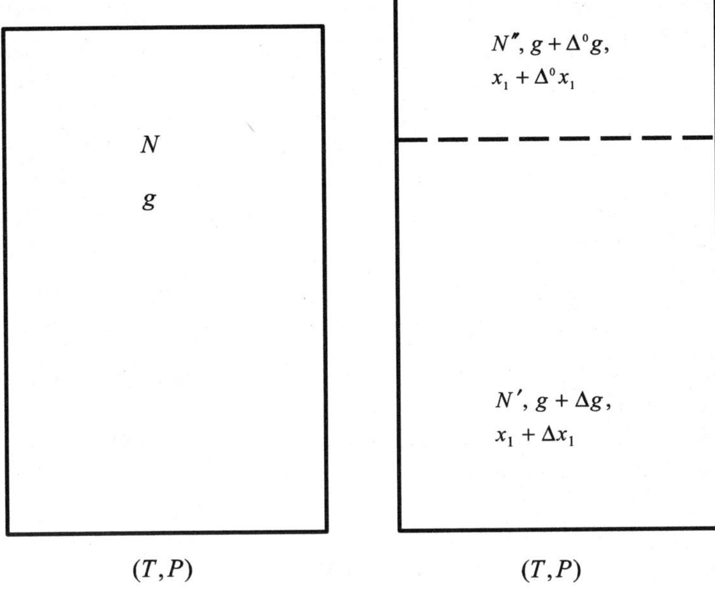

Figure 4.16   Stability of a two-component mixture at constant T, P and $\underline{N}$.

In Eq. (4.136), Δg is the molar Gibbs free energy difference between the primed phase and the original single phase, and $\Delta^0 g$ is the molar Gibbs free energy difference between the double-primed phase and the original phase. The constraints (see Fig. 4.16) are constant temperature and pressure and mole numbers of components 1 and 2; $N_i = N_i' + N_i''$, $i = 1, 2$ ($N = N_1 + N_2$, $N' = N_1' + N_2'$, and $N'' = N_1'' + N_2''$). From the mole constraints,

$$N'\Delta x_1 + N''\Delta^0 x_1 = 0. \tag{4.137}$$

The molar Gibbs free energy $g = g(T, P, x_1)$ and since T and P are held constant, then

$$\Delta g = (\partial g/\partial x_1)\Delta x_1 + (1/2)(\partial^2 g/\partial x_1^2)(\Delta x_1)^2 + \cdots \tag{4.138}$$

and

$$\Delta^0 g = (\partial g/\partial x_1)\Delta^0 x_1 + (1/2)(\partial^2 g/\partial x_1^2)(\Delta^0 x_1)^2 + \cdots, \tag{4.139}$$

where Δx_1 and $\Delta^0 x_1$ are assumed to be small. Combining Eqs. (4.135) to (4.139),

$$\Delta G = (N'N/2N'')(\partial^2 g/\partial x_1^2)(\Delta x_1)^2 + \cdots. \tag{4.140}$$

The stability of the single-phase state in comparison to that of the two-phase state requires $\Delta G > 0$. Since the multiplier of $(\partial^2 g/\partial x_1^2)$ is positive, then

$$\boxed{(\partial^2 g/\partial x_1^2)_{T,P} > 0} \tag{4.141}$$

is the stability condition of the original single phase. Note that when the second-order variation is zero, then

$$(\partial^4 g/\partial x_1^4)_{T,P} > 0. \tag{4.142}$$

If fourth-order variations are zero, then the higher-order even variations should be positive.

Equation (4.141) can be transformed into the chemical potential form. Since $(\partial g/\partial x_1)_{T,P} = \mu_1 - \mu_2$, then

$$(\partial^2 g/\partial x_1^2)_{T,P} = (\partial \mu_1/\partial x_1)_{T,P} - (\partial \mu_2/\partial x_1)_{T,P} \tag{4.143}$$

From the Gibbs-Duhem relation at constant T and P, $\sum_{i=1}^2 x_i(\partial \mu_i/\partial x_1)_{T,P} = 0$. Therefore, Eq. (4.141) transforms to $(\partial \mu_1/\partial x_1)_{T,P}/x_2$ and, since x_2 is positive,

$$\boxed{(\partial \mu_1/\partial x_1)_{T,P} > 0}. \tag{4.144}$$

Equation (4.144) is an alternative form of Eq. (4.141) for the stability criteria. Figure 4.17 shows a plot of μ_1 vs. x_1 at constant temperature and pressure in a binary mixture. For a system to be stable, $(\partial \mu_1/\partial x_1)$ should be positive. At T_1 and P_1, the system is in the stable single-phase state. At T_2 and P_2 only the points from A to B and from C to D are stable. At B and C, the limits of stability are defined by

$$\boxed{(\partial \mu_1/\partial x_1)_{T,P} = 0}. \tag{4.145}$$

We will show in Example 4.6 that the following inequalities hold for a stable binary system. At constant pressure,

$$\boxed{(\partial^2 g/\partial T^2)_{P,x_1} < 0} \tag{4.146}$$

and at constant temperature,

$$\boxed{(\partial^2 g/\partial P^2)_{T,x_1} < 0}. \tag{4.147}$$

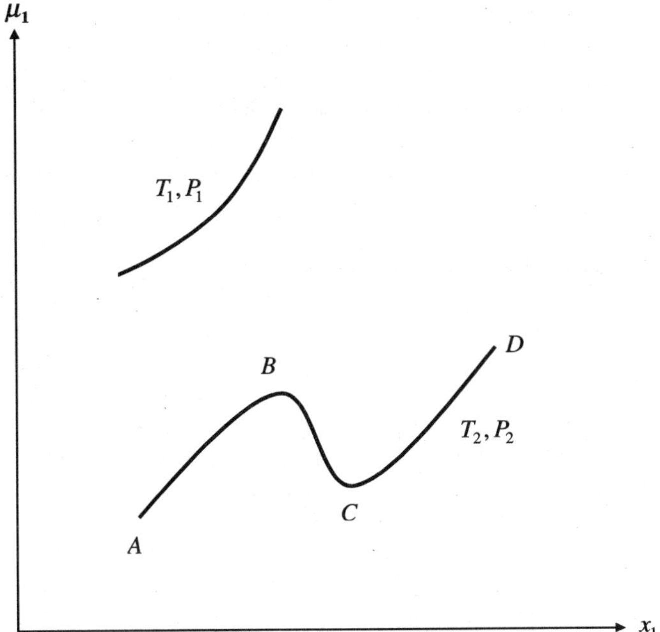

Figure 4.17   Plot of μ_1 *vs.* x_1 in a binary system at constant T and P.

Stability analysis for multicomponent mixtures

The stability criteria for single-component and binary systems are given above; those criteria are general, but restrictive assumptions were made in their derivations. These assumptions can be removed and the same results can be obtained. The generalized derivation relies on the use of the sum-of-the-squares expression, and Legendre and Jacobian transformations. The derivations of stability criteria presented above provide a physical sense of the problem using only the Taylor's series expansion. In the following, a more elaborate derivation for the general problem of multicomponent systems will be presented. An alternative derivation discussed in Tester and Modell (1996) relies more on Legendre transform.

Consider N moles of a multicomponent system consisting of c components. We wish to study the stability of the single-phase state in relation to the perturbed two-phase state. The perturbed two-phase state consists of N' moles of the primed phase and N'' moles of the double-primed phase. We assume that $N'' \ll N'$ solely for the purpose of determining whether the internal energy of a fixed mass could be reduced by introducing a new phase (that is, the double-primed phase) of infinitesimal amount while keeping the volume and entropy fixed. We can use the criterion of the U minimum (as was just mentioned) or G minimum or the criterion of the maximum of S to study stability. Let us use the U-minimum principle. The Legendre transform can be used to transform the results from U to any other basic thermodynamic function, such as H, A, or G.

The internal energy of the original homogeneous phase is a function of $(S, V, \underline{N})$. Figure 4.18 shows the state of the system before and after perturbation.

The change in the internal energy from a single phase to a two phase-state is

$$\Delta U = \Delta U' + \Delta U'', \tag{4.148}$$

where $\Delta U'$ denotes the internal energy change of N' moles from the original single-phase state to the primed phase, and $\Delta U''$ represents the internal energy change of N'' moles from the original single-phase state to the double-primed phase. The restrictions of the internal energy change are

$$\Delta S = \Delta S' + \Delta S'' = 0 \tag{4.149}$$

$$\Delta V = \Delta V' + \Delta V'' = 0 \tag{4.150}$$

$$\Delta N_i = \Delta N_i' + \Delta N_i'' = 0 \qquad i = 1, \dots, c. \tag{4.151}$$

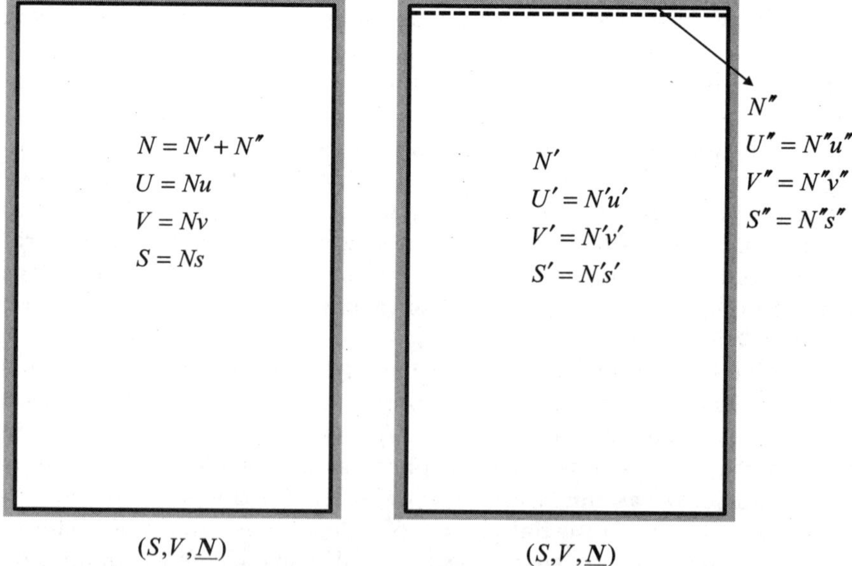

Figure 4.18   Stability of a multicomponent mixture at constant S, V, and $\underline{N}$.

In Eq. (4.149), $\Delta S'$ represents the entropy change of N' moles from the single-phase state to the primed phase, and $\Delta V'$ is the volume change of N' moles from the single-phase state to the primed phase. $\Delta S''$ and $\Delta V''$ have similar definitions. The expressions for various variations are given by

$$\Delta U' = N'(u' - u), \qquad \Delta V' = N'(v' - v), \qquad \Delta S' = N'(s' - s),$$
$$\Delta U'' = N''(u'' - u), \qquad \Delta V'' = N''(v'' - v), \qquad \Delta S'' = N''(s'' - s), \tag{4.152}$$

where u, u', and u'' are the molar internal energies of the original single-phase, primed and double-primed phases, respectively; s, s', and s'', and v, v', and v'' have similar definitions. Note that in the above equation, there is no restriction on the magnitude of the molar property changes.

The expressions for $\Delta U'$ and $\Delta U''$ are provided by the Taylor's series expansions, which assume that the amount of the double-primed phase is small.

$$\begin{aligned}
\Delta U' = {}& (\partial U/\partial S)'\Delta S' + (\partial U/\partial V)'\Delta V' + \sum_{i=1}^{c}(\partial U/\partial N_i)'\Delta N_i' \\
& + (1/2)\{(\partial^2 U/\partial S^2)'(\Delta S')^2 + (\partial^2 U/\partial V^2)'(\Delta V')^2 \\
& + 2(\partial^2 U/\partial S\partial V)'(\Delta S'\Delta V') \\
& + \sum_{i=1}^{c}\sum_{j=1}^{c}(\partial^2 U/\partial N_i\partial N_j)'(\Delta N_i'\Delta N_j') \\
& + 2\sum_{i=1}^{c}[(\partial U^2/\partial N_i\partial V)'(\Delta N_i'\Delta V') \\
& + (\partial^2 U/\partial N_i\partial S)'(\Delta N_i'\Delta S')]\} + \textit{higher-order terms}
\end{aligned} \tag{4.153}$$

and

$$\begin{aligned}
\Delta U'' = {}& (\partial U/\partial S)''\Delta S'' + (\partial U/\partial V)''\Delta V'' + \sum_{i=1}^{c}(\partial U/\partial N_i)''\Delta N_i'' \\
& + (1/2)\{(\partial^2 U/\partial S^2)''(\Delta S'')^2 + (\partial^2 U/\partial V^2)''(\Delta V'')^2 \\
& + 2(\partial^2 U/\partial S\partial V)''(\Delta S''\Delta V'') \\
& + \sum_{i=1}^{c}\sum_{j=1}^{c}(\partial U^2/\partial N_i\partial S)''(\Delta N_i''\Delta N_j'') \\
& + 2\sum_{i=1}^{c}[(\partial^2 U/\partial N_i\partial V)''(\Delta N_i''\Delta N_j''). \\
& + (\partial^2 U/\partial N_i\partial S)''(\Delta N_i''\Delta S'')]\} + \textit{higher-order terms.}
\end{aligned} \tag{4.154}$$

In Eq. (4.153), $(\partial U/\partial S)'$ represents the variation of the internal energy of N' moles with respect to the variation of the entropy of N' moles at the original state of the single phase. Similar representations for

$(\partial U/\partial S)''$, $(\partial^2 U/\partial S^2)'$, etc. follow. From $dU = TdS - PdV + \sum_{i=1}^{c} \mu_i dN_i$, at the original state,

$$(\partial U/\partial S)' = (\partial U/\partial S)'' = T$$

$$(\partial U/\partial V)' = (\partial U/\partial V)'' = -P \qquad (155)$$

$$(\partial U/\partial N_i)' = (\partial U/\partial N_i)'' = \mu_i \qquad i = 1, \dots, c.$$

The first derivatives are intensive variables. However, the second and higher-order derivatives at the original state depend on mass. The second derivatives are related by

$$N'(\partial^2 U/\partial S^2)' = N''(\partial^2 U/\partial S^2)''$$

$$N'(\partial^2 U/\partial V^2)' = N''(\partial^2 U/\partial V^2)''$$

$$N'(\partial^2 U/\partial S\partial V)' = N''(\partial^2 U/\partial S\partial V)''$$

$$N'(\partial^2 U/\partial S\partial V)' = N''(\partial^2 U/\partial N_i\partial S)'', \qquad i = 1, \dots, c$$

$$N'(\partial^2 U/\partial N_i\partial V)' = N''(\partial^2 U/\partial N_i\partial V)'', \qquad i = 1, \dots, c$$

$$N'(\partial^2 U/\partial N_i\partial N_j)' = N''(\partial^2 U/\partial N_i\partial N_j)'' \qquad i = 1, \dots, c, j = 1, \dots, c.$$

$$(156)$$

To prove the first relationship above, we write

$$(\partial^2 U/\partial S^2)' = \partial/\partial S'(\partial U/\partial S)' = (\partial T/\partial S)'$$

$$(\partial^2 U/\partial S^2)'' = \partial/\partial S''(\partial U/\partial S)'' = (\partial T/\partial S)''. \qquad (157)$$

But $S' = N's$ and $S'' = N''s$ at the original state, and therefore

$$(\partial^2 U/\partial S^2)' = (1/N')\partial T/\partial s,$$

$$(\partial^2 U/\partial S^2)'' = (1/N'')\partial T/\partial s. \qquad (4.158)$$

From the above two equations, the first relationship in Eq. (4.156) is established. (The last expression in Eq. (4.156) and some higher-order derivatives are derived in Example 4.4.) Combining Eqs (4.148) to (4.156), we obtain,

$$\Delta U = (1/2)(N/N'')\Big\{(\partial^2 U/\partial S^2)'(\Delta S')^2 + (\partial^2 U/\partial V^2)'(\Delta V')^2.$$

$$+ 2(\partial^2 U/\partial S\partial V)'(\Delta S'\Delta V') + \sum_{i=1}^{c}\sum_{j=1}^{c}(\partial^2 U/\partial N_i\partial N_j)'(\Delta N_i'\Delta N_j')$$

$$+ 2\sum_{i=1}^{c}(\partial^2 U/\partial N_i\partial V)'(\Delta N_i'\Delta V') + 2\sum_{i=1}^{c}(\partial^2 U/\partial N_i\partial S)'(\Delta N_i'\Delta S')\Big\}$$

$$+ \textit{higher-order terms}. \qquad (4.159)$$

Equation (4.159) can be written as

$$\Delta U = 1/2![1 + N'/N'']d^2U + 1/3![1 - (N'/N'')^2]d^3U$$
$$+ 1/4![1 + (N'/N'')^3]dU^4 + \cdots, \qquad (4.160)$$

where d^2U is the term in { } in Eq. (4.159); it represents the second-order variations of U. Note that there is no term to represent first-order variations of U in the above equations. The first-order variations drop out because of the equilibrium condition. ΔU in Eq. (4.160) is positive if the quadratic term representing second-order variations is positive for every small variation of variables ΔN_i, ΔS, and ΔV. (Note that the coefficient of d^2U is positive.) When the quadratic term is positive, the original homogeneous phase is stable and there is no need to examine higher-order terms. If the quadratic term is zero, then the cubic form must also be zero, and the fourth-order term must be positive for ΔU to be positive. In other words, the first nonvanishing even-order term must be positive for ΔU to be positive and odd-order terms below that must be zero. The expression for d^2U in a compact form can be written as

$$d^2U = \sum_{j=1}^{c+2}\sum_{k=1}^{c+2}(\partial^2U/\partial X_j\partial X_k)\Delta X_j\Delta X_k \qquad (4.161a)$$

Equation (4.161a) can also be written as

$$d^2U = \sum_{j=1}^{c+2}\sum_{k=1}^{c+2}U_{jk}\Delta X_j\Delta X_k, \qquad (4.161b)$$

where $U_{jk} \equiv (\partial^2U/\partial X_j\partial X_k)$.

Linear algebra provides a perfect shorthand for writing Eq. (4161b) because of the symmetry of U_{jk}: $U_{jk} = U_{kj}$. This equation can be written as the inner product of the perturbation vector $\Delta \underline{X}^T$ and the vector that results from the Hessian matrix of U on the perturbation vector $\Delta \underline{X}$. The Hessian matrix is the second-order-derivative matrix; the entries of the Hessian matrix of U are U_{jk}. Therefore,

$$d^2U = \Delta \underline{X}^T H_U \Delta \underline{X}, \qquad (4.162)$$

where the Hessian matrix, H_U, is given by

$$H_U = \begin{pmatrix} U_{1,1} & U_{1,2} & \cdots & U_{1,c+2} \\ U_{2,1} & U_{2,2} & \cdots & U_{2,c+2} \\ \vdots & \vdots & & \vdots \\ U_{c+2,1} & U_{c+2,2} & \cdots & U_{c+2,c+2} \end{pmatrix} \qquad (4.163)$$

and $\Delta \underline{X}^T = (\Delta X_1\ \Delta X_2\ \ldots\ \Delta X_{c+2})$; $\Delta \underline{X}^T$ is the transpose of the column vector $\Delta \underline{X} = (\Delta X_1, \Delta X_2, \ldots, \Delta X_{c+2})$. Note that we have commas between the entries of the column vector $\Delta \underline{X}$ to write it on a horizontal line! There is no comma between the entries of the row vector ΔX^T. The Hessian matrix, which has a pure quadratic form, is used to study the problem of maxima and minima for functions of many variables. From the Taylor's series expansion of $U(\underline{X} + \Delta \underline{X})$ (see Example 4.1),

$$U(\underline{X} + \Delta \underline{X}) = U(\underline{X}) + \Delta \underline{X}^T(\nabla U) + (1/2)\Delta \underline{X}^T H_U \Delta \underline{X}$$
$$+\ higher\text{-}order\ terms, \tag{4.164}$$

where ∇U represents the gradient of U. The gradient of U when set to zero provides the stationary point of $U(\underline{X})$; at a stationary point, $\nabla U = (\partial U/\partial X_1, \partial U/\partial X_2, \ldots, \partial U/\partial X_{c+2})$ is a vector of zeros: $\partial U/\partial X_1 = 0, \partial U/\partial X_2 = 0 \ldots, \partial U/\partial X_{c+2} = 0$ (see Strang, 1988). However, a stationary point can be either a maximum, a minimum, or a saddle point. The maxima and minima are, therefore, decided on the sign of $\Delta \underline{X}^T H_U \Delta \underline{X}$. U has a minimum when $\Delta \underline{X}^T H_U \Delta \underline{X}$ is positive. The Hessian matrix H_U should be positive definite in order to have $\Delta \underline{X}^T H_U \Delta \underline{X} > 0$. There are various tests for positive definiteness of the symmetric matrix H_U. One such test is that all the pivots of H_U should be positive (see Strang, 1988). Another test is that a positive-definite matrix has positive eigenvalues (also see Strang, 1988). For a single-component system, the derivations for positive and negative definiteness were straightforward because of dependence of S or U on two variables. When the number of variables exceeds two, a knowledge of the positive and negative definiteness of the Hessian matrix becomes helpful.

Let us return to Eq. (4.161b) and write it in the form of the sum-of-the-squares,

$$d^2 U = \sum_{j=1}^{c+2} y_{jj}^{(j-1)}(\Delta Z_j)^2. \tag{4.165}$$

The expression for ΔZ_j and $y_{jj}^{(j-1)}$ will be derived next. Equation (4.165) is often written without full proof. The criteria of stability and criticality rely on the use of intermediate steps in the expressions for the sum-of-the-squares in our derivations.

The double summation can be expanded as

$$d^2 U = \sum_{j=1}^{c+2} \sum_{k=1}^{c+2} U_{jk}\Delta X_j \Delta X_k$$
$$= \sum_{j=1}^{c+2} [U_{j1}\Delta X_j \Delta X_1 + U_{j2}\Delta X_j \Delta X_2 + U_{j3}\Delta X_j \Delta X_3 + \cdots], \tag{4.166}$$

which can be rearranged to

$$d^2U = \sum_{j=1}^{c+2}\left[U_{j1}\Delta X_j\Delta X_1\right] + \sum_{j=1}^{c+2}\left[U_{j2}\Delta X_j\Delta X_2 + U_{j3}\Delta X_j\Delta X_3 + \cdots\right]$$

$$= U_{11}(\Delta X_1)^2 + \sum_{j=2}^{c+2}(U_{j1}\Delta X_j\Delta X_1) + \left[U_{12}\Delta X_1\Delta X_2 + U_{13}\Delta X_1\Delta X_3 + \cdots\right]$$

$$+ \left[\sum_{j=2}^{c+2} U_{j2}\Delta X_j\Delta X_2 + U_{j3}\Delta X_j\Delta X_3 + \cdots\right]. \tag{4.167}$$

Use of $U_{j1} = U_{1j}$ and further rearrangement gives

$$d^2U = U_{11}(\Delta X_1)^2 + 2\sum_{j=2}^{c+2} U_{1j}\Delta X_1\Delta X_j + \sum_{j=2}^{c+2}\sum_{k=2}^{c+2} U_{jk}\Delta X_j\Delta X_k. \tag{4.168}$$

The above equation can be expressed as

$$d^2U = U_{11}\left[\Delta X_1 + (1/U_{11})\left(\sum_{j=2}^{c+2} U_{1j}\Delta X_j\right)^2\right] - (1/U_{11})\left(\sum_{j=2}^{c+2} U_{1j}\Delta X_j\right)^2$$

$$+ \sum_{j=2}^{c+2}\sum_{k=2}^{c+2} U_{jk}\Delta X_j\Delta X_k, \tag{4.169}$$

provided $U_{11} \neq 0$. We have already shown that $U_{11} > 0$ for a single-component system to be stable; that also applies when there is no change in composition or when composition is held constant for multi-component systems (see Eq. (4.125)).

The squared single summation $(\sum_{j=2}^{c+2} U_{1j}\Delta X_j)^2$ in Eq. (4.169) in terms of double summation is

$$\sum_{j=2}^{c+2}(U_{1j}\Delta X_j)^2 = \sum_{j=2}^{c+2}\sum_{k=2}^{c+2} U_{1j}U_{1k}\Delta X_j\Delta X_k \tag{4.170}$$

Equation (4.170) is derived by opening the double summation:

$$\sum_{j=2}^{c+2}\sum_{k=2}^{c+2} U_{1j}U_{1k}\Delta X_j\Delta X_k = \sum_{k=2}^{c+2}[U_{12}U_{1k}\Delta X_2\Delta X_k + U_{13}U_{1k}\Delta X_3\Delta X_k + \cdots]$$

$$= \{[U_{12}^2(\Delta X_2)^2 + U_{12}U_{13}\Delta X_2\Delta X_3 + \cdots]$$

$$+ [U_{12}U_{13}\Delta X_2\Delta X_3 + U_{13}^2(\Delta X_3)^2 + \cdots] + \cdots\}$$

$$= \sum_{j=2}^{c+2}(U_{1j}\Delta X_j)^2. \tag{4.171}$$

Defining ΔZ_1 and $U_{jk}^{(1)}$ as

$$\Delta Z_1 = \Delta X_1 + (1/U_{11}) \sum_{j=2}^{c+2} (U_{1j}\Delta X_j) \tag{4.172}$$

$$U_{jk}^{(1)} = U_{jk} - (1/U_{11})U_{1j}U_{1k} \qquad j,k \geq 2. \tag{4.173}$$

Combining Eqs. (4.169), (4.170), and the above definitions,

$$d^2 U = U_{11}(\Delta Z_1)^2 + \sum_{j=2}^{c+2} \sum_{k=2}^{c+2} U_{jk}^{(1)}\Delta X_j \Delta X_k. \tag{4.174}$$

The double summation in the above equation can be written as (similarly to Eq. (4.169))

$$\sum_{j=2}^{c+2} \sum_{k=2}^{c+2} U_{jk}^{(1)}\Delta X_j \Delta X_k = U_{22}^{(1)}\left[\Delta X_2 + (1/U_{22}^{(1)}) \sum_{j=3}^{c+2} U_{j2}^{(1)}\Delta X_j^2\right]^2$$

$$- (1/U_{22}^{(1)})\left[\sum_{j=3}^{c+2} U_{j2}\Delta X_j\right]^2 + \sum_{j=3}^{c+2} \sum_{k=3}^{c+2} U_{jk}^{(1)}\Delta X_j \Delta X_k \tag{4.175}$$

provided $U_{22}^{(1)} \neq 0$. The squared single summation can be written as (similarly to Eq. (4.170))

$$\left(\sum_{j=3}^{c+2} U_{j2}\Delta X_j\right)^2 = \sum_{j=3}^{c+2} \sum_{k=3}^{c+2} (U_{j2}U_{k2})(\Delta X_j \Delta X_k). \tag{4.176}$$

Define ΔZ_2 and $U_{jk}^{(2)}$ as

$$\Delta Z_2 = \Delta X_2 + (1/U_{22}^{(1)}) \sum_{j=3}^{c+2} U_{j2}^{(1)}\Delta Z_j \tag{4.177}$$

$$U_{jk}^{(2)} = \left[U_{jk}^{(1)} - (1/U_{22}^{(1)})(U_{j2}U_{k2})\right]. \qquad j,k \geq 3 \tag{4.178}$$

From Eqs. (4.174) and (4.178),

$$d^2 U = U_{11}(\Delta Z_1)^2 + U_{22}^{(1)}(\Delta Z_2)^2 + \sum_{j=3}^{c+2} \sum_{k=3}^{c+2} U_{jk}^{(2)}\Delta X_j \Delta X_k. \tag{4.179}$$

The above procedure can be continued to obtain the final expressions:

$$
\begin{aligned}
U_{jk}^{(0)} &= U_{jk} & j, k &= 1, \ldots, c+2 \\
U_{jk}^{(l)} &= U_{jk}^{(l-1)} - (1/U_{ll}^{(l-1)})[U_{lj}^{(l-1)} U_{lk}^{(l-1)}] & j, k &\geq l \\
& & j, k &= 2, \ldots, c+2 \\
& & l &= 1, \ldots, c+2 \\[2mm]
\Delta Z_j &= \Delta X_j + (1/U_{jj}^{(j-1)}) \sum_{k=j+1}^{c+2} U_{kj}^{(j-1)} \Delta X_k & j &= 1, \ldots, c+2 \\[2mm]
d^2 U &= U_{11}(\Delta Z_1)^2 + U_{22}^{(1)}(\Delta Z_2)^2 + \cdots + U_{c+2,c+2}^{(c+1)}(\Delta Z_{c+2})^2 &
\end{aligned}
$$

$$(4.180)$$

Let us define

$$
y_{11}^{(0)} = U_{11}
$$

$$
y_{jj}^{(j-1)} = U_{jj}^{(j-1)} \qquad j = 2, \ldots, c+2. \tag{4.181}
$$

The $y_{11}^{(0)}$, $y_{jj}^{(j-1)}$, and ΔZ_j in Eqs. (4.180) and (4.181) define the variables of Eq. (4.165). The purpose of introducing y in place of U is to emphasize that Eq. (4.165) represents not only U but other thermodynamic functions, H, A, and G.

Alternative expression for $d^2 U$ Equation (4.161b) can also be written as

$$
d^2 U = U_{11}(\Delta Z_1)^2 + \frac{\begin{vmatrix} U_{11} & U_{12} \\ U_{21} & U_{22} \end{vmatrix}}{U_{11}}(\Delta Z_2)^2 + \frac{\begin{vmatrix} U_{11} & U_{12} & U_{13} \\ U_{21} & U_{22} & U_{23} \\ U_{31} & U_{32} & U_{33} \end{vmatrix}}{\begin{vmatrix} U_{11} & U_{12} \\ U_{21} & U_{22} \end{vmatrix}}(\Delta Z_3)^2 + \cdots .
$$

$$(4.182)$$

This form of $d^2 U$ is very useful for the derivation of the criteria of stability. The key steps in the derivation of Eq. (4.182) are provided below. The coefficient of $(\Delta Z_2)^2$ is $U_{22}^{(1)}$. From Eq. (4.180),

$$
U_{22}^{(1)} = U_{22} - (1/U_{11})(U_{12} U_{12}) = \begin{vmatrix} U_{11} & U_{12} \\ U_{21} & U_{22} \end{vmatrix} \Big/ U_{11}. \tag{4.183}
$$

Similarly, the coefficient of $(\Delta Z_3)^2$, $U_{33}^{(2)}$, is

$$
U_{33}^{(2)} = \left[U_{33}^{(1)} U_{22}^{(1)} - U_{23}^{(1)} U_{23}^{(1)} \right] \Big/ U_{22}^{(1)}. \tag{4.184}
$$

The expressions for $U_{33}^{(1)}$, $U_{22}^{(1)}$, and $U_{23}^{(1)}$ from Eq. (4.180) when substituted into Eq. (4.184) result in

$$U_{33}^{(2)} = \frac{\begin{vmatrix} U_{11} & U_{12} & U_{13} \\ U_{21} & U_{22} & U_{23} \\ U_{31} & U_{32} & U_{33} \end{vmatrix}}{U_{22}^{(1)}}.$$

(4.185)

Other higher-order expressions are obtained in the same manner. The algebra is tedious, but the method is straightforward. Another compact form of Eq. (4.182) is

$$d^2U = \sum_{j=1}^{c+2} (\lambda^{(j)}/\lambda^{(j-1)})(\Delta Z_j)^2,$$

(4.186)

where $\lambda^{(0)} = 1$, $\lambda^{(1)} = U_{11}$, and

$$\lambda^{(j)} = \begin{vmatrix} U_{11} & U_{12} & \cdots & U_{1j} \\ U_{21} & U_{22} & \cdots & U_{2j} \\ \vdots & \vdots & & \vdots \\ U_{j1} & U_{j2} & \cdots & U_{jj} \end{vmatrix} \qquad j = 2, \ldots, c+2.$$

(4.187)

Since $y_{c+2,c+2}^{(c+1)} = 0$ (see Eq. (1.185) of Chapter 1), then Eqs. (4.165) and (4.186) can be written as

$$d^2U = \sum_{j=1}^{c+1} y_{jj}^{(j-1)}(\Delta Z_j)^2$$

(4.188)

and

$$d^2U = \sum_{j=1}^{c+1} (\lambda^{(j)}/\lambda^{(j-1)})(\Delta Z_j)^2.$$

(4.189)

Now we are ready to establish the criteria of stability. From the set of expressions in Eqs. (4.180) and (4.181),

$$y_{jj}^{(j-1)} = y_{jj}^{(j-2)} - \left[y_{j,j-1}^{(j-2)} \right]^2 / y_{j-1,j-1}^{(j-2)} \qquad j = 2, \ldots, c+1.$$

(4.190)

In this equation, if $y_{j-1,j-1}^{(j-2)}$ and $y_{jj}^{(j-1)}$ approach zero, $y_{jj}^{(j-1)}$ approaches zero first and $y_{j-1,j-1}^{(j-2)}$ approaches zero later (provided $(y_{j,j-1}^{(j-2)} \neq 0)$). Therefore, the positivity of d^2U which is related to the positivity of the coefficients of $(\Delta Z_j)^2$ simplifies considerably; in place of $y_{jj}^{(j-1)} > 0$ for $j = 1, \ldots, c + 1$, one can demand only that

$$y_{c+1,c+1}^{(c)} > 0. \tag{4.191}$$

The above simple relationship provides the criteria of stability for multicomponent systems. When

$$\boxed{y_{c+1,c+1}^{(c)} = 0}, \tag{4.192}$$

the multicomponent system is said to be at the limit of the stability; $y_{c+1,c+1}^{(c)}$ provides the locus of the limit of stability (or in fact the limit of metastability) discussed in Chapter 3. Examples are provided at the end of the chapter to give a clear picture of stability criteria and their significance. Let us transform Eqs. (4.191) and (4.192) in terms of more familiar variables. From Eq. (4.181),

$$y^{(0)}(X_1, X_2, \ldots, X_{c+2}) = U(S, V, N_1, N_2, \ldots, N_c), \tag{4.193}$$

where $y_{c+1,c+1}^{(c)}$ is simply

$$y_{c+1,c+1}^{(c)} = \left(\frac{\partial^2 y^{(c)}}{\partial X_{c+1}^2}\right)_{C_1,C_2,\ldots,C_c,X_{c+2}} = \left(\frac{\partial C_{c+1}}{\partial X_{c+1}}\right)_{C_1,C_2,\ldots,C_c,X_{c+2}} \tag{4.194}$$

From Eqs. (1.183) and (1.184) in Chapter 1,

$$C_1 = T, C_2 = -P, C_3 = \mu_3, \ldots, C_c = \mu_{c-2}, C_{c+1} = \mu_{c-1}$$

and

$$x_{c-1} = N_{c-1}/\sum_{i=1}^{c} N_i, \ y_{c+1,c+1}^{(c)} = (1 - x_{c-1})/\sum_{i=1}^{c} N_i \left(\frac{\partial \mu_{c-1}}{\partial x_{c-1}}\right)_{T,P,\mu_1,\mu_2,\ldots,\mu_{c-2},x_c}$$

Therefore, for a stable system,

$$\boxed{(\partial \mu_{c-1}/\partial x_{c-1})_{T,P,\mu_1,\mu_2,\ldots,\mu_{c-2}} > 0}. \tag{4.195}$$

The expression given by Eq. (4.195) implies that for a c-component system to be stable, $(\partial \mu_{c-1}/\partial x_{c-1})$ should be positive at constant $T, P, \mu_1, \ldots, \mu_{c-2}$, which is inconvenient in terms of the variables to which we are accustomed. Figure 4.19 is a plot of μ_{c-1} vs. x_{c-1}. The curve I has a positive slope and therefore is stable. The curve II shows the system can go through a two-phase region; between B and C, $(\partial \mu_{c-1}/\partial x_{c-1})$ is negative and the system is unstable. We will transform Eq. (4.195) in terms of the variables that are convenient for a pressure-explicit EOS in the examples at the end of this chapter.

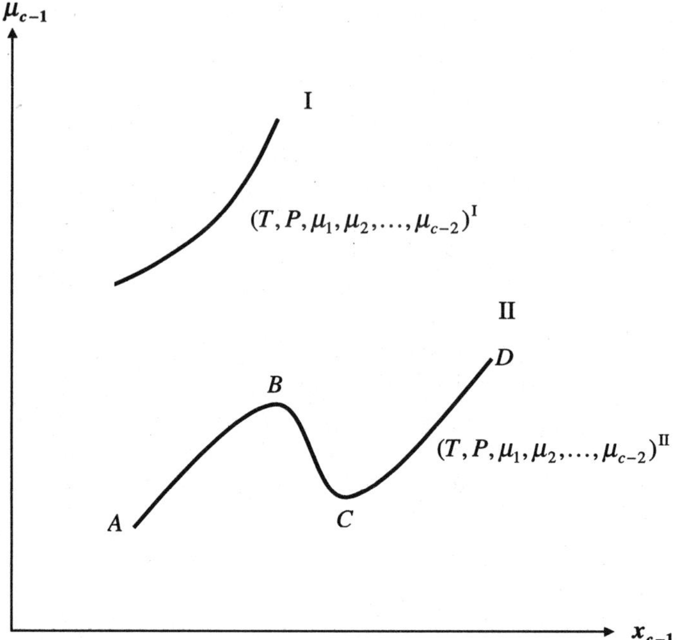

Figure 4.19   Plot of μ_{c-1} vs. x_{c-1}.

Criticality analysis

In the derivation of the criteria of criticality, we will first use a simple approach that relies on geometrical representation. Then an alternative approach will be presented which is more suitable for multicomponent systems with a large number of components.

Consider the vapor–liquid critical point of a single-component system by studying the $P\text{–}v$ graph shown in Fig. 4.20. In this figure, the spinodal curve is represented by the dashed curve and the binodal curve by the solid thick curve. The $P\text{–}v$ isotherms at four different temperatures $T_1, T_2, T_c,$ and T_3 are shown. Points B and C represent the limits of stability at temperature T_1. Points B' and C' represent the same limits at temperature T_2. According to the criteria of Eq. (4.115), the limits of stability at temperatures T_1 and T_2 are obtained from $(\partial P/\partial v)_{T_1} = 0$ and $(\partial P/\partial v)_{T_2} = 0$, respectively. Note that between points B and C, $(\partial P/\partial v)_{T_1} > 0$, and this is therefore the unstable segment of the isotherm. Also note that the curvature changes between B and C therefore, an inflection point where $(\partial P^2/\partial v^2)_{T_1} = 0$ exists between these two points.

There is another requirement for the inflection point; the third derivative (odd derivative) should be nonzero. At T_2 which is higher than T_1, points B' and C' are closer together than are points B and C, and the

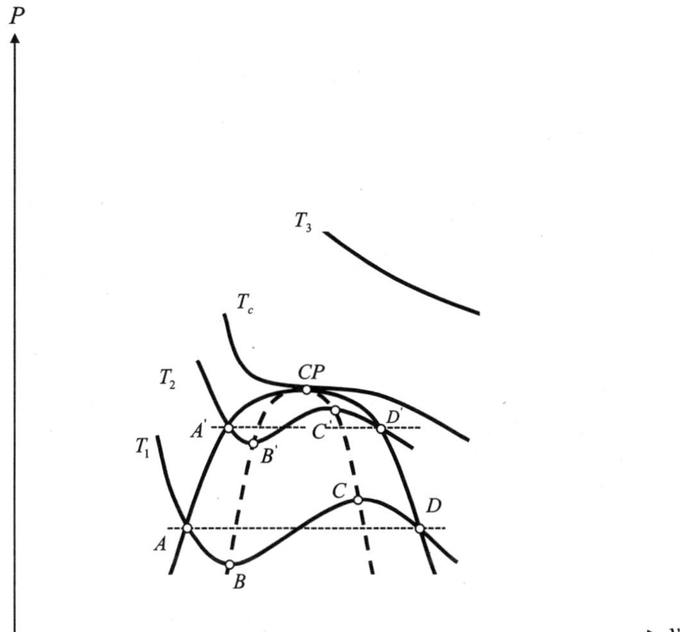

Figure 4.20 P–v plot for a single-component system.

inflection point $(\partial^2 P/\partial v^2)_{T_2} = 0$ is an unstable point. As the temperature approaches T_c, the two limits of stability and the inflection point coincide, and since the inflection point is now located on the binodal curve, the inflection point $(\partial^2 P/\partial v^2)_{T_c} = 0$ is a stable point. Points A and D and A' and D' represent equilibrium phases at T_1 and T_2, respectively. Toward the critical point, the points corresponding to A and A' and D and D' also coincide with the limit of stability. At the critical point, the gas and liquid phases can be transformed into each other without going through the two-phase region, which implies the continuity of gas and liquid states. The criteria for the critical point of a single-component system are, therefore,

$$\boxed{(\partial P/\partial v)_{T_c} = (\partial^2 P/\partial v^2)_{T_c} = 0, \qquad (\partial^3 P/\partial v^3)_{T_c} < 0}. \tag{4.196}$$

The last relationship in Eq. (4.196) implies that the critical point is neither a maximum nor a minimum. This inflection point from the algebraic standpoint has to have a nonvanishing odd-order derivative (see Example 4.3).

Now let us examine the graph of Helmholtz free energy, A vs. v, at constant temperature; $A(T, v)$ is a function of temperature and volume and a natural choice as the thermodynamic function for the criticality

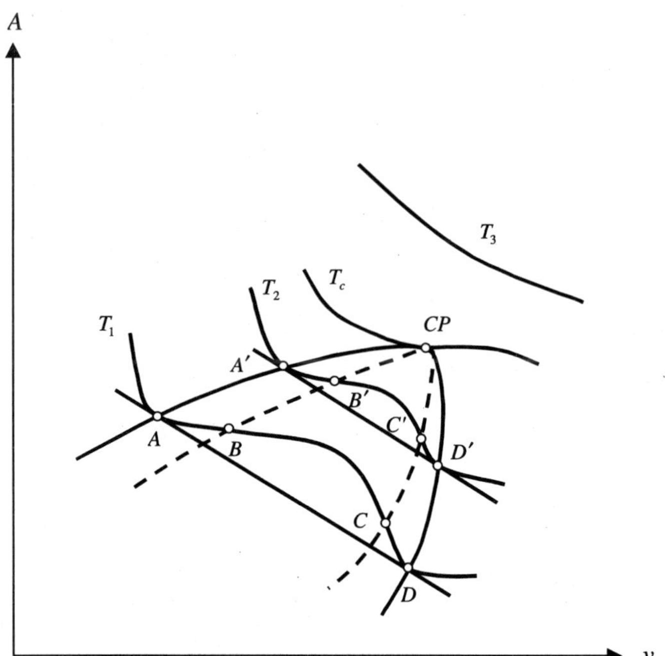

Figure 4.21   Plot of A vs. v for a single-component system.

criteria. Figure 4.21 shows a schematic diagram of the $A-v$ plot for a single-component system.

From the general criteria of stability given by Eq. (4.191), $y_{22}^{(1)} > 0$ for a single-component system. If $y^{(0)} = U$, then $y^{(1)} = A$ and $y_{22}^{(1)} = (\partial^2 A/\partial V^2)_{T,N}$ from $dA = -SdT - PdV + \mu dN$. The stability is, therefore, given by $(\partial^2 A/\partial V^2)_{T,N} > 0$ or $(\partial^2 A/\partial v^2)_T > 0$, and the limit of stability by $(\partial^2 A/\partial v^2)_T = 0$. Let us now examine the isotherm at T_1 in Fig. 4.21. In the stable segment, $(\partial^2 A/\partial v^2)_{T_1} > 0$, and in the segment between B and C, $(\partial^2 A/\partial v^2)_{T_1} < 0$, implying instability. Note that when $(\partial^2 A/\partial v^2)_{T_1} < 0$, the Helmholtz free energy curve is convex (like the surface of a ball as seen from inside). Points A and D are on the concave side, $(\partial^2 A/\partial v^2)_{T_1} > 0$, and are stable. Since these two points have a common tangent, $(\partial A/\partial v)_{T_1}|_{@A} = (\partial A/\partial v)_{T_1}|_{@D}$, they are at the same equilibrium pressure. Between B and C because of convexity, $(\partial^3 A/\partial v^3)_{T_1} > 0$. As the temperature rises, points B' and C' get closer. Finally, when $T = T_c$, points corresponding to B' and C' as well as A' and D' coincide. At the critical point,

$$\boxed{(\partial^2 A/\partial v^2)_{T_c} = (\partial^3 A/\partial v^3)_{T_c} = 0, \qquad (\partial^4 A/\partial v^4)_{T_c} > 0} \qquad (4.197)$$

At temperature, $T_3 > T_c$, $(\partial^2 A/\partial v^2) > 0$ and the system is stable.

Now let us examine the criteria of criticality for a multicomponent system. The criterion of stability is $y^{(c)}_{c+1,c+1} > 0$ (see Eq. (4.191)), and since $y^{(c)}_{c+1,c+1} = (\partial C_{c+1}/\partial X_{c+1})_{C_1,C_2,...,C_c,X_{c+2}}$, then for a phase to be stable,

$$(\partial C_{c+1}/\partial X_{c+1})_{C_1,C_2,...,C_c,X_{c+2}} > 0. \tag{4.198}$$

Let us examine the curve for C_{c+1} vs. X_{c+1} at fixed $C_1,\ldots,C_c,X_{c+2}$. Figure 4.22 shows three curves; for each curve, the $C_1,\ldots,C_c,X_{c+2}$ variables are fixed. The fixed variables for each curve are selected in such a way that curve I is in the two-phase region, curve II passes through the critical point, and curve III stays in the singe-phase region. For curve I, B and C are the stability limits; these two points are obtained from $(\partial C_{c+1}/\partial X_{c+1})_{C_1,C_2,...,C_c,X_{c+2}} = 0$. At the inflection point, E, between B and C, $(\partial^2 C_{c+1}/\partial X^2_{c+1})_{C_1,...,C_c,X_{c+2}} = 0$. Between B and C, $(C_{c+1}/\partial X_{c+1})_{C_1,...,C_c,X_{c+2}} < 0$, which implies instability. As X_{c+2} is varied, curve I may move upward and points B and C come closer together. When points B and C and inflection point E coincide,

$$\boxed{\begin{aligned}(\partial C_{c+1}/\partial X_{c+1})_{C_1,...,C_c,X_{c+2}} &= (\partial^2 C_{c+1}/\partial X^2_{c+1})_{C_1,...,C_c,X_{c+2}} = 0, \\ (\partial^3 C_{c+1}/\partial^3 X_{c+1})_{C_1,...,C_c,X_{c+2}} &> 0 \end{aligned}} \tag{4.199}$$

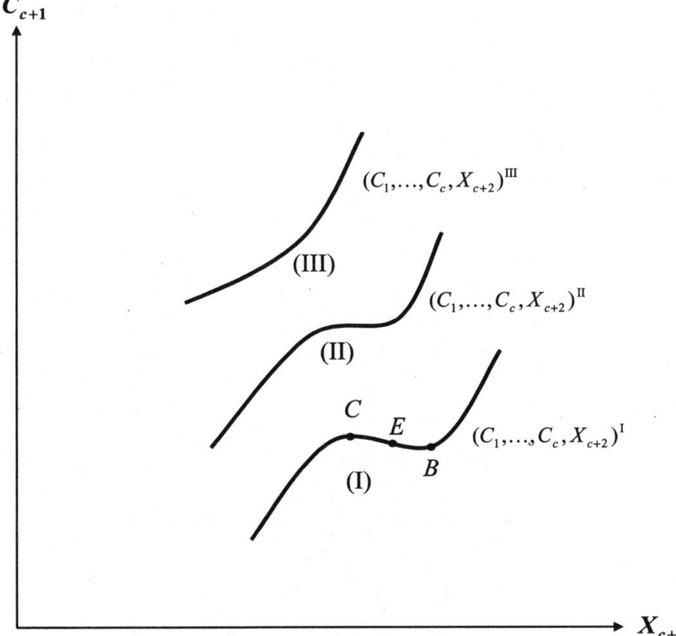

Figure 4.22 Plot of C_{c+1} vs. X_{c+1} for a c-component system.

In terms of $y^{(c)}$, the criteria of criticality are

$$y^{(c)}_{c+1,c+1} = 0, \qquad y^{(c)}_{c+1,c+1,c+1} = 0, \qquad y^{(c)}_{c+1,c+1,c+1,c+1,c+1} > 0. \quad (4.200)$$

The inequality provides the criterion that the critical point is an inflection point. If the third derivative vanishes, the next higher-order odd derivative should be positive. We will use the above criteria to establish the critical point for a pure component and for binary and multicomponent systems.

Single-component fluid. For a single component, $y^{(1)}_{22} = y^{(1)}_{222} = 0$ and $y^{(1)}_{2222} > 0$. From $y^{(0)} = U, dy^{(1)} = -SdT - PdV + \mu dN$ (see Eq. (1.175) of Chapter 1). Therefore, $y^{(1)}_2 = (\partial y^{(1)}/\partial V)_{T,N} = -P, y^{(1)}_{22} = -(\partial P/\partial V)_{T,N} = 0, y^{(1)}_{222} = (-\partial^2 P/\partial V^2)_{T,N} = 0,$ and $y^{(1)}_{2222} = (-\partial^3 P/\partial V^3)_{T,N} > 0$. Since $V = Nv$, then $(\partial P/\partial v)_T = (\partial^2 P/\partial v^2)_T = 0$ and $-(\partial^3 P/\partial v^3)_T > 0$ where $T = T_c$. These are the same expressions as Eqs. (4.196).

Two-component fluid From Eq. (4.200), $y^{(2)}_{33} = y^{(2)}_{333} = 0$ and $y^{(2)}_{3333} > 0$. Rewriting Eq. (1.176) of Chapter 1,

$$dy^{(2)} = dG = -SdT + VdP + \mu_1 dN_1 + \mu_2 dN_2, \qquad (4.201)$$

where $y^{(2)}_3 = (\partial y^{(2)}/\partial N_1)_{T,P,N_2} = \mu_1, y^{(2)}_{33} = (\partial \mu_1/\partial N_1)_{T,P,N_2} = 0, y^{(2)}_{333} = (\partial \mu_1^2/\partial N_1^2)_{T,P,N_2} = 0,$ and $y^{(2)}_{3333} = (\partial \mu_1^3/\partial N_1^3)_{T,P,N_2} > 0$. The above expressions in terms of mole fractions can be written as

$$\boxed{\begin{aligned} (\partial \mu_1/\partial x_1)_{T_c,P_c} &= (\partial \mu_1^2/\partial x_1^2)_{T_c,P_c} = 0 \\[4pt] (\partial \mu_1^3/\partial x_1^3)_{T_c,P_c} &> 0 \end{aligned}} , \qquad (4.202)$$

where $x_1 = N_1/(N_1 + N_2), (\partial \mu_1/\partial N_1)_{T,P,N_2} = (\partial \mu_1/\partial x_1)_{T,P}(\partial x_1/\partial N_1)_{T,P,N_2},$ and $(\partial x_1/\partial N_1)_{T,P,N_2} = (1 - x_1)/(N_1 + N_2)$.

In terms of fugacity, $(d\mu_1 = RTd\ln f_1)_T$, then

$$(\partial \mu_1/\partial x_1)_{T,P} = (RT/f_1)(\partial f_1/\partial x_1)_{T,P},$$

$$(\partial^2 \mu_1/\partial x_1^2)_{T,P} = RT[(-1/f_1^2)(\partial f_1/\partial x_1)^2_{T,P} + (1/f_1)(\partial^2 f_1/\partial^2 x_1)_{T,P}],$$

and for the third derivative a similar procedure is applied. The criteria of criticality are, therefore,

$$\boxed{(\partial f/\partial x_1)_{T_c,P_c} = 0, (\partial^2 f_1/\partial x_1^2)_{T_c,P_c} = 0, \text{ and } (\partial^3 f_1/\partial x_1^3)_{T_c,P_c} > 0} .$$

$$(4.203)$$

Multicomponent fluid   Equations (4.196) and (4.203) derived from the criteria of criticality are in a form that can readily be used to calculate the critical point of a single component and a binary mixture. For a three-component system, the criteria are of the form $y_{44}^{(3)} = 0$ and $y_{444}^{(3)} = 0$. Let us use $y^{(0)} = U(S, V, N_1, N_2, N_3)$. Then $dy^{(3)} = -SdT + VdP - N_1 d\mu_1 + \sum_{i=2}^{3} \mu_i dN_i$ (see Eqs. (1.21) and (1.181) of Chapter 1) and $y_4^{(3)} = \mu_2$. Therefore, at the critical point for a ternary mixture,

$$(\partial \mu_2 / \partial N_2)_{T,P,\mu_1,N_3} = 0 \tag{4.204}$$

$$(\partial^2 \mu_2 / \partial N_2^2)_{T,P,\mu_1,N_3} = 0. \tag{4.205}$$

The inequality constraint, $y_{4444}^{(3)} > 0$, results in $(\partial^3 \mu_2 / \partial N_2^3)_{T,P,\mu_1,N_3} > 0$, which is generally not tested. One may apply a stability analysis check on the critical point.

The above equations are not very practical to use! How can we keep μ_1 constant when a system is undergoing a change? An alternative procedure may be more convenient for the calculation of the critical point of mixtures with three or more components. Since we often use a pressure-explicit equation of state, $P = P(V, T, \underline{N})$, the following derivations are aimed towards the use of such equations. The first equation for criticality can be written as

$$y_{c+1,c+1}^{(c)} = \lambda^{(c+1)} / \lambda^{(c)}, \tag{4.206}$$

where λ's are defined in Eq. (4.187). Since $y_{c+1,c+1}^{(c)} = 0$ at the critical point, therefore

$$\lambda^{(c+1)} = 0. \tag{4.207}$$

The second equation for criticality is $y_{c+1,c+1,c+1}^{(c)} = 0$, which is $(\partial y_{c+1,c+1}^{(c)} / \partial X_{c+1}) = 0$. Taking the derivative of Eq. (4.206) with respect to X_{c+1},

$$y_{c+1,c+1,c+1}^{(c)} = \left[\lambda_{c+1}^{(c+1)} \lambda^{(c)} - \lambda_{c+1}^{(c)} \lambda^{(c+1)} \right] / \left[\lambda^{(c)} \right]^2 = 0. \tag{4.208}$$

Since $\lambda^{(c+1)} = 0$ at the critical point, therefore $\lambda_{c+1}^{(c+1)} = 0$, where the subscript $c + 1$ is used to denote the derivative of $\lambda^{(c+1)}$ with respect to variable X_{c+1}.

$$\lambda_{c+1}^{(c+1)} = \left[\partial \lambda^{(c+1)} / \partial X_{c+1} \right]_{C_1, \ldots, C_c, X_{c+2}} = 0. \tag{4.209}$$

Using the Jacobian transformations discussed in Chapter 1,

$$\lambda_{c+1}^{(c+1)} = \frac{\partial(C_1, \ldots, C_c, \lambda^{(c+1)}, X_{c+2})}{\partial(C_1, \ldots, C_c, X_{c+1}, X_{c+2})} = 0. \tag{4.210}$$

The above equation can be reduced to

$$\frac{\partial(C_2, \ldots, C_c, \lambda^{(c+1)})}{\partial(C_2, \ldots, C_c, X_{c+1})}\bigg|_{C_1, X_{c+2}} = 0. \tag{4.211}$$

The reduction is intended for the use the Helmholtz free energy $A = A(T, V, \underline{N})$. Dividing the numerator and denominator by $\partial(X_2, \ldots, X_{c+1})$ (see Eq. 1.185),

$$\frac{\partial(C_2, \ldots, C_c, \lambda^{(c+1)})/\partial(X_2, \ldots, X_c, X_{c+1})}{\partial(C_2, \ldots, C_c, \lambda_{c+1})/\partial(X_2, \ldots, X_c, X_{c+1})}\bigg|_{C_1, X_{c+2}} = 0. \tag{4.212}$$

In determinant form,

$$\frac{\begin{vmatrix} (\partial C_2/\partial X_2) & (\partial C_2/\partial X_3) & \cdots & (\partial C_2/\partial X_{c+1}) \\ \vdots & \vdots & \vdots & \vdots \\ (\partial C_c/\partial X_2) & (\partial C_c/\partial X_3) & \cdots & (\partial C_c/\partial X_{c+1}) \\ (\partial \lambda^{(c+1)}/\partial X_2) & (\partial \lambda^{(c+1)}/\partial X_3) & \cdots & (\partial \lambda^{(c+1)}/\partial X_{c+1}) \end{vmatrix}}{\begin{vmatrix} (\partial C_2/\partial X_2) & (\partial C_2/\partial X_3) & \cdots & (\partial C_2/\partial X_{c+1}) \\ \vdots & \vdots & \vdots & \vdots \\ (\partial C_c/\partial X_2) & (\partial C_c/\partial X_3) & \cdots & (\partial C_c/\partial X_{c+1}) \\ (\partial X_{c+1}/\partial X_2) & (\partial X_{c+1}/\partial X_3) & \cdots & (\partial X_{c+1}/\partial X_{c+1}) \end{vmatrix}} = 0. \tag{4.213}$$

Since $(\partial X_{c+1}/\partial X_i) = 0$ for $i = 2, \ldots, c$ and $= 1$ for $i = c + 1$, assuming that the resulting determinant in the denominator is not zero, then the numerator is zero. The $(\partial C_i/\partial X_j)$ elements are in the form $y_{ij}^{(1)}$, since for all of them C_1 and the Xs are held constant, i.e., $(\partial C_2/\partial X_2)_{C_1, X_3, \ldots, X_{c+2}} = y_{22}^{(1)}$ and $(\partial C_2/\partial X_3)_{C_1, X_2, X_4, \ldots, X_{c+2}} = y_{23}^{(1)}$ (see Eq. (1.183) of Chapter 1). From Eqs. (4.207) and (4.213), we can write the criticality conditions, respectively:

$$\lambda^{(c+1)} = \begin{vmatrix} y_{22}^{(1)} & y_{23}^{(1)} & \cdots & y_{2,c+1}^{(1)} \\ \vdots & \vdots & \vdots & \vdots \\ y_{c,2}^{(1)} & y_{c,3}^{(1)} & \cdots & y_{c,c+1}^{(1)} \\ y_{c+1,2}^{(1)} & y_{c+1,3}^{(1)} & \cdots & y_{c+1,c+1}^{(1)} \end{vmatrix} = 0 \tag{4.214}$$

and

$$
\begin{vmatrix}
y_{22}^{(1)} & y_{23}^{(1)} & \cdots & y_{2,c+1}^{(1)} \\
\vdots & \vdots & \vdots & \vdots \\
y_{c,2}^{(1)} & y_{c,3}^{(1)} & \cdots & y_{c,c+1}^{(1)} \\
\lambda_2^{(c+1)} & \lambda_3^{(c+1)} & \cdots & \lambda_{c+1}^{(c+1)}
\end{vmatrix} = 0 .
\tag{4.215}
$$

Since $y^{(1)} = A(T, V, N_1, \ldots, N_c)$, the elements of the above equations can be expressed as the derivatives of A with respect to V and $(N_1, N_2, \ldots, N_{c-1})$. The above set of equations were used by Baker and Luks (1980) to calculate the critical points of multicomponent systems.

Alternative approach for critical-point calculation

As we have already seen, the critical point can be calculated in a variety of related ways. However, the basic expressions that define the critical point involve setting two determinants equal to zero (that is, Eqs. (4.214) and (4.215)). Of the two determinants, one (that is, Eq. (4.215)) requires much effort, especially for multicomponent systems; it is necessary to evaluate the derivatives of certain determinants.

In 1980, Heidemann and Khalil introduced an alternative approach for the calculation of the critical point that is mathematically different from the expressions in Eqs. (4.214) and (4.215), but the concept is not different. As we have already seen, the critical point is a stable point at the limit of stability. Let us expand on this concept. Consider the P–v diagram of a multicomponent system of fixed composition sketched in Fig. 4.23. The thick solid curve on the left represents the bubblepoints and the thin solid curve on the right represents the dewpoints. The bubblepoints and dewpoints are stable equilibrium states; a perturbation in P, for example, results in a stable state. Critical point CP is the point at which bubblepoint and dewpoint converge and in addition to being a stable state, it is at the limit of stability. These two features were used by Gibbs in 1876 to derive the expressions for the critical point. We will repeat the use of stable and stability limit concepts in a more straightforward approach in this section.

Let us consider the Helmholtz free energy of the closed c-component system at constant T and V sketched in Fig. 4.24. The perturbed two-phase state consists of N' moles of the primed phase and N'' moles of the double-primed phase. We assume $N'' \ll N'$. Similar to the U

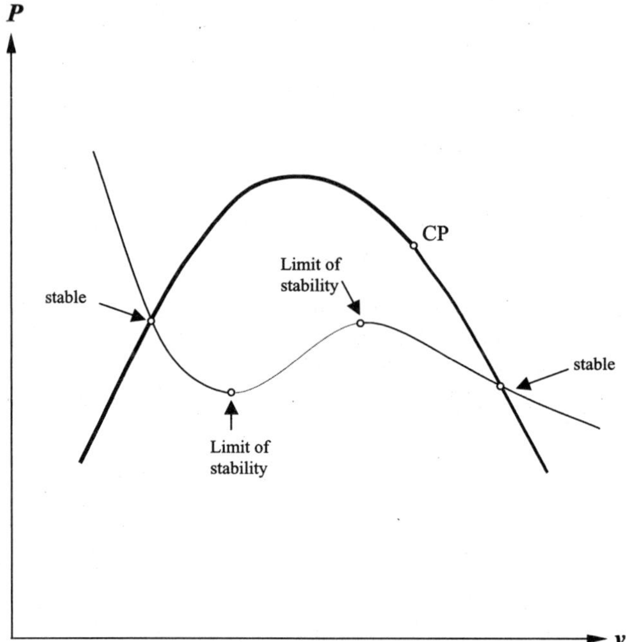

Figure 4.23   Critical point representation in a multicomponent system.

representation, the Helmholtz free energy change from single-phase to two-phase state is given by

$$\Delta A = \Delta A' + \Delta A'', \tag{4.216}$$

where $\Delta A'$ and $\Delta A''$ are the free energy changes of N' moles from the original single-phase state to the primed phase, and the free energy change of N'' moles from the original single-phase state to the double-primed phase, respectively. The restrictions of the Helmholtz free energy change are

$$\Delta V = \Delta V' + \Delta V'' = 0 \tag{4.217}$$

$$\Delta N_i = \Delta N_i' + \Delta N_i'' = 0 \qquad i = 1, \ldots, c \tag{4.218}$$

$$T = constant. \tag{4.219}$$

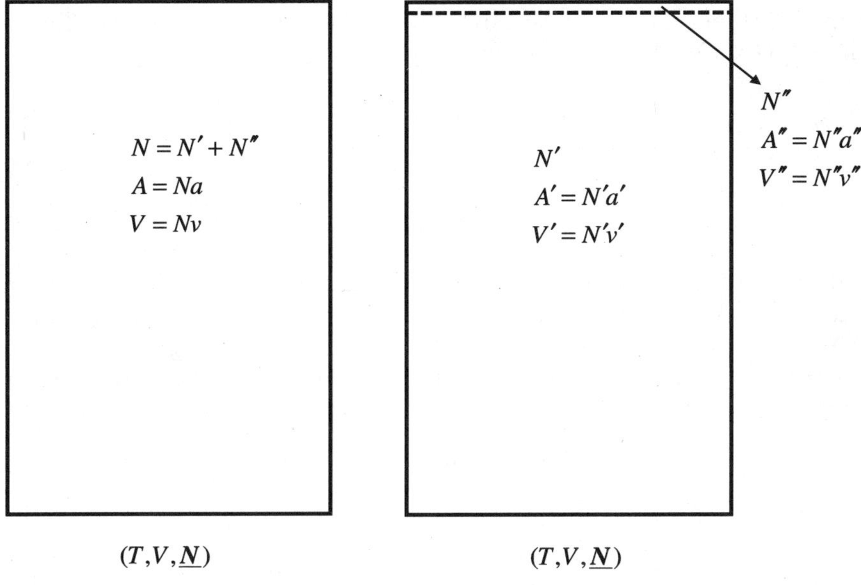

Figure 4.24 Stability of a multicomponent mixture at constant T, V, and $\underline{N}$.

Variables $\Delta V', \Delta V'', \Delta N_i', \Delta N_i''$ are defined in Eqs. (4.150) and (4.151). Manipulations similar to those for Eqs. (4.152) to (4.158) result in the following expression for ΔA:

$$\Delta A = 1/2![1 + (N'/N'')]\left\{ \sum_{i=1}^{c}\sum_{j=1}^{c}(\partial^2 A/\partial N_i \Delta N_j)(\Delta N_i \Delta N_j) + (\partial^2 A/\partial V^2)(\Delta V)^2 \right.$$

$$\left. + \sum_{i=1}^{c}(\partial^2 V/\partial N_i \partial V)(\Delta N_i \partial V) \right\}$$

$$+ 1/3![1 - (N'/N'')^2]\left\{ \sum_{i=1}^{c}\sum_{j=1}^{c}\sum_{k=1}^{c}(\partial^3 A/\partial N_i \partial N_j \partial N_k) \right.$$

$$\times (\Delta N_i \Delta N_j \Delta N_k) + (\partial^3 A/\partial V^3)(\Delta V)^3$$

$$\left. + \sum_{i=1}^{c}\sum_{j=1}^{c}(\partial^3 V/\partial V \partial N_i \partial N_j)(\Delta N_i \Delta N_j \Delta V) \right\} + \textit{higher-order terms.}$$

$$(4.220)$$

Let us represent the second- and third-order variation by $d^2 A$ and $d^3 A$ without the coefficients $[1 + (N'/N'')]$ and $[1 - (N'/N'')^2]$, respectively. For the original state to be stable, $\Delta A > 0$, which requires $d^2 A > 0$. If

$d^2A = 0$ then d^3A should be zero and $d^4A > 0$. Note that the coefficient of d^4A, $[1 + (N'/N'')^3]$, is positive. Therefore, at the critical point

$$d^2A = 0 \tag{4.221}$$

$$d^3A = 0 \tag{4.222}$$

$$d^4A > 0. \tag{4.223}$$

If $d^4A = 0$, d^5A should be zero, and the next higher-order even term, that is, d^6A should be positive, and so on. $d^2A = \Delta \underline{\mathbf{X}}^T A_H \Delta \underline{\mathbf{X}}$ where $\Delta \underline{\mathbf{X}}^T$ is the row vector $\Delta \underline{\mathbf{X}}^T = (\Delta V \ \Delta N_1 \ \Delta N_2 \ldots \Delta N_c)$, $\Delta \underline{\mathbf{X}}$ is the column vector $\Delta \underline{\mathbf{X}} = (\Delta V, \Delta N_1, \Delta N_2, \ldots, \Delta N_c)$, and A_H is the Hessian matrix A given by

$$A_H = \begin{pmatrix} A_{V,V} & A_{V,N_1} & \cdots & A_{V,N_c} \\ \vdots & & & \\ A_{N_c,V} & A_{N_c,N_1} & \cdots & A_{N_c,N_c} \end{pmatrix}. \tag{4.224}$$

For $d^2A > 0$, the matrix A_H should be positive definite. The matrix A_H is, positive definite if the principal submatrix B_H is positive definite (see Strang, 1988). The principal submatrix B_H is obtained by removing any ith row and ith column of A_H, $i = 1, \ldots, (c+1)$. Therefore, B_H is an $n \times n$ matrix, whereas A_H is an $(n+1)(n+1)$ matrix. Heidemann and Khalil (1980) based on physical grounds argue that when $\Delta V/V = (\Delta N_i/N_i) = k$ for all $i = 1, \ldots c$ (k is a constant), such variation does not qualify as a variation in the phase; the Helmholtz free energy A is simply multiplied by the constant k. Therefore, the quadratic, cubic, and higher-order terms can be contracted to eliminate the case of $\Delta V/V = \Delta N_i/N_i = k, i = 1, \ldots, c$. The criticality in the contracted (by removing volume variable) form can be written as

$$\sum_{i=1}^{c} \sum_{j=1}^{c} (\partial^2 A/\partial N_i \partial N_j)(\Delta N_i \Delta N_j)$$

$$= (\Delta N_1 \Delta N_2 \ldots \Delta N_c) \begin{pmatrix} A_{N_1,N_1} & \cdots & A_{N_1,N_c} \\ \vdots & & \vdots \\ A_{N_c,N_1} & \cdots & A_{N_c,N_c} \end{pmatrix} \begin{pmatrix} \Delta N_1 \\ \vdots \\ \Delta N_C \end{pmatrix} = \begin{pmatrix} 0 \\ \vdots \\ 0 \end{pmatrix} \tag{4.225}$$

and

$$\boxed{\sum_{i=1}^{c} \sum_{j=1}^{c} \sum_{k=1}^{c} (\partial^3 A/\partial N_i \partial N_j \partial N_k)(\Delta N_i \Delta N_j \Delta N_k) = 0}. \tag{4.226}$$

The inequality constraint given by Eq. (4.223) is generally not tested. Heidemann and Khalil (1980) have proposed the use Eqs. (4.225) and (4.226) to calculate the critical point. One distinct feature of the criticality criteria in terms of Eqs. (4.225) and (4.226) in comparison with Eqs. (4.214) and (4.215) is that the vector $\Delta \underline{N}$ should be evaluated. Equations (4.214) and (4.215) bypass this step. At the limit of stability, the quadratic term represented by Eq. (4.225) is positive semidefinite; there will be some variations in the phase for which the quadratic is positive, but for some other variations it will be exactly zero. These variations will make the cubic form in Eq. (4.226) zero.

From Eq. (4.225), it follows that the determinant

$$\begin{vmatrix} A_{N_1,N_1} & \cdots & A_{N_1,N_c} \\ \vdots & & \vdots \\ A_{N_c,N_1} & \cdots & A_{N_c,N_c} \end{vmatrix} = 0. \tag{4.227}$$

When the above determinant is zero, then from Eq. (4.225),

$$\begin{pmatrix} A_{N_1,N_1} & \cdots & A_{N_1,N_c} \\ \vdots & & \vdots \\ A_{N_c,N_1} & \cdots & A_{N_c,N_c} \end{pmatrix} \begin{pmatrix} \Delta N_1 \\ \vdots \\ \Delta N_c \end{pmatrix} = \begin{pmatrix} 0 \\ \vdots \\ 0 \end{pmatrix} \tag{4.228}$$

which can have a nonzero solution. By fixing one of the ΔN_is ($i = 1, \ldots, c$), other ΔN_is can be found from Eq. (4.228). With vector $\Delta \underline{N} = (\Delta N_1, \Delta N_2, \ldots, \Delta N_c)$, the cubic form of Eq. (4.226) can be solved. In other words, Eqs. (4.225) and (4.226) can provide two unknowns such as critical pressure and critical temperature or any other two unknowns. Note that Eq. (4.227), which provides the limit of stability, is the same as the result from Eq. (4.214), when instead of the vector $(V, \Delta N_1, \ldots, \Delta N_{c-1})$, the vector $(\Delta N_1, \ldots, \Delta N_c)$ is selected. Also note that in Eqs. (4.225) to (4.228), we could have selected the vector $(\Delta V, \Delta N_1, \ldots, \Delta N_{c-1})$. There are certain advantages in the vector of $\Delta N = (\Delta N_1, \ldots, \Delta N_c)$ (see Problem 4.11).

In the following, we will further discuss the use of Eqs. (4.225), (4.227), and (4.228) for the calculation of the critical point of a single-component system and binary and ternary mixtures. We will also use Eq. (4.215) for the binary mixture to show that more computational effort is required in the calculation of the critical point when the cubic Eq. (4.226) is not used.

Single-component fluid For a pure substance from Eq. (4.225), $d^2 A = A_{N,N}(\Delta N)^2 = 0$, and from Eq. (4.226) $d^3 A = A_{N,N,N}(\Delta N)^3 = 0$ are established at the critical point. Assign a nonzero value to ΔN, then

$A_{N,N} = 0$, and $A_{N,N,N} = 0$. From $dA = -SdT - PdV + \mu dN$, one obtains $(\partial \mu / \partial N)_{T,V} = 0$ and $(\partial^2 \mu / \partial N^2)_{T,V} = 0$. If ΔV is selected instead of ΔN, one obtains $(\partial P / \partial V)_{T,N} = 0$ and $(\partial^2 P / \partial V^2)_{T,N} = 0$; these are the same expressions as in Eq. (4.196).

Two-component fluid.   For a binary mixture from Eq. (4.227),

$$A_{N_1,N_1} A_{N_2,N_2} - A_{N_1,N_2}^2 = 0 \tag{4.229}$$

From Eq. (4.228),

$$A_{N_1,N_1} \Delta N_1 + A_{N_1,N_2} \Delta N_2 = 0 \tag{4.230}$$

$$A_{N_1,N_2} \Delta N_1 + A_{N_2,N_2} \Delta N_2 = 0 \tag{4.231}$$

Using Eq. (4.229), both Eqs. (4.230) and (4.231) collapse into the following equation,

$$A_{N_1,N_2} \Delta N_1 + A_{N_2,N_2} \Delta N_2 = 0 \tag{4.232}$$

We can set $\Delta N_1 = A_{N_1,N_2}$, then $\Delta N_2 = -A_{N_1,N_1}$. From Eq. (4.226),

$$A_{N_1,N_1,N_1}(\Delta N_1)^3 + A_{N_2,N_2,N_2}(\Delta N_2)^3 + 3A_{N_1,N_1,N_2}(\Delta N_1)^2(\Delta N_2)$$
$$+ 3A_{N_1,N_2,N_2}(\Delta N_1)(\Delta N_2)^2 = 0 \tag{4.233}$$

Substituting for ΔN_1 and ΔN_2 in Eq. (4.233), we obtain

$$A_{N_1,N_1,N_1}(A_{N_1,N_2})^3 - A_{N_2,N_2,N_2}(A_{N_1,N_1})^3 - 3A_{N_1,N_1,N_2}(A_{N_1,N_2})^2(A_{N_1,N_1})$$
$$+ 3A_{N_1,N_2,N_2}(A_{N_1,N_2})(A_{N_1,N_1})^2 = 0 \tag{4.234}$$

Substituting for A_{N_1,N_2}^2 from Eq. (4.229) in the first term of the above equation results in

$$A_{N_1,N_1,N_1}(A_{N_1,N_2} A_{N_2,N_2})^3 - A_{N_2,N_2,N_2}(A_{N_1,N_1})^2 - 3A_{N_1,N_1,N_2}(A_{N_1,N_2})^2$$
$$+ 3A_{N_1,N_2,N_2}(A_{N_1,N_2} A_{N_1,N_1}) = 0. \tag{4.235}$$

Equation (4.235) provides the same expression as Eq. (4.214) when ΔN_1 and ΔN_2 are selected in place of ΔV and ΔN_1. However, Eq. (4.214) requires much additional work. We need to find the derivative of the determinant in Eq. (4.214) with respect to N_1;

$$A_{N_1,N_1,N_2} A_{N_2,N_2} + A_{N_1,N_2,N_2} A_{N_1,N_1} - 2A_{N_1,N_2} A_{N_1,N_1,N_2}$$

and N_2;

$$A_{N_1,N_1,N_2} A_{N_2,N_2} + A_{N_1,N_1} A_{N_2,N_2,N_2} - 2A_{N_1,N_2} A_{N_1,N_2,N_2}.$$

Then these derivatives should be substituted in appropriate entries of the determinant in Eq. (4.215). We also need to combine the result with Eq. (4.229) to obtain Eq. (4.235).

Three-component fluid. For a three-component system, Eq. (4.227) results in

$$A_{N_1,N_1}(A_{N_2,N_2}A_{N_3,N_3} - A^2_{N_3,N_2}) - A_{N_1,N_2}(A_{N_1,N_2}A_{N_3,N_3} - A_{N_2,N_3}A_{N_1,N_3})$$

$$+ A_{N_1,N_3}(A_{N_1,N_2}A_{N_2,N_3} - A_{N_1,N_3}A_{N_2,N_2}) = 0. \tag{4.236}$$

From Eq. (4.228),

$$\begin{pmatrix} A_{N_1,N_1} & A_{N_1,N_2} & A_{N_1,N_3} \\ A_{N_2,N_1} & A_{N_2,N_2} & A_{N_2,N_3} \\ A_{N_3,N_1} & A_{N_3,N_2} & A_{N_3,N_3} \end{pmatrix} \begin{pmatrix} \Delta N_1 \\ \Delta N_2 \\ \Delta N_3 \end{pmatrix} = \begin{pmatrix} 0 \\ 0 \\ 0 \end{pmatrix}, \tag{4.237}$$

and from Eq. (4.226)

$$A_{N_1,N_1,N_1}(\Delta N_1)^3 + A_{N_2,N_2,N_2}(\Delta N_2)^3 + A_{N_3,N_3,N_3}(\Delta N_3)^2$$

$$+ 3A_{N_1,N_1,N_2}(\Delta N_1)^2(\Delta N_2)$$

$$+ 3A_{N_1,N_1,N_3}(\Delta N_1)^2(\Delta N_3) + 3A_{N_1,N_2,N_2}(\Delta N_1)(\Delta N_2)^2$$

$$+ 6A_{N_1,N_2,N_3}(\Delta N_1\Delta N_2\Delta N_3)$$

$$+ 3A_{N_1,N_3,N_3}(\Delta N_1)(\Delta N_3)^2 + 3A_{N_2,N_2,N_3}(\Delta N_2)^2(\Delta N_3)$$

$$+ 3A_{N_2,N_3,N_3}(\Delta N_2)(\Delta N_3)^2 = 0. \tag{4.238}$$

There are different ways to solve the above system of nonlinear equations (Eqs. (4.236) to (4.238)) to obtain two unknowns at the critical point. Heidemann and Khalil used Eq. (4.236) to solve for the critical temperature at a given critical volume, and then used the result to obtain ΔN_2 and ΔN_3 in Eq. (4.238). Note that one ΔN_i, e.g., ΔN_1, can be fixed. With values of $\underline{\Delta N}$ known, Eq. (4.238) is used to estimate critical volume. With known critical volume and critical temperature, the EOS is used to calculate critical pressure. Use of the nested calculations was successful for every mixture that had a vapor–liquid critical point, including mixtures with more than 40 components (Heidemann and Khalil, 1980). It should be pointed out that some mixtures may have more than one critical point for any given composition and some may have none.

Michelsen (1982b) and Michelsen and Heidemann (1981) have proposed methods for computational variation of the critical-point calculation. Michelsen's method relies on numerical differentiation of a single variable to obtain the cubic form of Eq. (4.226).

Examples and theory extension

Example 4.1   Write the expression for the Taylor's series expansion of the internal energy $U = U(\underline{X}) = U(X_1, X_2, \ldots, X_{c+2})$ around point $\underline{X}$ in vector representation.

Solution   The Taylor's series expansion of U around the perturbed point $\underline{X}$ can be written as

$$U(\underline{X} + \Delta\underline{X}) = U(\underline{X}) + \sum_{i=1}^{c+2} U_i \Delta X_i + \frac{1}{2!} \sum_{i=1}^{c+2}\sum_{j=1}^{c+2} U_{ij}\Delta X_i \Delta X_j + \textit{higher-order terms,}$$

where $U_i = (\partial U/\partial X_i)$ and $U_{ij} = (\partial^2 U/\partial X_i \partial X_j)$. The second term on the right side of the above equation can be written as the inner product or the dot product of the transpose vector $\Delta\underline{X}^T = (\Delta X_1 \Delta X_2 \ldots \Delta X_{c+2})$ and the gradient vector $\nabla U = (\partial U/\partial X_1, \partial U/\partial X_2 \ldots \partial U/\partial X_{c+2})$.

$$\Delta\underline{X}^T \nabla U = (\Delta X_1 \Delta X_2 \ldots \Delta X_{c+2}) \begin{pmatrix} \partial U/\partial X_1 \\ \partial U/\partial X_2 \\ \vdots \\ \partial U/\partial X_{c+2} \end{pmatrix} = \sum_{i=1}^{c+2} U_i \Delta X_i$$

Matrix multiplication is used to derive this equation. We also use matrix multiplication to show that

$$\Delta\underline{X}^T H_U \Delta\underline{X} = (\Delta X_1 \Delta X_2 \ldots \Delta X_{c+2}) \begin{pmatrix} U_{11} & U_{12} & \cdots & U_{1,c+2} \\ U_{21} & U_{22} & \cdots & U_{2,c+2} \\ \vdots & \vdots & & \vdots \\ U_{c+2,1} & U_{c+2,2} & \cdots & U_{c+2,c+2} \end{pmatrix} \begin{pmatrix} \Delta X_1 \\ \Delta X_2 \\ \vdots \\ \Delta X_{c+2} \end{pmatrix}$$

$$= \left(\sum_{i=1}^{c+2} U_{i1}\Delta X_i \sum_{i=1}^{c+2} U_{i2}\Delta X_i \ldots \sum_{i=1}^{c+2} U_{i,c+2}\Delta X_i \right) \begin{pmatrix} \Delta X_1 \\ \Delta X_2 \\ \vdots \\ \Delta X_{c+2} \end{pmatrix}$$

$$= \sum_{i=1}^{c+2}\sum_{j=1}^{c+2} U_{ij}\Delta X_i \Delta X_j$$

Example 4.2   The initial estimate of the K_i values for the calculation of three-phase vapor–liquid equilibria (VLLE) can be made from the following expressions.

$$K_i^V = y_i/x_i^I = \varphi_i^I(\boldsymbol{z})/\varphi_{pure\ i}^V$$

$$K_i^L = x_i^I/x_i^{II} = \varphi_{pure\ i}^L/\varphi_i^I(\boldsymbol{z})$$

In the above expressions x_i^I is the mole fraction of component i in the reference phase, which can be the light liquid phase. The definition of other parameters are: x_i^{II} is the mole fraction of component i in the heavy liquid phase (phase II); y_i the mole fraction of component i in the vapor phase; $\varphi_i^I(z)$ and $\varphi_i^{II}(z)$ are the fugacity coefficients of component i in the reference phase I and phase II, respectively with the overall composition z; $\varphi_{pure\ i}^V$ and $\varphi_{pure\ i}^L$ are the fugacity coefficients of pure component i in the vapor and liquid phases, respectively. Trebble (1989) suggested the above expressions for VLLE calculations. What is the basis of those two expressions?

Solution   Consider Eq. (4.70), and assume that the feed composition is the same as the light liquid phase composition $x_i^I = z_i$. If the amounts of the vapor and heavy liquid phases are very small, then from Eq. (4.70),

$$\ln Y_i = \ln x_i^I - \ln \varphi_i(y) + \ln \varphi_i(x^I)$$

$$\ln X_i^{II} = \ln x_i^I - \ln \varphi_i(x^{II}) + \ln \varphi_i(x^I)$$

Now assume that $\varphi_i(y) = \varphi_{pure\ i}^V$, and $\varphi_i(x^{II}) = \varphi_{pure\ i}^L$, then from the above two expressions,

$$K_i^{VL} = \frac{Y_i}{x_i^I} = \frac{\varphi_i(x^I)}{\varphi_{pure\ i}^V}$$

$$K_i^{LL} = \frac{x_i^I}{X_i^{II}} = \frac{\varphi_{pure\ i}^L}{\varphi_i(x^I)}$$

Note that $x_i^I = z_i$, and Y_i and X_i^{II} are close to normalized values y_i, and x_i^{II}, respectively.

Example 4.3   Derive the criteria for the maximum and minimum of a single-variable function.

Solution   Let us write the Taylor's series expansion of $f(x)$ around $x = 0$,

$$f(x) = \sum_{n=0}^{\infty} \frac{f^{(n)}(0)}{n!} x^n = f(0) + \sum_{n=1}^{\infty} \frac{f^{(n)}(0)}{n!} x^n,$$

where $f^{(0)}(0) = f(0)$, and $f^{(1)}, f^{(2)}, \ldots$, are the first-, second-, and higher-order derivatives of f, respectively. At the stationary point, $f^{(1)}(0) = 0$; to develop the criteria for a maximum or minimum, one needs to examine the higher-order derivatives at $x = 0$. Close to the point $x = 0$,

$$f(x) - f(0) \approx \frac{f^{(m)}(0)x^m}{m!},$$

where m is the first nonvanishing term of the Taylor's series expansion. If m is odd, the function $f^{(m)}(0)x^m/m!$ may look like the graph in Fig. 4.25a. The function $f(x)$ has an inflection point at $x = 0$. However, if m is even, depending on the sign of $f^{(m)}(0)x^m/m!$, one may have either a maximum or a minimum. If $f^{(m)}(0)$ is positive, the graph of $f(x) - f(0)$ vs. x may look like the sketch in Fig.

4.25b and if $f^{(m)}(0)$ is negative, $f^{(m)}(0)x^m/m!$ vs. x may look like the sketch in Fig. 4.25c. Therefore, for $f(x)$ or $f(x) - f(0)$ to have a maximum at $x = 0$, the first nonvanishing higher-order odd derivative of $f(x)$ should be negative (see Fig. 4.25b). Similarly, for $f(x)$ or $f(x) - f(0)$ to have a minimum at $x = 0$, the first nonvanishing higher-order even derivative of $f(x)$ should be positive (see Fig. 4.25c).

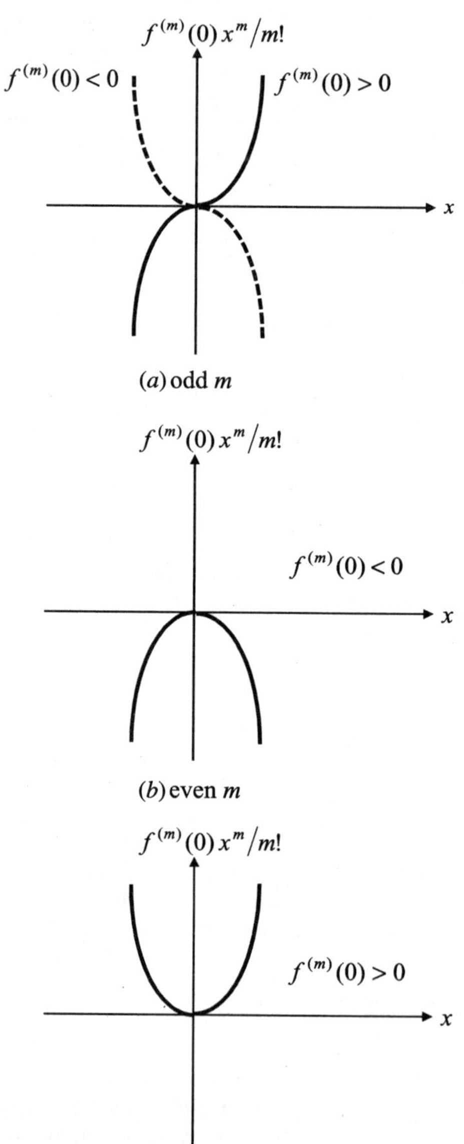

Figure 4.25 Plots of $f(x) - f(0) \approx f^{(m)}(0)x^m/m!$ vs. x.

Example 4.4   Derive the following expressions:

$$N'\left(\frac{\partial^2 U}{\partial N_i \partial N_j}\right)' = N''\left(\frac{\partial^2 U}{\partial N_i \partial N_j}\right)''$$

$$(N')^2\left(\frac{\partial^3 U}{\partial S^3}\right)' = (N'')^2\left(\frac{\partial^3 U}{\partial S^3}\right)''$$

$$(N')^2\left(\frac{\partial^3 U}{\partial N_i \partial N_j \partial N_k}\right) = (N'')^2\left(\frac{\partial^3 U}{\partial N_i \partial N_j \partial N_k}\right)''.$$

Solution   To derive the first expression, we write

$$(\partial^2 U/\partial N_i \partial N_j)' = \partial/\partial N_i'(\partial U/\partial N_j)',$$

where $(\partial U/\partial N_j)'$ is at the original state and is equal to μ_j. Similarly, $(\partial U/\partial N_j)''$ is at the original state and is equal to μ_j. Therefore,

$$(\partial^2 U/\partial N_i \partial N_j)' = (\partial \mu_j/\partial N_i').$$

Next we can write $\partial/\partial N_i' = (\partial/\partial x_i)(\partial x_i/\partial N_i')$, where x_i is the mole fraction of component i in the original single-phase state. Since $x_i = N_i'/\sum_{i=1}^c N_i' = N_i'/N'$, then

$$\frac{\partial}{\partial x_i}\left[\frac{\partial x_i}{\partial N_i'}\right] = \frac{\partial}{\partial x_i}\left[\frac{\partial}{\partial N_i'}\left(\frac{N_i'}{N'}\right)\right] = \frac{1}{N'}\frac{\partial}{\partial x_i}(1 - x_i)$$

and therefore,
$$\boxed{\frac{\partial}{\partial N_i'} = \frac{1}{N'}\frac{\partial}{\partial x_i}(1 - x_i)}.$$

Similarly,
$$\boxed{\frac{\partial}{\partial N_i''} = \frac{1}{N''}\frac{\partial}{\partial x_i}(1 - x_i)}.$$

Combining the above results would lead to the first expression. The derivation of the second expression is straightforward:

$$(\partial^3 U/\partial S^3)' = \partial/\partial S'[\partial^2 U/\partial S^2]' = (\partial/\partial S')(\partial T/\partial S)'$$

Since at the original state, $S' = N's$, then

$$(\partial^3 U/\partial S^3)' = (1/N')^2(\partial^2 T/\partial s^2).$$

Similarly,
$$(\partial^3 U/\partial S^3)'' = (1/N'')^2(\partial^2 T/\partial s^2).$$

Therefore,
$$(N')^2(\partial^3 U/\partial S^3)' = (N'')^2(\partial^3 U/\partial S^3)''.$$

The third expression is derived in a similar manner. Higher-order derivatives follow the same trend:

$$(N')^{l-1}(\partial^l U/\partial S^l)' = (N'')^{l-1}(\partial^l U/\partial S^l)''.$$

Example 4.5 *Criteria of criticality in terms of derivatives of G* Derive the criteria of criticality in terms of the derivatives of G.

Solution We can start from Eq. (4.210) and reduce it to

$$\left.\frac{\partial(C_3, \ldots, C_c, \lambda^{c+1})}{\partial(C_3, \ldots, C_c, X_{c+1})}\right|_{C_1, C_2, X_{c+2}} = 0.$$

Dividing the numerator and denominator by $\partial(X_3, \ldots, X_{c+1})$,

$$\left.\frac{\partial(C_3, \ldots, C_c, \lambda^{(c+1)})/\partial(X_3, \ldots, X_c, X_{c+1})}{\partial(C_3, \ldots, C_c, X_{c+1})/\partial(X_3, \ldots, X_c, X_{c+1})}\right|_{C_1, C_2, X_{c+2}} = 0.$$

Since C_1, C_2, and the Xs are held constant, $(\partial C_3/\partial X_3)_{C_1, C_2, X_4, \ldots, X_{c+2}} = y^{(2)}_{33}$, $(\partial C_3/\partial X_{c+1})_{C_1, C_2, \ldots, X_c, X_{c+2}} = y^{(2)}_{3,c+1}$, etc. (see Eq. (1.183) of Chapter 1). $\lambda^{(c+1)} = 0$ can be expressed as

$$\lambda^{(c+1)} = \begin{vmatrix} y^{(2)}_{33} & y^{(2)}_{34} & \cdots & y^{(2)}_{3,c+1} \\ \vdots & \vdots & \vdots & \vdots \\ y^{(2)}_{c,3} & y^{(2)}_{c,4} & \cdots & y^{(2)}_{c,c+1} \\ y^{(2)}_{c+1,3} & y^{(2)}_{c+1,4} & \cdots & y^{(2)}_{c+1,c+1} \end{vmatrix} = 0, \tag{E4.5.1}$$

and the second criterion becomes

$$\begin{vmatrix} y^{(2)}_{33} & y^{(2)}_{34} & \cdots & y^{(2)}_{3,c+1} \\ \vdots & \vdots & \vdots & \vdots \\ y^{(2)}_{c,3} & y^{(2)}_{c,4} & \cdots & y^{(2)}_{c,c+1} \\ \lambda^{(c+1)}_3 & \lambda^{(c+1)}_4 & \cdots & \lambda^{(c+1)}_{c+1} \end{vmatrix} = 0. \tag{E4.5.2}$$

From $y^{(2)} = G(T, P, \underline{N})$, one obtains $y^{(2)}_{33} = (\partial^2 G/\partial N_1)^2_{T, P, N_1}$ and all the other elements of the two determinants above. Therefore, only derivatives of G with respect to $\underline{N}$ are required. These are the set of equations that were used by Peng and Robinson (1977) to predict the critical points of multicomponent systems.

Example 4.6   Show that for a stable binary homogeneous phase, the following inequalities should hold:

$$(a)\quad (\partial^2 g/\partial T^2)_{P,x_1} < 0$$

$$(b)\quad (\partial^2 g/\partial P^2)_{T,x_1} < 0.$$

Solution   For a stable homogeneous phase, whether a pure component, a binary mixture, or a multicomponent system, the heat capacities c_P and c_V and the isothermal compressibility C_T should be positive. The derivations for a single-component system were presented in the text. Example 4.10 provides the derivation for C_T. The derivations for c_V and c_P are straightforward (see Problem 4.8).

Let us write the expression for dG of a multicomponent system,

$$dG = -SdT + VdP + \sum_{i=1}^{c} \mu_i dn_i.$$

At constant moles,

$$[dG = -SdT + VdP]_{\underline{n}}.$$

Then

$$(\partial G/\partial T)_{P,\underline{n}} = -S$$

and

$$(\partial^2 G/\partial T^2)_{P,\underline{n}} = -(\partial S/\partial T)_{P,\underline{n}}.$$

But $(\partial S/\partial T)_{P,\underline{n}} = nc_P/T$ (see Eq. (3.110)).
Since $c_P > 0$, then

$$(\partial^2 G/\partial T^2)_{P,\underline{n}} < 0$$

and for a binary system to be stable

$$(\partial^2 g/\partial T^2)_{P,x_1} < 0.$$

The second relationship, $(\partial^2 g/\partial P^2)_{T,x_1} < 0$, can be obtained from

$$(\partial G/\partial P)_{T,\underline{n}} = V$$

and

$$(\partial^2 G/\partial P^2)_{T,\underline{n}} = (\partial V/\partial P)_{T,\underline{n}}.$$

Since $(\partial V/\partial P)_{T,\underline{n}} = -VC_T$ and $C_T > 0$ for a stable system, then

$$(\partial^2 g/\partial P^2)_{T,x_1} < 0.$$

Example 4.7   *Stability of pure substances and binary mixtures*   The purpose of this example is to have an appreciation of supersaturation in pure hydrocarbons and binary mixtures. Supersaturation can be defined as the difference between the equilibrium pressure at the saturation point and the pressure at which an infinitesimal amount of a new phase is formed. The temperature is held constant in the process.

(a) Consider pure hydrocarbons nC_5, and nC_7. Calculate the limits of stability in the gas and liquid states. Plot the results on a $P-T$ diagram and compare them with the vapor pressure curve. Skripov and Ermakov (1964) have measured the limit of stability of liquid nC_5 and nC_7. Compare the results of calculation with the measured data of these authors.

(b) Compute the stability limits of a mixture of 70% C_1 and 30% nC_5, a mixture of 70% C_1 and 30% nC_7, and a mixture of 70% C_1 and 30% nC_{10} both in the gas and liquid states. Plot the results on a $P-T$ diagram. Show also the saturation pressures of the above systems on the $P-T$ plots.

Solution   (a) For a pure substance, the stability limit is given by the first expression of Eq. (4.196), $(\partial P/\partial v)_T = 0$. The PR-EOS can be used to evaluate $(\partial P/\partial v)_T = 0$. The results are presented in Fig. 4.26 for both the vapor and liquid phases. In the same figure, the experimental data of Skripov and Ermakov (1964) for the liquid-phase stability limit are also presented. Note that there is qualitative agreement between the experimental data and calculated results. The difference between the pressure at the limit of stability and the vapor pressure at a given temperature is the maximum supersaturation in pressure. Note that close to the critical point, supersaturation is small.

Figure 4.27 provides the calculated limit of stability for the gas and liquid phases, the vapor pressure, and the measured limit of stability for the liquid phase (Skripov and Ermakov, 1964) for nC_7. The PR-EOS was used in the stability limit calculations. The results presented in Fig. 4.27 for nC_7 are very similar to the results in Fig. 4.26 for nC_5. Similarly to the nC_5, the supersaturation in pressure for nC_7 (at a given temperature) is higher for the liquid phase than the gas phase. Note that because of supersaturation, the pressure in the liquid phase can be zero or even negative.

(b) The limit of stability for a binary mixture is calculated from $(\partial f_1/\partial x_1)_{T,P} = 0$ (the first expression in Eq. (4.203)). The results of the calculation for the C_1/nC_5 mixture are presented in Fig. 4.28a. The same figure also shows the saturation pressure. Note that on the bubblepoint side over a large range of pressure in the neighborhood of the critical point, the supersaturation in pressure is negligible.

Figures 4.28b and 4.28c show both the stability limits and saturation pressures for the C_1/nC_7 and C_1/nC_{10} mixtures. The results are very similar to those presented in Fig. 4.28a for the C_1/nC_5 mixture. The maximum supersaturation in pressure for the liquid state becomes more pronounced for the C_1/nC_{10} mixture than for the C_1/nC_7 and C_1/nC_5 mixtures.

Example 4.8   *Various derivatives of the Helmholtz free energy for the PR-EOS* Derive the expressions for the following derivatives using the PR-EOS: $(\partial A/\partial V)$, $(\partial^2 A/\partial V^2)$, $(\partial^3 A/\partial V^3)$, $(\partial A/\partial N_i)$, $(\partial^2 A/\partial V \partial N_i)$, $(\partial^2 A/\partial N_i \partial N_j)$, $(\partial^3 A/\partial V \partial N_i \partial N_j)$, and $(\partial^3 A/\partial N_i \partial N_j \partial N_k)$.

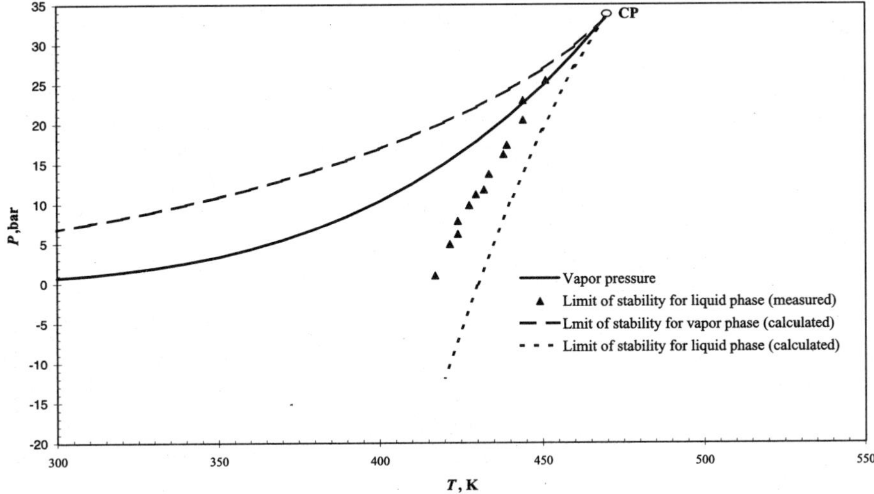

Figure 4.26   Limit of stability for nC_5 in the gas and liquid states.

Solution   Software such as Mathematica have made life easy for deriving the expressions for derivatives of the equations of state. Earlier, such softwares were not available and apparently Baker and Luks (1980) spent considerable effort to derive the above derivations for the SRK-EOS, although the derivations for this EOS are much simpler than those of the PR-EOS. In the following, the derivatives from Mathematica are presented.

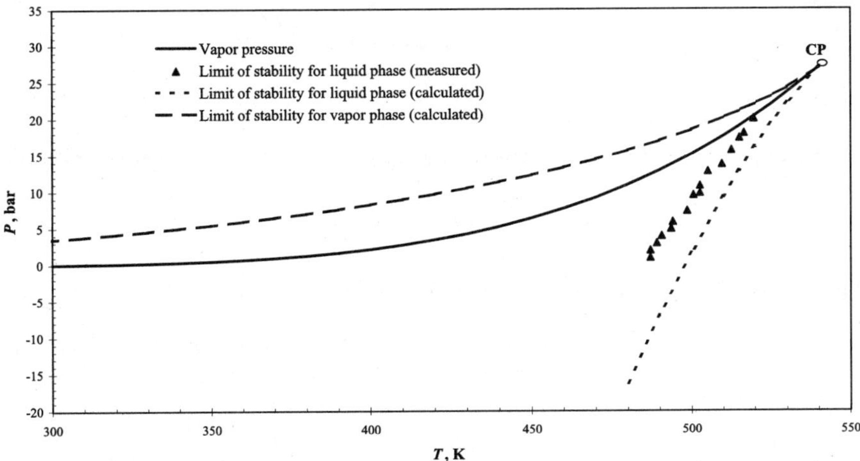

Figure 4.27   Limit of stability for nC_7 in the gas and liquid states.

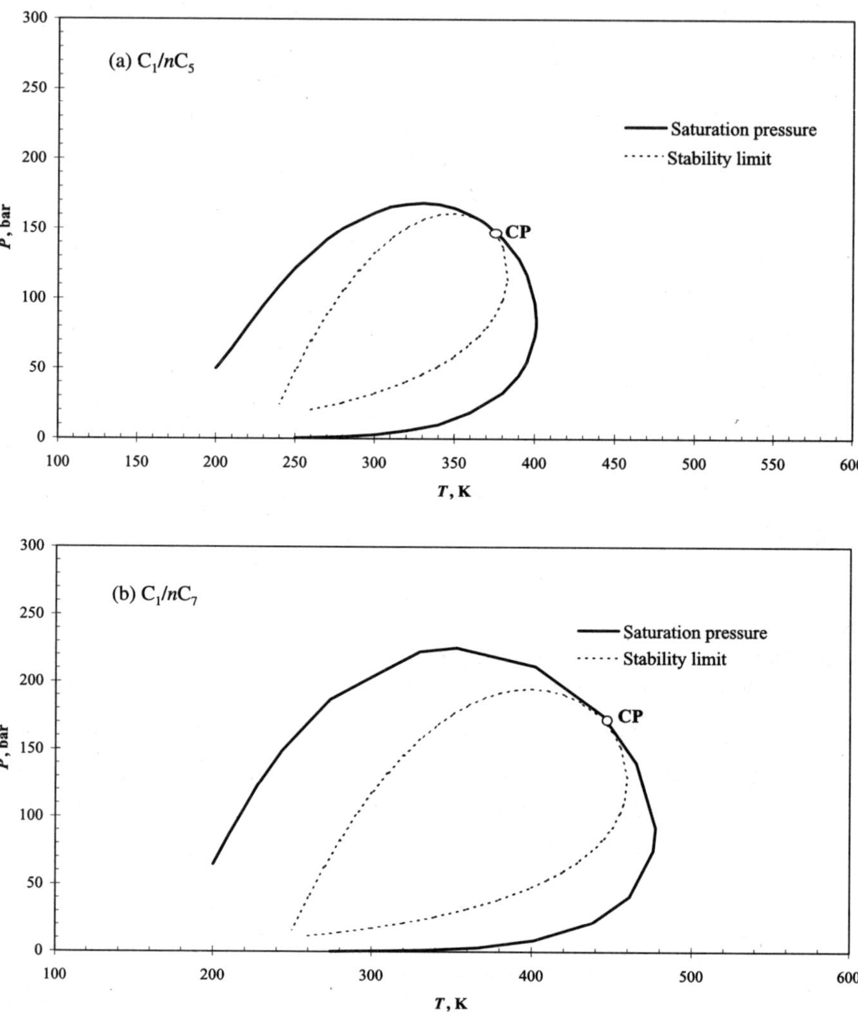

Figure 4.28 Calculated saturation pressures and stability limits for three binary hydrocarbon mixtures ($x_{c1} = 70\%$).

The PR-EOS can be written as (see Eq. 3.6)

$$P = \frac{NRT}{V - \mathscr{B}} - \frac{\mathscr{A}}{V(V + \mathscr{B}) + \mathscr{B}(V - \mathscr{B})},$$

where

$$\mathscr{A} = N^2 a = \sum_{i=1}^{c} \sum_{j=1}^{c} N_i N_j a_{ij}$$

$$\mathscr{B} = Nb = \sum_{i=1}^{c} N_i b_i.$$

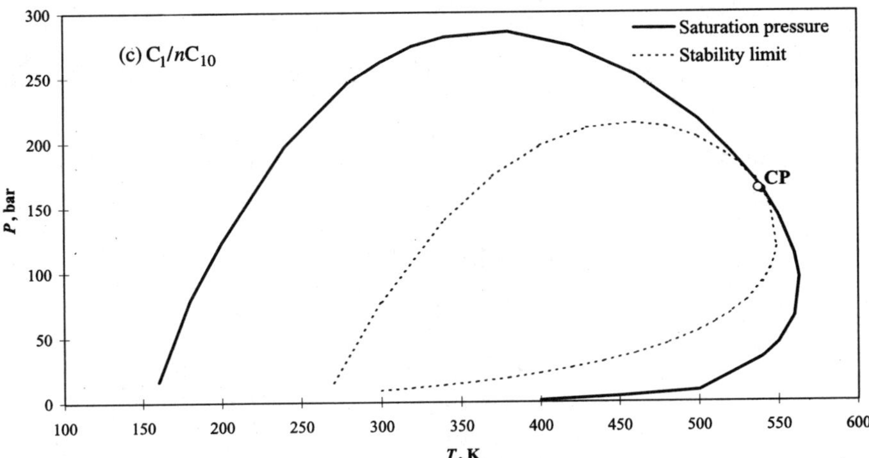

Figure 4.28   (*Continued*)

Introducing the above equation into the expression given in Example 1.5 of Chapter 1 yields (using Mathematica)

$$A = -NRT \ln \frac{V - \mathscr{B}}{RT} - \frac{\mathscr{A}}{2\sqrt{2}\mathscr{B}} \ln \frac{V + (\sqrt{2}+1)\mathscr{B}}{V - (\sqrt{2}-1)\mathscr{B}} + RT \sum_{i=1}^{c} N_i \ln N_i$$

$$+ \sum_{i=1}^{c} N_i(u_i^0 - Ts_i^0)$$

The various derivatives of A are also obtained by using Mathematica:

$$(\partial A/\partial V) = -\frac{NRT}{V - \mathscr{B}} + \frac{\mathscr{A}}{V^2 + 2V\mathscr{B} - \mathscr{B}^2}$$

$$(\partial^2 A/\partial V^2) = \frac{NRT}{(V - \mathscr{B})^2} - \frac{2\mathscr{A}(V + \mathscr{B})}{(V^2 + 2V\mathscr{B} - \mathscr{B}^2)^2}$$

$$(\partial^3 A/\partial V^3) = -\frac{2NRT}{(V - \mathscr{B})^3} + \frac{2\mathscr{A}(3V^2 + 6V\mathscr{B} + 5\mathscr{B}^2)}{(V^2 + 2V\mathscr{B} - \mathscr{B}^2)^3}$$

$$(\partial A/\partial N_i) = \mu_i = RT\left[1 + \ln N_i + \frac{Nb_i}{V - \mathscr{B}} - \ln \frac{V - \mathscr{B}}{RT}\right]$$

$$+ \left[\frac{AVb_i}{\mathscr{B}^2(\mathscr{B}^2 - 2\mathscr{B}V - V^2)} + \frac{(b_i A - A_i'\mathscr{B})}{2\sqrt{2}\mathscr{B}} \ln \frac{V + (\sqrt{2}+1)\mathscr{B}}{V - (\sqrt{2}-1)\mathscr{B}}\right]$$

$$+ (u_i^0 - Ts_i^0)$$

$$(\partial^2 A/\partial V \partial N_i) = -\frac{RT(V - \mathscr{B} + Nb_i)}{(V - \mathscr{B})^2} + \frac{(V^2 + 2V\mathscr{B} - \mathscr{B}^2)\mathscr{A}_i' + 2\mathscr{A}b_i(\mathscr{B} - V)}{(V^2 + 2V\mathscr{B} - \mathscr{B}^2)^2}$$

$$(\partial^3 A/\partial V^2 \partial N_i) = \frac{RT(2Nb_i + V - B)}{(V - \mathscr{B})^3} - \frac{2}{(V^2 + 2V\mathscr{B} - \mathscr{B}^2)^3}$$
$$\times \left[(V^3 + 3V^2\mathscr{B} + V\mathscr{B}^2 - \mathscr{B}^3)\mathscr{A}'_i + \mathscr{A}b_i(-3V^2 + 2V\mathscr{B} + 3\mathscr{B}^2)\right]$$

$$(\partial^2 A/\partial N_i \partial N_j) = \frac{RT\delta_{ij}}{N_j} + \frac{RT}{(V - \mathscr{B})^2}\left[(V - \mathscr{B})b_i + (V - \mathscr{B} + Nb_i)b_j\right]$$
$$+ \frac{1}{2\sqrt{2}\mathscr{B}^3}\left\{[\mathscr{B}(\mathscr{A}'_i b_j + \mathscr{A}'_j b_j - 2\mathscr{B}a_{ij}) - 2\mathscr{A}b_i b_j]\right.$$
$$\times \ln \frac{V + (\sqrt{2} + 1)\mathscr{B}}{V - (\sqrt{2} - 1)\mathscr{B}}$$
$$- \frac{2\sqrt{2}\mathscr{B}V}{V^2 + 2\mathscr{B}V - B^2}[\mathscr{B}(\mathscr{A}'_i b_j + \mathscr{A}'_j b_i) - 2\mathscr{A}b_i b_j]$$
$$\left.+ \frac{4\sqrt{2}\mathscr{A}\mathscr{B}^2 V(V - b)b_i b_j}{(V^2 + 2\mathscr{B}V - \mathscr{B}^2)^2}\right\}$$

$$(\partial^3 A/\partial V \partial N_i \partial N_j) = \frac{-RT}{(V - \mathscr{B})^3}\left[(V - b)(b_i + b_j) + 2Nb_i b_j\right]$$
$$+ \frac{2}{(V^2 + 2\mathscr{B}V - \mathscr{B}^2)^3}\left\{(V^2 + 2\mathscr{B}V - \mathscr{B}^2)[a_{ij}(V^2 + 2\mathscr{B}V - \mathscr{B}^2)\right.$$
$$+ \mathscr{A}'_i b_j(\mathscr{B} - V)] - b_i[\mathscr{A}'_j(V^3 + \mathscr{B}V^2 - 3\mathscr{B}^2 V + \mathscr{B}^3)$$
$$\left.+ \mathscr{A}b_j(-5V^2 + 6\mathscr{B}V - 3\mathscr{B}^2)]\right\}$$

$$(\partial^3 A/\partial N_i \partial N_j \partial N_k) = -\frac{RT\delta_{ij}\delta_{jk}}{N_k^2} + \frac{RT}{(V - \mathscr{B})^3}\left[(V - \mathscr{B})(b_i b_j + b_i b_k + b_j b_k)\right.$$
$$+ 2Nb_i b_j b_k] + \frac{1}{2\sqrt{2}\mathscr{B}^4}\left\{[\mathscr{B}^2(b_i a_{jk} + b_j a_{ik} + b_k a_{ij})\right.$$
$$- \mathscr{B}(\mathscr{A}'_i b_j b_k + \mathscr{A}'_j b_i b_k + \mathscr{A}'_k b_i b_j) + 3\mathscr{A}b_i b_j b_k]$$
$$\times \left[2\ln \frac{V + (\sqrt{2} + 1)\mathscr{B}}{V - (\sqrt{2} - 1)\mathscr{B}} - \frac{4\sqrt{2}\mathscr{B}V}{V^2 + 2\mathscr{B}V - \mathscr{B}^2}\right]$$
$$- \frac{4\sqrt{2}\mathscr{B}^2 V}{(V^2 + 2\mathscr{B}V - \mathscr{B}^2)^2}[\mathscr{B}(\mathscr{B} - V)$$
$$\times (\mathscr{A}'_i b_j b_k + \mathscr{A}'_j b_i b_k + \mathscr{A}'_k b_i b_j) + b_i b_j b_k \mathscr{A}(3V - 2\mathscr{B})]$$
$$\left.- \frac{16\sqrt{2}\mathscr{A}\mathscr{B}^3(V - \mathscr{B})^2 Vb_i b_j b_k}{(V^2 + 2\mathscr{B}V - \mathscr{B}^2)^3}\right\},$$

where $\mathscr{A}'_i = \partial\mathscr{A}/\partial N_i$, $\mathscr{A}'_j = \partial\mathscr{A}/\partial N_j$, $\mathscr{A}'_k = \partial\mathscr{A}/\partial N_k$, $\delta_{ij} = 1$ for $i = j$, and $\delta_{ij} = 0$ for $i \neq j$.

Example 4.9 *Critical-point calculation of multicomponent mixtures* Consider mixtures 23, and 25 of Peng and Robinson (1977) shown below. Calculate the critical pressure and critical temperature of these two mixtures using (1) Baker and Luks (1980), and (2) Heidemann and Khalil (1980) formulations.

Mixture 23: $x_{nC_5} = 0.2465$, $x_{nC_6} = 0.2176$, $x_{nC_7} = 0.1925$, $x_{nC_8} = 0.1779$, $x_{nC_9} = 0.1656$ (mole fraction)

Mixture 25: $x_{C_1} = 0.7057$, $x_{C_2} = 0.0669$, $x_{C_3} = 0.0413$, $x_{nC_4} = 0.0508$, $x_{nC_5} = 0.1353$ (mole fraction)

Solution The system of two equations (Eqs. (4.214) and (4.215)) and two unknowns (that is T_c and P_c) for the Baker and Luks formulation can be solved via the secant method. This is just a Newton-Raphson method on numerical derivatives. Computation of the residual in Eq. (4.214) is fairly straight forward, because it only requires expressions for the second derivatives of the Helmholtz free energy A in terms of V and N_i. These derivatives are provided in Example 4.8. Equation (4.215) is somewhat more complicated; its determinant requires derivatives of Eq. (4.214) with respect to V and N_i. The procedure presented in Problem 4.14 can be used to evaluate the determinant derivatives. After both residuals are computed, the system of two equations and two unknowns are solved by the Newton-Raphson method to convergence.

The Heidemann and Khalil formulation (1980) as was discussed in this chapter retains the equation that provides the stability test limit (Eq. (4.225)) but utilizes the fact that the critical point is a special minimum of the scalar function A in the multidimensional space of the mole numbers. It is special in the way that a perturbation along the most unstable direction has a vanishing third derivative in that direction. An interpretation is that at the stability limit, one set of eigenvalues becomes zero while all the others remain non-zero with the same sign. Along the eigenvector in this direction, the next higher order derivative must be zero at the critical point. In other words, the "triadic" product of the third derivative tensor with the limiting direction vector triple product must vanish. The numerical algorithm for implementing the Heidemann and Khalil formulation is coded as follows. First, at an initial guess of pressure and temperature, the temperature in Eq. (4.225) is determined via a 1D Newton-Raphson technique. At that point, the Hessian of A is singular, therefore, the eigenvector corresponding to this vanishing eigenvalue is computed. This is accomplished using a standard linear algebra packages such as Lapack (developed by the University of Tennessee and the Oak Ridge National Laboratory).

The result from the two formulations are the same: Mixture 23, $T_c = 540\text{K}$, $P_c = 31\,\text{bar}$, Mixture 25: $T_c = 309\,\text{K}$, $P_c = 152\,\text{bar}$. PR-EOS was used in the calculations. The Baker and Luks formulation was found to be more sensitive to the initial guess, requiring a value to be within several degrees or bars of the final computed temperature and pressure. The Heidemann and Khalil formulation was found to be sufficiently insensitive to initial guess of $v_c = 4b$ and $T_c = 1.5 \sum_{i=1}^{c} x_i T_{ci}$. The run times as suspected were faster for the Heidemann and Khalil formulation by a factor of three for the 5-component systems. When very small amounts of C_1, C_2, and C_3 (less that 0.00001 mole fractions) were introduced to mixture 23, the run time was 10 times faster for the Heidemann and Khalil formulation than for the Baker and Luks formulation.

Note: The criteria for stability and criticality can be expressed in terms of mole numbers N_i or mole fractions x_i. The mole-number derivatives are discussed in the text. The mole-fraction derivatives will be presented below.

Consider the criteria of stability in terms of derivatives of G in Example 4.5; $y^{(2)} = G(T, P, \underline{N})$ can be also expressed by

$$y^{(2)} = G(T, P, N_1, N_2, \ldots, N_{c-1}, N)$$

where $N = \sum_{i=1}^{c} N_i$. Therefore, (see Example 1.2, Chapter 1),

$$dy^{(2)} = dG = -SdT + VdP + \sum_{i=1}^{c-1}(\mu_i - \mu_c)dN_i + \mu_i dN.$$

The expression for $y_{33}^{(2)}$, for example, is given by

$$y_{33}^{(2)} = \left(\frac{\partial \mu_1}{\partial N_1}\right)_{T,P,N_2,\ldots,N_{c-1},N}$$

or

$$y_{33}^{(2)} = \frac{1}{N}\left(\frac{\partial \mu_1}{\partial x_1}\right)_{T,P,x_2,\ldots,x_{c-1}}.$$

In terms of fugacity, from $d\mu_i = (RTd\ln f_i)_T$,

$$y_{33}^{(2)} = \frac{RT}{N}\left(\frac{\partial \ln f_1}{\partial x_1}\right)_{T,P,x_2,\ldots,x_{c-1}}.$$

Similar expressions for $y_{34}^{(2)}$ etc., can be obtained. Therefore, there is no change in the criteria of stability and criticality where mole fractions are used.

Example 4.10   Show that for a stable multicomponent homogeneous fluid similar to a single-component fluid, the isothermal compressibility should be positive.

Solution   For a stable system the coefficient $y_{jj}^{(j-1)}$ of $(\Delta Z_j)^2$ in Eq. (4.165) should be positive, and since $y_{c+2,c+2}^{(c+1)} = 0$ (see Eq. (1.185) of Chapter 1),

$$y_{jj}^{(j-1)} > 0 \qquad j = 1, \ldots, c+1.$$

If $y^{(0)} = U(S, V, n_1, \ldots, n_c)$, then $y^{(1)} = A(T, V, n_1, \ldots, n_c)$, and $dy^{(1)} = -SdT - PdV + \sum_{i=1}^{c} \mu_i dn_i$ which gives $(-\partial P/\partial V)_{T,\underline{n}} > 0$.

Therefore the isothermal compressibility of a multicomponent stable system $-(1/V)(\partial V/\partial P)_{T,\underline{n}} > 0$.

Example 4.11   Consider a binary mixture of C_1 and nC_5 at $P = 100$ bar $T = 350$ K. Plot Δg_{mix} *vs.* x_2 (mole fraction of nC_5) over the whole range (that is, $0 \le x_2 \le 1$), assuming that the C_1/nC_5 mixture will stay in the hypothetical single-phase state. Then derive

$$\left(\frac{\partial^2 \Delta g_{mix}}{\partial x_2^2}\right)_{T,P} = -\frac{1}{x_2}\left(\frac{\partial \mu_1}{\partial x_2}\right)_{T,P}.$$

Based on stability criteria and the use of the above equation, show unstable, stable, and metastable parts of the Δg_{mix} curve. Relate the curvatures to the stability.

Solution　In order to plot Δg_{mix} (see Eq. (4.39)), the expression for $(\mu_i - \mu_i^0)$ is first evaluated (see Eq. (1.111) of Chapter 1):

$$\mu_i(T, P, x_2) - \mu_i^0(T, P) = RT \ln f_i(T, P, x_2)/f_i^0(T, P)$$

Using the above equation in Eq. (4.39),

$$(\Delta g_{mix}/RT) = \sum_{i=1}^{2} x_i \ln f_i/f_i^0.$$

The PR-EOS can be used to evaluate f_i and f_i^0 for the C_1/nC_5 mixture, and for C_1 and nC_5 pure components at 100 bar and 350 K, respectively. The results are presented in Fig. 4.29.

Taking the derivative of Eq. (4.41), one readily obtains

$$\left(\frac{\partial^2 \Delta g_{mix}}{\partial x_2^2}\right)_{T,P} = -\frac{1}{x_2}\left(\frac{\partial \mu_1}{\partial x_2}\right)_{T,P}.$$

For a two-component system to be stable, $(\partial \mu_1/\partial x_2)_{T,P}$ should be less than zero (see Eq. (4.144)). Therefore for stability of a two-component system,

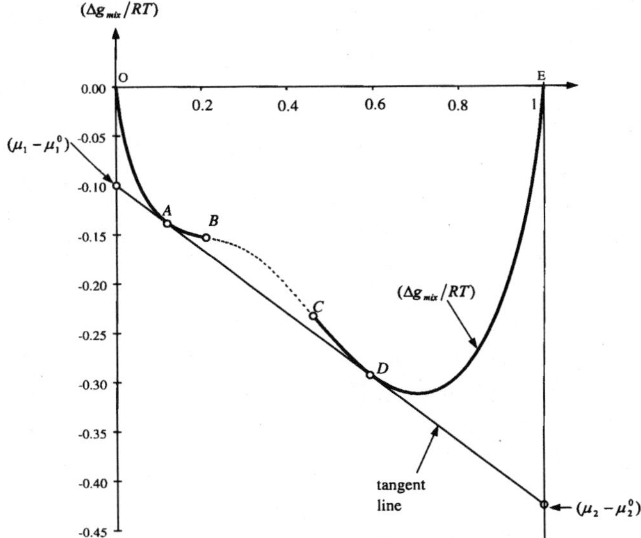

Figure 4.29　($\Delta g_{mix}/RT$) vs. x_2 for the C_1/nC_5 mixture at 350 K and 100 bar.

$(\partial^2 \Delta g_{mix}/\partial x_2^2)_{T,P}' > 0$. Points B and C in Fig. 4.29 represent $(\partial^2 \Delta g_{mix}/\partial x_2^2)_{T,P} = 0$. For points between B and C on the Δg_{mix} curve $(\partial^2 \Delta g_{mix}/\partial x_2^2)_{T,P} < 0$ and therefore are unstable; the curve is convex. In other words, at points B and C the curvature of the Δg_{mix} curve changes. All the points on the curves OAB and CDE are stable; the curves are concave. The points between A and B and between C and D are of metastable nature.

Problems

4.1   In the successive substitution method, sometimes because of a poor initial guess for K_i or other reasons, after a few iterations the vapor phase fraction α from Eq. (4.12) is calculated to be outside the range [0,1]. Consequently either $h(\alpha = 0) = \sum_{i=1}^{c} K_i z_i - 1 < 0$ or $h(\alpha = 1) = 1 - \sum_{i=1}^{c} z_i/K_i > 1$. The fact that $h(\alpha = 0) < 0$ or $h(\alpha = 1) > 1$ does not imply subcooled liquid or superheated vapor. In case $h(\alpha = 0) < 0$, we may estimate the vapor phase composition from $y_i = K_i z_i/\sum_{j=1}^{c} K_j z_j$ as we set $x_i = z_i$. How would you argue that the estimated vapor phase composition is reasonable? What would be the similar expression for the estimation of the liquid phase composition if $h(\alpha = 1) > 1$?

4.2   Consider the total Gibbs free energy of a multicomponent two-phase mixture given by $G = \sum_{i=1}^{c} \sum_{j=1}^{2} n_{ij}\mu_{ij}$, where n_{ij} is the number of moles of component i in phase j and μ_{ij} is the chemical potential of component i in phase j. Derive the basic equation of equilibrium $\mu_{i1} = \mu_{i2}(i = 1, \ldots, c)$ by taking the derivative of G with respect to either n_{i1} or n_{i2} at constant T and P. Note that $n_{i1} + n_{i2} = n_i$ $(i = 1, \ldots, c)$ is a constant.
   Hint: Use the Gibbs-Duhem equation in your derivation.

4.3   Show that at the critical point of a pure substance,

$$(\partial^3 \mu/\partial n^3)_{T,V} \geq 0.$$

Note: The above expression is equivalent to the criterion expressed by $(\partial^3 P/\partial V^3)_{T,N} \leq 0$.

4.4   Gibbs presented one of the alternative sets of the criteria for criticality in the following form:

$$(\partial \mu_c/\partial N_c)_{T,V,\mu_1,\ldots,\mu_{c-1}} = 0$$
$$(\partial^2 \mu_c/\partial N_c^2)_{T,V,\mu_1,\ldots,\mu_{c-1}} = 0$$
$$(\partial^3 \mu_c/\partial N_c^3)_{T,V,\mu_1,\ldots,\mu_{c-1}} \geq 0$$

Derive the above set of equations.

4.5   The molar Gibbs free energy of mixing of a binary mixture, Δg_{mix}, is plotted vs. x_2 (mole fraction of component 2) in Fig. 4.29 at constant temperature and pressure. Show that the intercepts of Δg_{mix} with $x_2 = 0$ and $x_2 = 1$ axes are

$\mu_1 - \mu_1^0$ and $\mu_2 - \mu_2^0$, respectively; μ_1 and μ_2 are the chemical potentials at points A and D, respectively.

4.6 The stability test for a homogeneous phase can be expressed in terms of A and G for the following inequalities to hold for every possible variation around the initial state of the homogeneous phase:

$$\sum_{i=1}^{c}\sum_{j=1}^{c}(\partial^2 A/\partial N_i \partial N_j)_{T,V}(\Delta N_i \Delta N_j) + 2\sum_{i=1}^{c}(\partial^2 A/\partial N_i \partial V)_{T,V}(\Delta N_i \Delta V)$$

$$+ (\partial^2 A/\partial V^2)_{T,V} + \textit{higher-order terms} > 0$$

and $\qquad \sum_{i=1}^{c}\sum_{j=1}^{c}(\partial^2 G/\partial N_i \partial N_j)_{T,P}\Delta N_i \Delta N_j + \textit{higher-order terms} > 0.$

Derive the above inequalities in the same manner that Eq. (4.159) was derived. Note that the derivatives are evaluated at constant mole number of the components other than those that are varied.

4.7 Show that Maxwell's equal-area rule for multicomponent systems takes the following form: a plot of $-\mu_{c-1}$ vs. N_{c-1} at constant $T, P, \mu_1, \ldots, \mu_{c-2}, N_c$ in the two-phase region gives two equal areas shown by the dotted region of Fig. 4.30.

4.8 Show that for a stable multicomponent homogeneous mixture, similar to a single-component fluid, c_P and c_V are positive. Can these parameters be positive at the limit of stability? If the answer is yes, why?

4.9 The evaluation of the cubic in Eq. (4.226) can be performed using terms involving at most double sums (Heidemann, 1994). The triple sums in Eq. (4.226) can be reduced as follows:

$$\sum_{i=1}^{c}\sum_{j=1}^{c}\sum_{k=1}^{c}\Delta N_i \Delta N_j \Delta N_k = \left(\sum_{i=1}^{c}\Delta N_i\right)^3$$

$$\sum_{i=1}^{c}\sum_{j=1}^{c}\sum_{k=1}^{c}\beta_i \Delta N_i \Delta N_j \Delta N_k = \left(\sum_{i=1}^{c}\beta_i\Delta N_i\right)\left(\sum_{i=1}^{c}\Delta N_i\right)^2$$

$$\sum_{i=1}^{c}\sum_{j=1}^{c}\sum_{k=1}^{c}\gamma_{ij} \Delta N_i \Delta N_j \Delta N_k = \left(\sum_{i=1}^{c}\sum_{j=1}^{c}\gamma_{ij}\Delta N_i\Delta N_j\right)\left(\sum_{i=1}^{c}\Delta N_i\right)$$

Derive the above relationships and relate β_i and γ_{ij} to the coefficients in Eq. (4.226).

4.10 Show that c different matrices of the form given by Eq. (4.225) can be composed. Why do permutations of the component index not produce a new matrix?

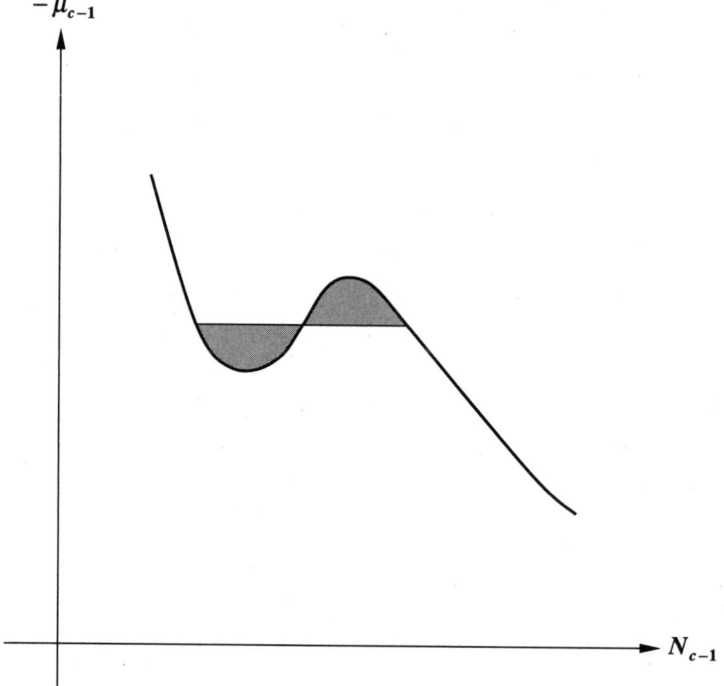

Figure 4.30   Maxwell's equal-area rule for multicomponent mixtures.

4.11   The first criterion for the limit of stability (at the critical point) in terms of the Helmholtz free energy can be written as

$$
\begin{vmatrix}
A_{V,V} & A_{V,N_1} & \cdots & A_{V,N_{c-1}} \\
\vdots & & & \\
A_{N_{c-1},V} & A_{N_{c-1},N_1} & \cdots & A_{N_{c-1},N_{c-1}}
\end{vmatrix} = 0,
$$

where $A_{V,V} = (\partial^2 A/\partial V^2)_{T,N_1,\dots,N_{c-1}}$ and $A_{V,N_1} = (\partial^2 A/\partial V \partial N_1)_{T,N_2,\dots,N_{c-1}}$, etc. The same criterion can also be written as

$$
\begin{vmatrix}
A_{N_1,N_1} & A_{N_1,N_2} & \cdots & A_{N_1,N_c} \\
\vdots & & & \vdots \\
A_{N_c,N_1} & A_{N_c,N_2} & \cdots & A_{N_c,N_c}
\end{vmatrix} = 0,
$$

where $A_{N_1,N_1} = (\partial^2 A/\partial N_1^2)_{T,V,N_2,\dots,N_c}$, $A_{N_1,N_2} = (\partial^2 A/\partial N_1 \partial N_2)_{T,V,N_3\dots N_c}$, etc. In the last expression all the derivatives of the Helmholtz free expression are taken at constant T and V. What is the advantage of the second equation over the first equation? Is the second equation simpler than the first for using pressure-explicit equations of state?

4.12   Use Gibbs free energy representation of the criticality criteria from Eqs. (4.225) and (4.226) to show that for a binary mixture $(\partial \mu_1/\partial N_1)_{T,P,N_2} = 0$ and $(\partial \mu_1^2/\partial N_1^2)_{T,P,N_2} = 0$. Note that these two expressions are the same expressions as those of Eq. (4.202).

4.13   Show that at the critical point of a c-component mixture, the following determinant is zero:

$$\begin{vmatrix} (\partial \mu_1/\partial y_1) & (\partial \mu_1/\partial y_2) & \cdots & (\partial \mu_1/\partial y_{c-1}) \\ (\partial \mu_2/\partial y_1) & (\partial \mu_2/\partial y_2) & \cdots & (\partial \mu_1/\partial y_{c-1}) \\ \vdots & \vdots & \ddots & \vdots \\ (\partial \mu_{c-1}/\partial y_1) & (\partial \mu_{c-1}/\partial y_2) & \cdots & (\partial \mu_{c-1}/\partial y_{c-1}) \end{vmatrix} = 0.$$

Note that the above is the determinant of the matrix in Eq. (2.19b) of Chapter 2. *Hint*: You may use the results from Examples 4.5 in your derivation.

4.14   Use the second criterion of criticality expressed by Eq. (4.215) for a three-component system with $y^{(1)} = A(T, V, N_1, N_2, N_3)$, with index 2 for V, index 3 for N_1, and index 4 for N_2. Show that the derivative with respect to V for the determinant given by Eq. (4.214) is

$$\begin{aligned} \lambda_2^{(4)} &= A_{222}(A_{33}A_{44} - A_{43}^2) + A_{332}(A_{22}A_{44} - A_{42}^2) + A_{442}(A_{22}A_{33} - A_{32}^2) \\ &\quad + 2A_{223}(A_{42}A_{43} - A_{32}A_{44}) + 2A_{422}(A_{32}A_{43} - A_{33}A_{42}) \\ &\quad + 2A_{432}(A_{32}A_{42} - A_{22}A_{43}), \end{aligned}$$

and the derivative of the same determinant with respect to N_1 is given by

$$\begin{aligned} \lambda_3^{(4)} &= A_{223}(A_{33}A_{44} - A_{43}^2) + A_{333}(A_{22}A_{44} - A_{42}^2) + A_{443}(A_{22}A_{33} - A_{32}^2) \\ &\quad + 2A_{323}(A_{42}A_{43} - A_{32}A_{44}) + 2A_{423}(A_{32}A_{43} - A_{33}A_{42}) \\ &\quad + 2A_{433}(A_{32}A_{42} - A_{22}A_{43}). \end{aligned}$$

Then derive the final form of the expression from Eq. (4.215) given below.

$$\begin{aligned} & A_{222}(A_{33}A_{44}A_{23}A_{34} - A_{33}A_{44}A_{24}A_{33} - A_{23}A_{34}^3 + A_{24}A_{33}A_{34}^2) \\ &\quad + A_{233}(3A_{22}A_{23}A_{34}A_{44} - 2A_{22}A_{34}^2A_{24} - A_{22}A_{24}A_{33}A_{44} + A_{23}A_{34}A_{24}^2 \\ &\quad - 2A_{23}^2A_{24}A_{44} + A_{24}^3A_{33}) + A_{223}(-A_{22}A_{33}A_{34}A_{44} + A_{22}A_{34}^3 \\ &\quad + A_{23}A_{24}A_{33}A_{44} - 2A_{24}^2A_{33}A_{34} + A_{23}A_{34}^2A_{24}) + A_{333}(-A_{22}^2A_{34}A_{44} \\ &\quad + A_{22}A_{23}A_{24}A_{44} + A_{22}A_{34}A_{24}^2 - A_{23}A_{24}^3) + A_{244}(3A_{22}A_{23}A_{33}A_{34} \\ &\quad - 3A_{22}A_{24}A_{33}^2 - 3A_{23}^3A_{34} + 3A_{23}^2A_{24}A_{33}) + A_{224}(A_{22}A_{33}^2A_{44} - A_{22}A_{33}A_{34}^2 \\ &\quad + 3A_{23}^2A_{34}^2 - A_{23}^2A_{33}A_{44} - 4A_{23}A_{33}A_{24}A_{34} + 2A_{24}^2A_{33}^2) + A_{234}(-4A_{22}A_{23}A_{34}^2 \end{aligned}$$

$$+ 6A_{22}A_{24}A_{33}A_{34} - 2A_{22}A_{23}A_{33}A_{44} + 2A_{23}^3A_{44} + 2A_{23}^2A_{24}A_{34} - 4A_{23}A_{24}^2A_{33})$$

$$+ A_{344}(-3A_{22}^2A_{33}A_{34} + 3A_{22}A_{23}^2A_{34} + 3A_{22}A_{23}A_{24}A_{33} - 3A_{23}^3A_{24})$$

$$+ A_{334}(A_{22}^2A_{33}A_{44} + 2A_{22}^2A_{34}^2 - A_{22}A_{23}^2A_{44} - 4A_{22}A_{23}A_{24}A_{34}$$

$$- A_{22}A_{33}A_{24}^2 + 3A_{23}^2A_{24}^2) + A_{444}(A_{22}^2A_{33}^2 - 2A_{22}A_{33}A_{32}^2 + A_{23}^4) = 0$$

Hint: In order to obtain the derivative of determinant $\lambda^{(4)}$ with respect to V or any other variable, the following steps may facilitate the algebra: (1) take the derivative of each element of determinant $\lambda^{(4)}$ with respect to V or any other variable; call the matrix with derivation elements $\lambda_*^{(4)}$, (2) form the cofactor matrix (see Strang, 1988) of the determinant $\lambda^{(4)}$, and (3) the sum of the element by element products of the cofactor matrix and the matrix $\lambda_*^{(4)}$ is the determinant of $\lambda^{(4)}$ with respect to V or any other variable.

4.15   Show that the isothermal compressibility of a stable two-phase multicomponent mixture is positive.

References

1. Baker, L.E., and K.D. Luks: "Critical Point and Saturation Pressure Calculations for Multicomponent Systems," *Soc. Pet. Eng. J.* p. 15, Feb. 1980.
2. Baker, L.E., A.C. Pierce, and K.D. Luks: "Gibbs Energy Analysis of Phase Equilibria," *Soc. Pet. Eng. J.* p. 731, Oct. 1982.
3. Chang, Y.-B.: "Development and Application of an Equation of State Compositional Simulator," Ph.D. Dissertation, University of Texas-Austin, Austin, TX, Aug. 1990.
4. Corana, A., M. Marchesi, C. Martini, and S. Ridella: "Minimizing Multimodal Functions of Continuous Variables with the Simulated Annealing Algorithm," ACM Transactions on Mathematical Software, vol. 13, p. 262, 1987.
5. Crowe, A.M., and M. Nishio: "Convergence Promotion in the Simulation of Chemical Processes — the General Dominant Eigenvalue Method," *AIChE J.* vol. 21, p. 528, 1975.
6. Gibbs, J.W.: "Transaction of the Connecticut Academy (1876). As Reprinted in the Scientific Papers of J. Willard Gibbs," Vol. I.
7. Goffe, W.L., G.D. Ferrier, and J. Rogers: "Global Optimization of Statistical Functions with Simulated Annealing," *J. Econometrics*, vol. 60, p. 65, 1994.
8. Haase, R.: "Thermodynamik der Mischphasen," *Springer-Verlag*, Berlin-Goettingen-Heidelberg, 1956.
9. Heidemann, R.A., and A.M. Khalil: "The Calculation of Critical Points," *AIChE J.*, vol. 26, p. 769, 1980.
10. Heidemann, R.A.: *Thermodynamics of Aqueous Systems with Industrial Applications*, S.A. Newman, (ed.), ACS Symposium Series No. 133, American Chemical Society, Washington, D.C. p. 379, 1980.
11. Heidemann, R.A.: "The Classical Theory of Critical Points," *Supercritical Fluids, Fundamentals for Application,* E. Kiran and J.M.H. Levelt Sengers, (editors), NATO Advanced Study Institute Series E, vol. 273, Kluwer Academic Publishers: Dordrecht, p. 39, 1994.
12. Hendriks, E.M., and A.R.D. Van Bergen: "Application of a Reduction Method to Phase Equilibria Calculations," *Fluid Phase Equilibria*, vol. 74, p. 17, 1992.
13. Michelsen, M.L, and R.A. Heidemann: "Calculation of Critical Points from Cubic Two-Constant Equations of State," *AIChE J.,* vol. 27, p. 521, 1981.
14. Michelsen, M.L.: "The Isothermal Flash Problem, Part 1: Stability," *Fluid Phase Equilibria*, vol. 9, p. 1, 1982a.
15. Michelsen, M.L.: "The Isothermal Flash Problem, Part II: Phase-Split Calculation," *Fluid Phase Equilibria*, vol. 9, p. 21, 1982b.

16. Michelsen, M.: "Simplified Flash Calculations for Cubic Equations of State," *Ind. Eng. Chem Process Des. Dev.*, vol. 25, p. 184, 1986.
17. Pan, H., and A. Firoozabadi: "Complex Multiphase Equilibrium Calculations by Direct Minimization of Gibbs Free Energy Using Simulated Annealing," *SPE Res. Eval. and Engl.*, p. 36, Feb. 1998.
18. Peng, D.Y., and D.B. Robinson: "A Rigorous Method for Predicting the Critical Properties of Multicomponent Systems from an Equation of State," *AIChE J.* vol. 23, p. 137, 1977.
19. Perschke, D.R.: "Equation of State Phase Behavior Modeling for Compositional Simulation," Ph.D. Dissertation, University of Texas-Austin, Austin, TX, December 1988.
20. Powell, M.J.D.: "A Hybrid Method for Nonlinear Equations," *Numerical Methods for Nonlinear Algebraic Equations,* P. Rabinowitz, (ed), Gordon and Breach, London, p. 87, 1970.
21. Press, W.H., S.A. Teukolsky, W.T. Vetterling, and B.P. Flannery: *Numerical Recipes: The Art of Scientific Computing,* 2nd ed., Cambridge University Press, New York, Chap. 10, 1992.
22. Rachford, H.H., Jr. and J.D. Rice: "Procedure for Use of Electronic Digital Computers in Calculating Flash Vaporization Hydrocarbon Equilibrium," *Trans. Soc. Pet. Eng.* vol. 195 p. 327, 1952.
23. Robinson, D.B., H. Kalara, and H. Rempis: "The Equilibrium Phase Properties of a Synthetic Sour Gas Mixture and a Simulated Natural Gas Mixture," *GPA Research Report No. RR-31,* Tulsa, OK., May 1978.
24. Shelton, J.L., and L. Yarborough: "Multiple Phase Behavior in Porous Media during CO_2 or Rich-Gas Flooding," *J. Pet. Tech.* p. 1171, Sept. 1977 .
25. Skripov, V.P., and G.V. Ermakov: "Pressure Dependence of the Ultimate Superheating of a Liquid," *Russian J. Phys. Chem.* vol. 38, p. 396, 1964.
26. Strang, G.: *Introduction to Applied Mathematics,* Wellesley Cambridge Press, Wellesley, MA, 1986.
27. Strang, G.: *Linear Algebra and Its Applications,* 3d ed., Harcourt Brace Jovanovich College Publishers, New York, 1988.
28. Tester, J.W., and M. Modell: "Thermodynamics and Its Applications," Third Edition, Princeton-Hall PTR, Upper Saddle River, NJ, 1996.
29. Trangenstein, J.A.: "Customized Minimization Techniques for Phase Equilibrium Computations in Reservoir Simulation," *Chem. Eng. Science*, vol. 12, p. 2847, 1987.
30. Trebble, M. A.: "A Preliminary Evaluation of Two and Three Phase Flash Initiation Procedures," *Fluid Phase Equilibria*, vol. 53, p. 113, 1989.
31. Wilson, G.M: "A Modified Redlich-Kwong Equation of State, Application to General Physical Data Calculation," paper No. 15C presented at the *1969 AIChE 65th National Meeting,* Cleveland, Ohio, May 4–7, 1969.

5

Thermodynamics of wax and asphaltene precipitation

Wax and asphaltene precipitation are serious problems in production from some hydrocarbon reservoirs. Wax, which is a solid precipitate, can occur in the well, in the production facilities, and in the pipelines. There is a field case in which wax has been reported in a natural state in the reservoir. Wax precipitation can occur for gas-condensate, light-oil and heavy-oil fluids at temperatures as high as 150°F. Asphaltene precipitation may occur in the reservoir, in the production facilities, and in the pipelines. Asphaltenes may precipitate from some light oils, but there may be no precipitation from some heavy oils. Asphaltene precipitation has not been reported from gas condensate reservoirs.

As we will see in this chapter, the modeling of wax and asphaltene precipitation requires different approaches because they are fundamentally different. The effect of pressure, composition, and temperature on wax and asphaltene precipitation is also fundamentally different.

Temperature has a strong effect on wax precipitation from both gas condensates and crude oils. The pressure effect on wax precipitation from gas condensates is often different from crude oils. In general, when pressure increases isothermally, wax precipitates from crude oil. On the other hand, with pressure increase at constant temperature wax can dissolve in natural gases and prevent precipitation. Asphaltene precipitation may not be sensitive to temperature, and temperature increase may increase or may decrease the onset of asphaltene precipitation. Pressure can have an important effect on asphaltene precipitation; pressure decrease often causes asphaltene precipitation. The effect of composition is also very different on wax and asphaltene precipitation.

Wax and asphaltene precipitation are old problems, but only recently have attempts been made to develop a thermodynamic description for these processes. The asphaltene thermodynamic models have just begun to evolve. In this chapter, we will first present the thermodynamics of wax precipitation and discuss the effects of pressure, temperature, and composition. Then we will present a thermodynamic micellization model for asphaltene precipitation, and study the pressure and composition effects. All the calculations for the equilibrium between the precipitated phase and the crude oil will be based on direct minimization of the Gibbs free energy of the total system.

Wax precipitation

Components in a gas-condensate fluid contain hydrocarbons from methane, C_1, ethane, C_2, and other hydrocarbons as heavy as C_{40} or C_{50} or even heavier. Reservoir crudes may contain hydrocarbons as heavy as C_{100}. At room temperature (75°F) and atmospheric pressure, C_1, C_2, C_3, and C_4 are in the gas state, nC_5 to nC_{15} are in the liquid state, and normal alkanes heavier than nC_{15} are in the solid state. The broad volatility and melting-point range of these hydrocarbon components found in petroleum fluids cause formations of gas, liquid, and solid phases in response to changes in pressure, temperature, or composition. Let us consider a mixture of two hydrocarbons—nC_5 and nC_{28}. The melting-point temperature of nC_{28} is 57°C at atmospheric pressure. The solubility of nC_{28} in nC_5 at atmospheric pressure is 0.5 mole percent at 14°C. At 40°C and atmospheric pressure, the solubility of nC_{28} in nC_5 increases to 12 mole percent. It is therefore natural that when the temperature falls, heavy hydrocarbons in a crude or even a gas condensate may precipitate as wax crystals. In the petroleum industry, wax precipitation is undesirable because it may plug the pipeline and processing equipment.

Currently, there is no equation of state that can describe the volumetric behavior of the solid phase. However, one can relate the chemical potential of the solid to the chemical potential of the liquid phase in terms of certain melting properties. Let $\mu^S_{pure\ i}(P, T)$ represent the chemical potential of pure solid-component i at pressure P and temperature T, and $\mu^L_{pure\ i}(P, T)$, the chemical potential of the pure subcooled-liquid component i at the same temperature and pressure. We are interested in finding an expression for $\Delta\mu_i = \mu^L_{pure\ i}(P, T) - \mu^S_{pure\ i}(P, T)$. Let us examine the diagram shown in Fig. 5.1. In this diagram, $\Delta\mu_i$ represents the chemical potential difference of pure component i from point 1 to point 6. Since temperature is constant from 1 to 6,

$$\Delta\mu_i = \Delta h_i - T\Delta s_i, \tag{5.1}$$

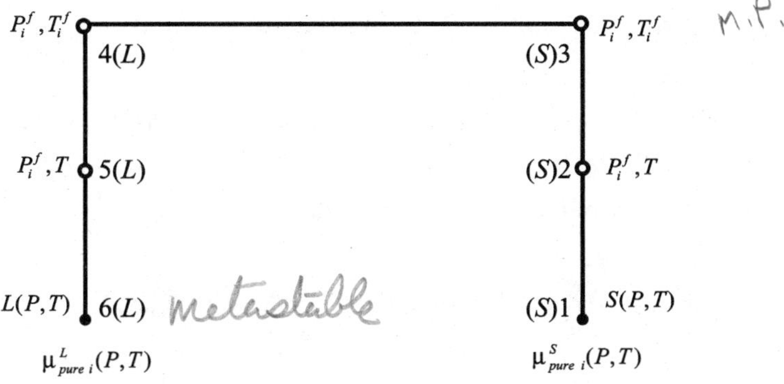

Figure 5.1   Diagram relating $\mu_{pure\,i}^S(P,\,T)$ to $\mu_{pure\,i}^L(P,\,T)$.

where Δh_i, similar to $\Delta \mu_i$, represents $\Delta h_i = h_{pure\,i}^L(P,\,T) - h_{pure\,i}^S(P,\,T)$. The same is true for Δs_i. The next step is the calculation of Δh_i and Δs_i where we will follow the path from 1 to 2, to 3, ... to 6. From 1 to 2, according to Eq. (3.113) of Chapter 3,

$$\Delta h_i^{2\to 1} = \int_P^{P_i^f} [v_i^s - T(\partial v_i^S/\partial T)_P]dP, \tag{5.2}$$

where P_i^f is the pressure at the melting-point temperature T_i^f. From 2 to 3, using Eq. (3.113) of Chapter 3,

$$\Delta h_i^{3\to 2} = \int_T^{T_i^f} c_{Pi}^S dT. \tag{5.3}$$

From 3 to 4, Δh_i represents the heat of fusion at T_i^f and P_i^f, which is shown by Δh_i^f, $\Delta h_i^{4\to 3} = \Delta h_i^f$. From 4 to 6, the pure component i is in the subcooled-liquid state. Similarly to the previous steps, one can write

$$\Delta h_i^{5\to 4} = \int_{T_i^f}^{T} c_{Pi}^L dT \tag{5.4}$$

$$\Delta h_i^{6\to 5} = \int_{P_i^f}^{P} [v_i^L - T(\partial v_i^L/\partial T)_P]dP. \tag{5.5}$$

Then,

$$\Delta h_i = \Delta h_i^{2\to 1} + \Delta h_i^{3\to 2} + \Delta h_i^{4\to 3} + \Delta h_i^{5\to 4} + \Delta h_i^{6\to 5}. \tag{5.6}$$

Let $\Delta c_{Pi} = c_{Pi}^L - c_{Pi}^S$ and $\Delta v_i = v_i^L - v_i^S$. Using the above equations,

$$\Delta h_i = \Delta h_i^f - \int_T^{T_i^f} \Delta c_{Pi}\,dT + \int_P^{P_i^f} [-\Delta v_i + T(\partial \Delta v_i/\partial T)_P]\,dP. \qquad (5.7)$$

Assuming negligible temperature effect on Δc_{Pi},

$$\Delta h_i = \Delta h_i^f + \Delta c_{Pi}(T - T_i^f) - \int_P^{P_i^f} \Delta v_i\,dP + T\int_P^{P_i^f}(\partial \Delta v_i/\partial T)_P\,dP. \qquad (5.8)$$

Similarly, one can derive the expression for Δs_i,

$$\Delta s_i = \frac{\Delta h_i^f}{T_i^f} + \Delta c_{Pi} \ln\frac{T}{T_i^f} + \int_P^{P_i^f}(\partial \Delta v_i/\partial T)_P\,dP. \qquad (5.9)$$

Substituting Eqs. (5.8) and (5.9) into Eq. (5.1) results in

$$\mu_{pure\,i}^L(P, T) - \mu_{pure\,i}^S(P, T) = \Delta h_i^f\left(1 - \frac{T}{T_i^f}\right) + \Delta c_{Pi}\left[(T - T_i^f) - T\ln\frac{T}{T_i^f}\right]$$
$$-\int_P^{P_i^f} \Delta v_i\,dP. \qquad (5.10)$$

Equation (5.10) can be written in dimensionless form when divided by RT.

$$\boxed{\begin{aligned}
\frac{\mu_{pure\,i}^L(P, T) - \mu_{pure\,i}^S(P, T)}{RT} &= \frac{\Delta h_i^f}{RT_i^f}\left(\frac{T_i^f}{T} - 1\right) - \frac{\Delta c_{Pi}}{R}\left(\frac{T_i^f}{T} - 1\right)\\
&\quad + \frac{\Delta c_{Pi}}{R}\left[\ln\frac{T_i^f}{T}\right] + \frac{1}{RT}\int_{P_i^f}^{P} \Delta v_i\,dP
\end{aligned}} \qquad (5.11)$$

Equation (5.11) allows the calculation of the chemical potential of a pure solid in terms of the chemical potential of a pure subcooled liquid component. Data necessary to calculate $\mu_{pure\,i}^S(P, T)$ are Δh_i^f, melting-point enthalpy, T_i^f, melting-point temperature, Δc_{Pi}, heat capacity of fusion, and Δv_i which may be negligible.

Now suppose that, in a multicomponent mixture, the solid solution and the liquid solution are in equilibrium. Then one can write

$$\mu_i^L(P, T, \mathbf{x}^L) = \mu_i^S(P, T, \mathbf{x}^S). \qquad (5.12)$$

From $(d\mu_i = RTd \ln f_i)_T$,

$$\mu_i^L(P, T, \mathbf{x}^L) = \mu_{pure\ i}^L(P, T) + RT \ln \frac{f_i^L(P, T, \mathbf{x}^L)}{f_{pure\ i}^L(P, T)} \tag{5.13}$$

and

$$\mu_i^S(P, T, \mathbf{x}^S) = \mu_{pure\ i}^S(P, T) + RT \ln \frac{f_i^S(P, T, \mathbf{x}^S)}{f_{pure\ i}^S(P, T)}. \tag{5.14}$$

Combining Eqs. (5.12) to (5.14),

$$f_{pure\ i}^S(P, T) = f_{pure\ i}^L(P, T) \exp\left[-\frac{\mu_{pure\ i}^L(P, T) - \mu_{pure\ i}^S(P, T)}{RT}\right]. \tag{5.15}$$

Note that $f_i^L(P, T, \mathbf{x}^L) = f_i^S(P, T, \mathbf{x}^S)$, but the subcooled pure liquid may not be in equilibrium with the solid at temperature T and pressure P.

If we neglect the Poynting correction term, that is, the last term on the right side of Eq. (5.11), and substitute the remaining terms in Eq. (5.15), we obtain the expression for the pure-solid component fugacity:

$$\boxed{\begin{aligned} f_{pure\ i}^S(P, T) &= f_{pure\ i}^L(P, T) \\ &\times \exp\left[\frac{\Delta h_i^f}{RT_i^f}\left(1 - \frac{T_i^f}{T}\right) - \frac{\Delta c_{Pi}}{R}\left(1 - \frac{T_i^f}{T}\right)\right. \\ &\quad \left. - \frac{\Delta c_{Pi}}{R} \ln \frac{T_i^f}{T}\right] \end{aligned}} \tag{5.16}$$

An equation of state can be used to calculate $f_{pure\ i}^L(P, T)$. The expression for the PR-EOS for the calculation of $f_{pure\ i}^L(P, T)$ is provided by Eq. (3.22) of Chapter 3. Then with Δh_i^f, T_i^f, and Δc_{Pi}, one can calculate $f_{pure\ i}^S(T, P)$. In order to proceed with the wax-precipitation calculations, one needs to assume a proper solid model. Currently, there are two types of solid models. One is the solid-solution model, and the other is the multisolid-phase model. These models are presented below.

Solid-solution model.   If one assumes that the precipitation forms a solid solution, then

$$f_i^S(P, T, \mathbf{x}^S) = \gamma_i^S(P, T, \mathbf{x}^S) x_i^S f_{pure\ i}^S(P, T), \tag{5.17}$$

which requires determining $\gamma_i^S(P, T, \boldsymbol{x}^S)$. Figure 5.2 shows the sketch for the solid-solution model. For an ideal-solid solution, $\gamma_i^S(P, T, \boldsymbol{x}^S) = 1$, and therefore

$$f_i^S(P, T, \boldsymbol{x}^S) = x_i^S f_{pure\ i}^S(P, T), \text{ ideal solid solution.} \tag{5.18}$$

The fugacities in the vapor and liquid phases may be obtained from an EOS. Alternatively, an activity-coefficient model can be used to describe the liquid phase to estimate γ_i^L. However, as discussed in Chapter 1, activity coefficient models, in general, may not be suitable for reservoir fluids because they are based on the assumption of no change in volume due to mixing. At equilibrium, when liquid and solid phases are present,

$$f_i^L(P, T, \boldsymbol{x}^L) = f_i^S(P, T, \boldsymbol{x}^S). \tag{5.19}$$

With an activity coefficient model for the liquid phase,

$$f_i^L(P, T, \boldsymbol{x}^L) = \gamma_i^L(P, T, x)x_i^L f_i^L(P, T). \tag{5.20}$$

Combining Eqs. (5.17) and (5.20), the solid-to-liquid equilibrium ratio for component i is given by

$$K_i^{SL} = \frac{x_i^S}{x_i^L} = \frac{f_i^L(P, T)\gamma_i^L(P, T, \boldsymbol{x}^L)}{f_i^S(P, T)\gamma_i^S(P, T, \boldsymbol{x}^S)}. \tag{5.21}$$

Equation (5.16) provides the expression for $f_i^L(P, T)/f_i^S(P, T)$.

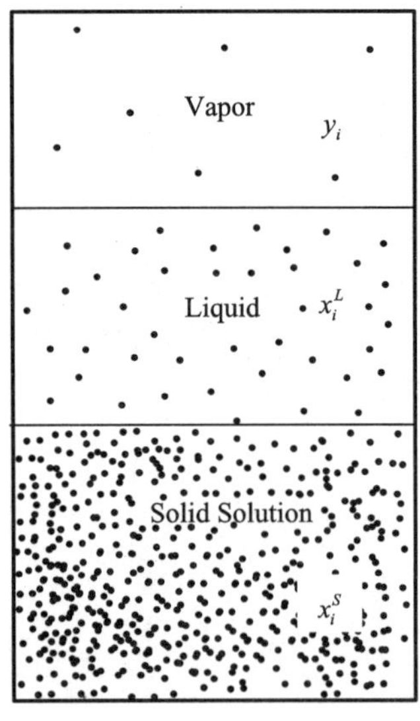

Figure 5.2   Solid-solution model for wax precipitation.

Multisolid-phase model. A number of studies show that when binary normal alkane mixtures are cooled, the precipitation is unstable and segregates into two solid phases, provided the chain-length difference between the two alkanes exceeds a certain value. As an example, Dorset (1990) reports that the segregated solid phases of binary n-alkanes consist predominantly of pure components. Snyder *et al.* (1992, 1993, 1994) have also studied the kinetics of the segregation of binary normal-alkane mixtures using spectroscopy, calorimetry, and electron diffraction. They observed that the rate of the segregation is very sensitive to the chain-length difference. Hansen *et al.* (1991) observed phase transitions of the precipitated wax from North Sea crude oils. Based on these observations, Lira-Galeana, Firoozabadi, and Prausnitz (1996) developed a thermodynamic multisolid-wax model (see Fig. 5.3). In this model, each solid phase is described as a pure component which does not mix with other solid phases. The calculations for the liquid/multisolid phases become very simple once the stability analysis from Chapter 4 is used. From stability analysis considerations (see Eq. (4.50) of Chapter 4), a component may exist as a pure solid if

$$\boxed{f_i(P, T, \boldsymbol{z}) - f^S_{pure\ i}(P, T) \geq 0 \qquad i = 1, \ldots, c} \, , \qquad (5.22)$$

where $f_i\,(P, T, \boldsymbol{z})$ is the fugacity of component i with feed composition $\boldsymbol{z}$. The mixture components that fulfill the above expression will precipitate. Note that the above simple expression is only applicable for precipitation from a single phase. Precipitation from a vapor–liquid system is formulated in Example 5.4 at the end of this chapter.

Here, we provide both the equilibrium and the material balance equations for wax precipitation calculations for solid–liquid equilibria. At fixed temperature and pressure, for every component i, the multisolid-phase model must satisfy

$$f^L_i(P, T, \boldsymbol{x}^L) = f^S_{pure\ i}(P, T) \qquad i = (c - c_S + 1), \ldots, c, \qquad (5.23)$$

where c_S is the number of solid phases determined from Eq. (5.22). The material balances for the nonprecipitating components are

$$z_i - x^L_i \left[1 - \sum_{j=(c-c_S+1)}^{c} n^S_j / F \right] = 0 \qquad i = 1, \ldots, (c - c_S), \qquad (5.24)$$

where n^S_j is the moles of solid phase j and F is the feed mole numbers.

For the precipitating components where all solid phases are pure,

$$z_i - x^L_i \left[1 - \sum_{j=(c-c_S+1)}^{c} n^S_j / F \right] - n^S_i / F = 0$$

$$i = (c - c_S + 1), \ldots, c - 1, c_S > 1. \qquad (5.25)$$

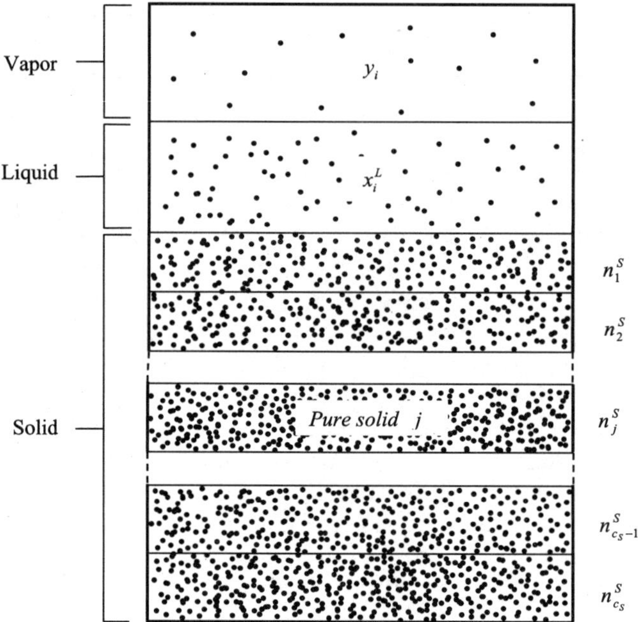

Figure 5.3 Multisolid-phase model for wax precipitation.

The constraint equation for component i in the liquid phase is

$$\sum_{i=1}^{c} x_i^L = 1.\tag{5.26}$$

There are $(c + c_S)$ equations and $(c + c_S)$ unknowns. The unknowns are $\underline{x}^L$ and $n_j^S(j = 1, \ldots, c_S)$.

When a vapor phase is present, some of the above equations should be modified (see Example 5.4). One can use the direct minimization of Gibbs free energy to solve the general problem of vapor–liquid-multisolid phase equilibria (see also Example 5.4). Simulated annealing can be employed to provide the global minimization of the Gibbs free energy (Pan and Firoozabadi, 1998a). Equation (5.16) requires the melting-point properties of the precipitating components. Normally wax consists of hydrocarbons heavier than C_{15}; there is a significant difference in the melting-point properties of heavy paraffins, naphthenes, and aromatics (PNA). Therefore, there is a need to provide the melting-point and physical properties of PNA groups.

Melting-point properties. The precipitated wax consists mainly of paraffins and some naphthenes; the aromatics are absent. Lighter hydrocarbons from C_1 to C_{15} are also absent in the wax (Philp, 1994). As shown by Eq. (5.16), the fugacity of solid-component i depends on the melting-point properties of component i. In the following, the melting-point properties of the components which appear in the wax are provided.

Melting-point temperature T_i^f.   Won (1986) has given a correlation for the melting point of pure n-alkanes:

$$T_i^f = 374.5 + 0.02617M_i - 20172/M_i, \qquad (5.27)$$

where T^f is in degrees Kelvin and M is the molecular weight. In some n-alkanes, there may exist one, two, or more solid-solid transitions below the melting point (see Example 5.3). Eq. 5.27 and Eq. 5.29 (to be discussed later) provide average values. The melting-point temperatures for naphthenes, aromatics, and iso-paraffins are given by (Pan, Firoozabadi, and Fotland, 1997b)

$$T_i^f = 333.45 - 419 \exp[-0.00855M_i], \qquad (5.28)$$

where T^f is in degrees Kelvin. Figure 5.4 provides a plot of Eqs. (5.27) and (5.28). This figure clearly shows that n-alkanes would form wax in preference to other species in the gas condensates and crude oils.

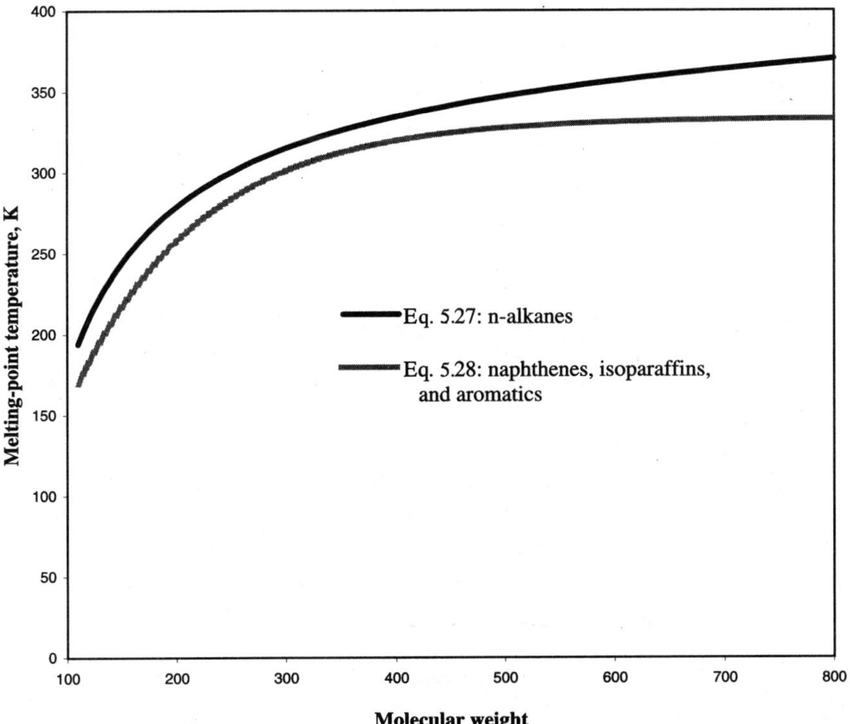

Figure 5.4   Melting-point temperature of hydrocarbons.

Enthalpy of fusion Δh_i^f.   Won (1986) developed a correlation for calculating the melting-point enthalpies (enthalpy of fusion) of paraffin hydrocarbons:

$$\Delta h_i^f = 0.1426 M_i T_i^f, \tag{5.29}$$

where T_i^f is in degrees Kelvin and Δh_i^f is in calories/gmole.

The enthalpy of fusion for naphthenes and iso-paraffins is given by (Lira-Galena *et al.*, 1996),

$$\Delta h_i^f = 0.0527 M_i T_i^f. \tag{5.30}$$

The enthalpy of fusion for aromatic species is given by (Pan *et al.*, 1997b):

$$\Delta h_i^f = 11.2 T_i^f. \tag{5.31}$$

Note that from the above correlation the entropy of fusion, $\Delta h_i^f / T_i^f$, for aromatics is independent of the molecular weight. Figure 5.5 shows the entropy of fusion using Eqs. (5.29) to (5.31); there is a substantial difference between the entropies of fusion of paraffins and naphthenes.

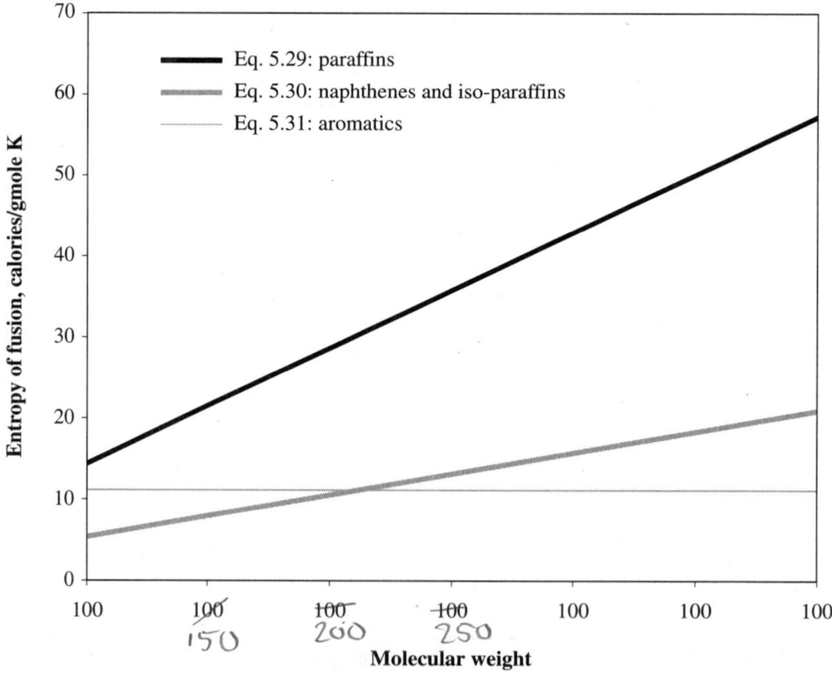

Figure 5.5   Entropies of fusion of hydrocarbons.

Heat capacity of fusion Δc_{Pi}. The correlation proposed by K. S. Pederson *et al.* (1991) can be used to calculate Δc_{Pi} for all, P, N, and A species:

$$\Delta c_{Pi} = 0.3033 M_i - 4.635 \times 10^{-4} M_i T, \tag{5.32}$$

where Δc_{Pi} is in calories/(gmole·K) and T is in degrees K. For the limitation of the above equation, the discussions by Lira-Galeana *et al.* (1996) may be useful.

Critical properties and acentric factor. The correlations for critical properties discussed in Chapter 3 are based on the properties of light hydrocarbons ($<C_{20}$). Those correlations do not differentiate between paraffins, naphthenes and aromatics. Riazi and Al-Sahhaf (1995) provide a general correlation of critical properties and acentric factors for PNA species. Pan, *et al.* (1997b) have modified the correlation of Riazi and Al-Sahhaf for the critical pressure of PNA species with a molecular weight of more than 300 g/gmole:

$$P_c = A - Be^{-CM}, \tag{5.33}$$

where M is the molecular weight. Coefficients A, B, and C are listed in Table 5.1. P_c, A, and B have the units of bar, and C has the unit of gmole/g. The correlation of Riazi and Al-Sahhaf for T_c is satisfactory:

$$\ln(\theta_\infty - \theta) = a - bM^c \tag{5.34}$$

Coefficients θ_∞, a, b, and c for various PNA species are given in Table 5.2 for both $\theta = T_b$ (boiling-point temperature) and $\theta = T_b/T_c$ (reduced boiling-point temperature).

Pan *et al.* (1997b) proposed the following expression for estimating ω of aromatics:

$$\ln \omega = -36.1544 + 30.94 M_i^{0.026261} \ (M \le 800). \tag{5.35}$$

When $M > 800$, $\omega = 2.0$. Riazi and Al-Sahhaf's expressions for the acentric factor of paraffins and naphthenes are given by Eq. (5.34) with the coefficients given in Table 5.2.

TABLE 5.1   Coefficients of Eq. (5.33) (from Pan *et al.*, 1997b)

Coefficient	Paraffins	Naphthenes	Aromatics
A	0.679091	2.58854	4.85196
B	−22.1796	−27.6292	−42.9311
C	0.00284174	0.00449506	0.00561927

TABLE 5.2   Constants of Eq. (5.34) (from Riazi and Al-Sahhaf, 1995)

θ	Paraffins	Naphthenes	Aromatics
$\theta = T_b$			
θ_∞	1070	1028	1015
a	6.9829	6.9565	6.91062
b	0.02013	0.02239	0.02247
c	2/3	2/3	2/3
$\theta = T_b/T_c$			
θ_∞	1.15	1.2	1.03
a	-0.41966	0.06765	-0.29875
b	0.02436	0.13763	0.06814
c	0.580	0.35	0.5
$\theta = \omega$			
θ_∞	0.3	0.3	
a	-3.06826	-8.25682	
b	-1.04987	-5.33934	
c	0.20	0.08	

Let us now use the multisolid model described above to study (1) wax composition, (2) the effect of pressure, and (3) the effect of composition on wax precipitation.

Wax composition.   Consider a model fluid consisting of normal pentane and PNA C_{10}, C_{15}, C_{20}, C_{25}, C_{30}, C_{35}, C_{40}, and C_{45} with the compositions listed in Table 5.3. Figure 5.6 shows the wax precipitation process at 1 atm as the temperature decreases. This figure was obtained using the multisolid-wax model (Pan *et al.*, 1997b). The results reveal that (1) the precipitated wax does not contain aromatics, (2) the normal paraffins with the same carbon number as the naphthenes precipitate first, (3) at high temperatures, n-paraffins constitute the wax, and (4) the lightest component that can be found in the wax phase is P–C_{15} which precipitates around 237 K. The multiphase nature of precipitation and other features were also observed experimentally by Rønningsen *et al.* (1991), and by Coutinho and Ruffier-Meray (1997). The model results and experimental data indicate that if a light crude or a gas condensate has high concentration of paraffins, wax precipitation may be encountered in production and transfer in pipelines at temperatures as high as 150°F.

Now let us examine the effect of pressure on wax precipitation.

Effect of pressure on wax precipitation.   Pressure increase causes the cloudpoint temperature (which is the temperature at which wax first begins to precipitate) to increase for a liquid mixture of fixed composition.

TABLE 5.3   Composition and Molecular Weight of the Model-Synthetic Oil (from Pan *et al.*, 1997b)

Comp.*	Mole%	M, g/mole
P-C5	40.0	72
P-C10	5.0	144
N-C10	5.0	142
A-C10	5.0	136
P-C15	5.0	212
N-C15	5.0	210
A-C15	5.0	204
P-C20	2.0	284
N-C20	3.0	282
A-C20	2.0	276
P-C25	1.0	352
N-C25	3.5	350
A-C25	2.5	344
P-C30	1.0	424
N-C30	2.0	422
A-C30	2.0	416
P-C35	1.0	492
N-C35	2.5	490
A-C35	1.5	484
P-C40	0.5	562
N-C40	1.0	560
A-C40	1.5	554
P-C45	0.5	632
N-C45	1.0	630
A-C45	1.5	624

*P-normal paraffins; N-iso-paraffin and naphthenes; A-aromatics

The solubility of the first precipitating component in the liquid, which is expected to be a heavy normal paraffin at the cloudpoint temperature (CPT), can be expressed as (see Example 5.1)

$$x_i^L \approx F(T) \exp\left[-\frac{P(v_i^L - v_i^S)}{RT} \right], \tag{5.36}$$

where x_i^L is the mole fraction (solubility) in the liquid, $F(T)$ is a parameter that is a function of temperature, and v_i^L, and v_i^S are the molar volumes of component i in the liquid and solid phases, respectively. The liquid molar volume is a weak function of pressure, and the effect of pressure on solid molar volume is negligible. According to Eq. (5.36), a pressure increase results in a decrease in the solubility of the precipi-

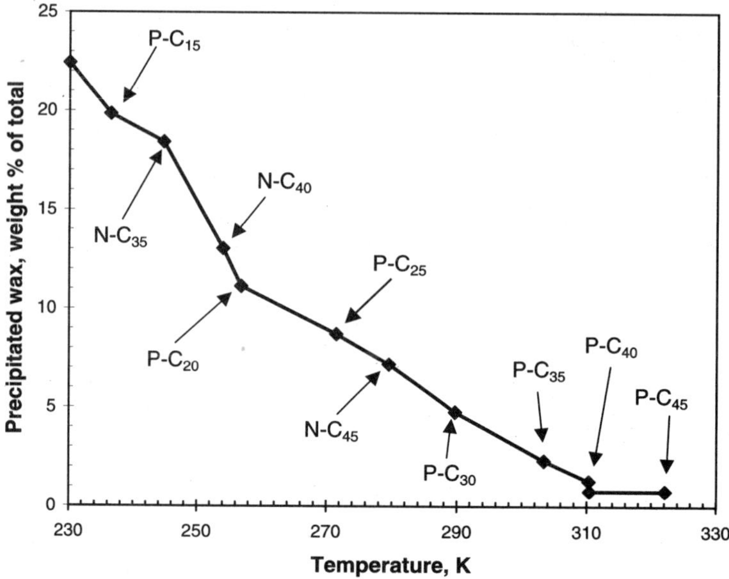

Figure 5.6 Calculated wax amount and precipitating species for the model-synthetic oil: $P = 1$ atm (adapted from Pan *et al.*, 1997b).

tating component in the liquid phase. Consequently, the cloudpoint temperature (CPT) may increase as a result of pressure increase.

Figure 5.7 shows the effect of pressure on the cloudpoint temperature for three different stock tank oils. Note that these results are based on calculations from the multisolid model. For all three oils, the CPT increases as pressure increases. However, oil 2 is more sensitive to pressure than the other two oils. As we will see in Problems 5.7 and 5.8, the effect of pressure on wax precipitation in gases is different than in liquids.

Effect of composition on wax precipitation. Wax precipitation is often measured using the stock-tank oil at atmospheric pressure. When some gas is dissolved into the crude oil, the cloudpoint temperature may decrease, which is desirable. Low cloudpoint temperature implies that one may be less concerned about wax precipitation.

A very clear effect of composition on CPT can be seen from the data presented in Table 5.4. The composition of the light crude oil (that is, oil 4) is shown in Table 5.5. The CPT at 38.3 bar is 318.9 K for the stock-tank oil. When 30 mole% C_1 is dissolved in the crude oil as a result of pressure increase, the CPT is 316.8 K. The change in CPT is the result of two effects: (1) pressure, and (2) composition. The pressure increase from 38.3 to 106.2 bar raises the CPT say by 0.5 K. The net effect of

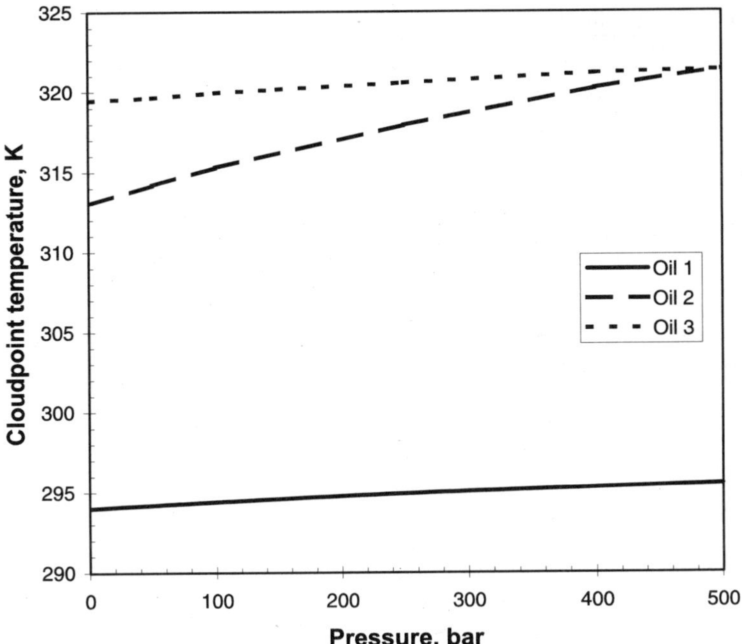

Figure 5.7   Calculated cloudpoint temperature vs. pressure for three stock-tank oils (adapted from Pan *et al.*, 1997b).

TABLE 5.4   Measured results for solvent effect on cloudpoint temperature of oil 4 (from Pan *et al.*, 1997b)

Solvent, mole%	Pressure, bar	CPT, K
0	38.3	318.9
C_1, 30	106.2	316.8
C_3, 30	73.5	316.8
nC_5, 30	37.2	314.4
nC_7, 30	37.6	314.2
nC_{10}, 30	36.2	316.0
nC_{12}, 30	35.5	322.0
nC_{15}, 30	35.5	323.0

composition change is, therefore, 1.6 K. The effect of nC_5, and nC_7 is more pronounced than the effect of C_1; 30 mole% nC_7 lowers the CPT to 314.2 K, which is 4.7 K lower than the CPT of the stock-tank oil. For normal alkanes nC_{12} and nC_{15} the trend reverses and the CPT increases.

Madsen and Boistelle (1976, 1979) have measured the solubilities of nC_{28}, nC_{32}, and nC_{36} in normal-alkane solvents from nC_5 to nC_{12} at atmo-

TABLE 5.5   Oil 4 Composition Data (from Pan _et al._, 1997b)

Comp.	Mole%	M, g/mole;
C_2	0.0041	30
C_3	0.0375	44
iC_4	0.0752	58
nC_4	0.1245	58
iC_5	0.3270	72
nC_5	0.2831	72
C_6	0.3637	86
C_7	3.2913	100
C_8	8.2920	114
C_9	10.6557	128
C_{10}	11.3986	142
C_{11}	10.1595	156
C_{12}	8.7254	170
C_{13}	8.5434	184
C_{14}	6.7661	198
C_{15}	5.4968	212
C_{16}	3.5481	226
C_{17}	3.2366	240
C_{18}	2.1652	254
C_{19}	1.8098	268
C_{20}	1.4525	282
C_{21}	1.2406	296
C_{22}	1.1081	310
C_{23}	0.9890	324
C_{24}	0.7886	338
C_{25}	0.7625	352
C_{26}	0.6506	366
C_{27}	0.5625	380
C_{28}	0.5203	394
C_{29}	0.4891	408
C_{30}	0.3918	422
C_{31}	0.3173	436
C_{32}	0.2598	450
C_{33}	0.2251	464
C_{34}	0.2029	478
C_{35}	0.1570	492
C_{36}	0.1461	506
C_{37}	0.1230	520
C_{38}	0.1093	534
C_{39}	0.1007	548
C_{40+}	3.0994	700

spheric pressure. The results reveal that the solubility of a heavy hydrocarbon in a solvent first increases with the carbon number and then decreases. As an example, at 298 K, the solubility of nC_{36} increases from nC_5 to nC_7, then decreases as the solvent carbon number increases. Crude oils may have a similar behavior, as the data in Table 5.4 show.

<u>Errata</u>

<u>Thermodynamics of Hydrocarbon Reservoirs</u>

1 – p. 35 – Replace μ_1 with μ_i in Eq. (1.175)

2 – p. 74 – The order of symbols in Fig. 2.14 description are $\circ, \square, \triangle$, and $\bullet$.

3 – p. 79

I – Eq. (2.78) – Replace ∇v with $\nabla \vec{v}$.

II – Eq. (2.80) – Replace J_i with $\vec{J}_i$.

4 – p. 80

I – Eq. (2.81) – Multiply $\displaystyle\sum_{i=1}^{c} \tilde{S}_i \vec{J}_i$ by T.

II – Eq. (2.82) – Multiply $\displaystyle\sum_{i=1}^{c} \tilde{S}_i \vec{J}_i$ by T.

III – Eq. (2.83) should read

$$\tau_{ii} = -\frac{2}{3}\mu \frac{\partial v_i}{\partial x_i} + \frac{2}{3}\mu (\nabla \cdot \vec{v}) \quad i = 1,2,3 \,.$$

5 – p. 105

I – Part (b), add g to the right side of the expression below the line that ends with …above equations,

II – Part (c)

a – First line, add "molar" after the first word. It should read "The molar specific volume…"

b – Correct the right side of the two expressions after "="; replace "0.03157 ft³/lbm" with "1.39 ft³/lbmole" and "0.02283 ft³/lbm" with "2.28 ft³/lbmole".

c – "nC₅" in line 5 should read "nC₇".

d – The right side of the third equation, replace "1.2" with "1.026".

6 – p. 121 – Eq. of Problem 2.2 should read

$$P = P^0 \exp\left[-\frac{Mg}{RT} z \right] \,.$$

7 – p. 125 – In the Hint section, $\vec{J}_i = \rho_i (\vec{v}_i - \vec{v})$ and in the second line $\vec{v}_i$ is correct.

8 – p. 143 – Replace " $(a_{ij})^{1/2}$ " by " a_{ij} " in Eq. (3.29).

9 – p. 143 – The first multiplier of the last term in Eq. (3.32) should read $-\dfrac{A}{2\sqrt{2}B}$.

10 – p. 175 – The right side of Eqs. (3.103) and (3.104) should be divided by "M".

11 – p. 182

    I – The s on the left side of Eq. (3.131) should be lower case.

    II – Add M in the denominator of Eq. (3.133) next to C_T.

12 – p. 194 – Remove "-" sign in the exponential term in the fourth equation from the top.

13 – p. 199 – Delete "z" in "nzC_7" in the figure title

14 – p. 206 – Ref. 13, add ", 1986." at the end.

15 – p. 213- Eq. 4.16 – Replace $\ln(P/P_{ci})$ by $\ln(P_{ci}/P)$.

16 – p. 214 – Line below Eq. (4.20) – Replace $h(0)$ and $h(1)$ with α .

17 – p. 216 – Eq. (4.30) – Insert "=" between the last two terms on the right.

18 – p. 243 – Line 9 – It should read at $T_1 > T_c$ ...

19 – p. 255

    I – Eq. (4.169) – Delete power "2" from the "( )" on the left and add to "[   ]"

    II – Eq. (4.170) – Remove "(" and add to the left of single sum.

    III – Eq. (4.171) – Remove "(" from bottom line and add before the sum.

20 – p. 266 – Eq. (4.212) – Replace λ_{c+1} by X_{c+1} .

21 – p. 269 – Eq. (4.220) – Multiply the term in the second line of the equation by 2.

22 – p. 288 – Modify $h(\alpha = 0) > 0$ and $h(\alpha = 1) < 0$.

23 – p. 291 – Remove "s" from Examples in the Hint to problem (4.13).

Asphaltene precipitation

A variety of substances of diverse chemical nature comprise a crude oil. In order to study asphaltene precipitation, these substances can be classified into two groups (1) nonpolar hydrocarbons such as paraffins, naphthenes, and aromatics of moderate molecular weight, and (2) polar polyaromatic materials. The polar aromatics, which may contain metals and nitrogen, are part of the heavy-nonvolatile end of the crude oil and can be subdivided into resins and asphaltenes; resins are less polar than asphaltenes. Under certain conditions, resins and asphaltenes precipitate from the crude oil.

The chemical structure and physicochemical properties of asphaltenes and resins are not well understood. The operational definitions of asphaltenes and resins are based on their solubility in different diluents. Asphaltenes are defined as the fraction of crude oil insoluble in excess normal alkanes such as *n*-pentane but soluble in excess benzene and toluene at room temperature. Resins are defined as the fraction of crude oil insoluble in excess liquid propane at room temperature. Resins are adsorbed on silica, alumina, or other surface-active material. Figure 5.8 shows the precipitation when a bitumen oil is mixed with various diluents. The normal alkanes used are *n*-pentane, *n*-hexane, *n*-heptane, *n*-octane, *n*-nonane, and *n*-decane. According to the operational definition stated above, the asphaltene content of the bitumen

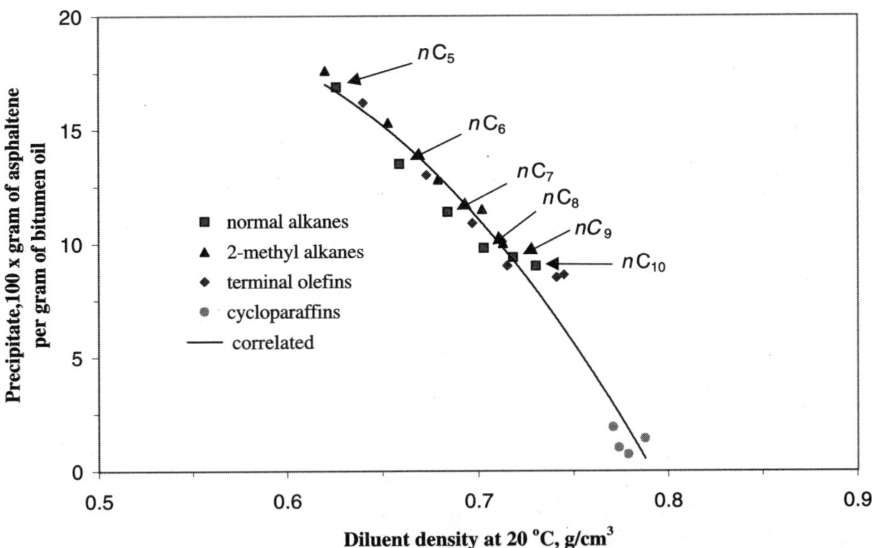

Figure 5.8 Effect of various diluents on precipitation from bitumen oil at 20°C: 40 cm^3 diluent/cm^3 bitumen oil (adapted from Wu, 1998; data of Mitchell and Speight, 1973).

oil from Fig. 5.8, is about 17% (wt). Note that cycloparaffins result in precipitation of a very small amount. The cycloparaffins used in Fig. 5.8 include cyclopentane and decalin. Since the density of toluene and benzene at 20°C is about 0.88 g/cm^3, according to Fig. 5.8 there will be no precipitation. This figure also implies that C_2 to nC_4 should result in greater precipitation. One reason for greater precipitation is that C_2 to nC_4 may cause precipitation of resins in addition to asphaltenes; C_2 diluent may also precipitate components other than asphaltenes and resins.

Since asphaltenes and resins are polar, they may associate. Association between asphaltene molecules in different diluents results in different molecular weights ranging from 800 to 50,000 or even higher. Small-angle X-ray scattering (SAXS) measurements suggest that all asphaltene molecules are of similar size and molecular weight (Lin *et al.,* 1991). (We will later discuss the molecular weight of asphaltenes and resins.) Association of different species in a crude is schematically shown in Fig. 5.9. This sketch is based on comprehensive studies of the asphaltene aggregation in crude oils using SAXS, small-angel neutron scattering (SANS), rheology, and specific-heat measurements (Storm and Sheu, 1995; Storm *et al.*, 1995; Wiehe and Liang, 1996). Storm *et al.* (1995) suggest that the asphaltene micelles have spherical shape. Other shapes have also been assumed. The asphalt-free species in Fig.

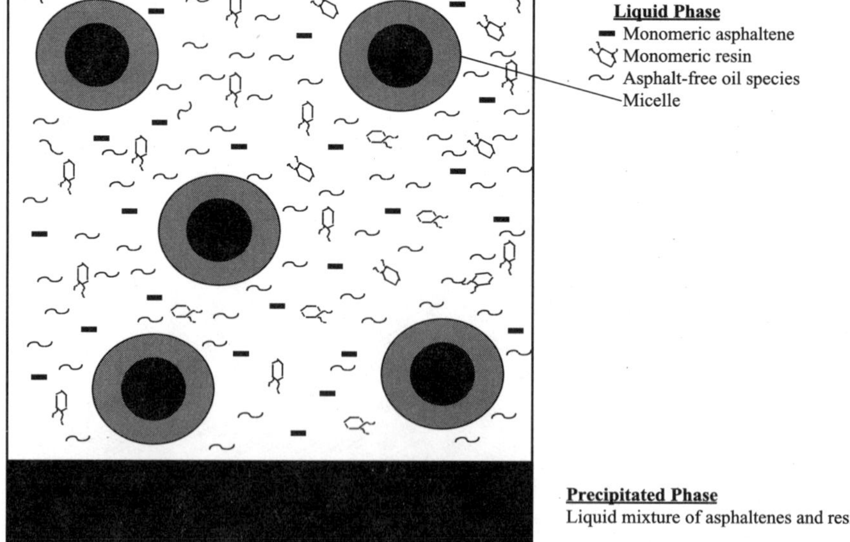

Figure 5.9   Schematic representation of a crude/precipitate system (adapted from Pan and Firoozabadi, 1998b).

5.9 are all the species except asphaltenes and resins. The asphaltene molecules can be found as monomers (that is, single molecules) in the bulk phase and in the micellar core. The resins are also found both in the monomeric state in the bulk phase and in the micellar shell. The nonpolar asphalt-free species are also found both in the bulk phase and in the micellar shell. Figure 5.10 portrays a single micelle in the crude oil. This sketch shows that the micellar core is comprised only of asphaltene molecules. The shell surrounding the asphaltene core contains the resins and the asphalt-free oil components. The sketch in Fig. 5.10 is very similar to the micellar structure suggested by Wiehe and Liang (1996) with the exception that those authors assume a high concentration of aromatics in the shell. Figure 5.11 depicts the sketch for the shell when a crude is diluted with an aromatic. Note that the aromatic solvent does not adsorb onto the asphaltene core.

As a result of change in pressure, composition, and temperature to a lesser degree, the micellar size shown in Figs. 5.10 and 5.11 may change (Nielsen, *et al.*, 1994). The equilibrium between the precipitated phase and the bulk-liquid phase will be effected as a result of the micellar size. This is one main reason that cubic equations such as the PR-EOS, which have been used so successfully in vapor–liquid equilibria, may not be suitable for asphaltene-precipitation and asphaltene-precipitation inhibition calculations. This point will become clear later.

Various methods have been used for asphaltene-precipitation calculations. Most models in the literature are based on the classical Flory-Huggins polymer-solution theory (see Chapter 1) coupled with the Hildebrand regular solution (see Chapter 1) (Hirschberg *et al.*, 1984; Burke *et al.*, 1990; Kokal *et al.*, 1992; Kawanaka *et al.*, 1991; MacMillan

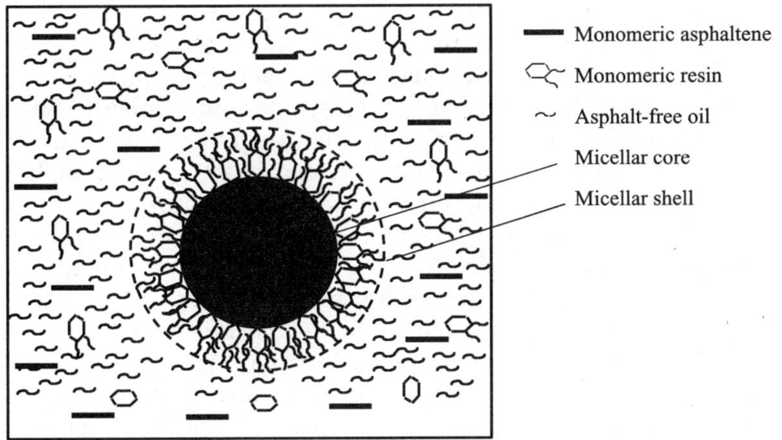

Figure 5.10   Schematic representation of a micelle in crude (adapted from Pan and Firoozabadi, 1998b).

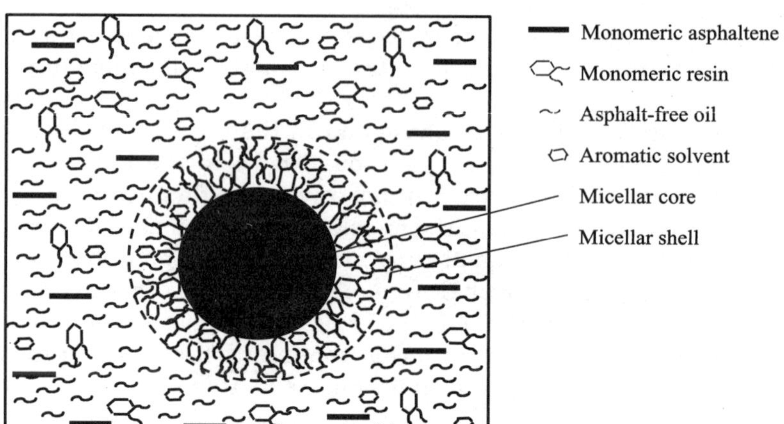

Figure 5.11   Schematic representation of a micelle in a mixture of the crude/aromatic solvent (adapted from Pan and Firoozabadi, 1998c).

et al., 1995). Those models do not consider (1) the change in association between asphaltene molecules, and (2) the peptizing effect of resin molecules (the term peptizing means the dispersion of asphaltenes in the crude by a substance such as resins and amphiphiles (Gonzales and Middea, 1991)). The EOS-approach has also been used to model asphaltene precipitation (Nghiem *et al.*, 1993). Leontaritis and Mansoori (1987) proposed a colloidal model that is based on the assumption that the insoluble solid asphaltene particles are suspended in the crude oil; the suspended asphaltene particles are stabilized by the adsorbing resins on their surface. In this model, resins are necessary for the asphaltenes to exist in solution. The above models have not been tested for the effects of pressure, temperature, and composition including aromatics and polar material when added to the crude oil.

It was stated earlier that wax and asphaltene precipitation are different processes; the effects of composition, pressure, and temperature on wax precipitation can be in the opposite direction to those on asphaltene precipitation. As an example, an increase in pressure often increases the CPT in a crude oil; that is, pressure increase enhances wax precipitation from crude oils (see Fig. 5.7). On the other hand, as we will see soon, pressure increase can inhibit asphaltene precipitation. Wax precipitation is strongly effected by temperature; temperature may weakly effect asphaltene precipitation, and may enhance or inhibit it. The composition effect is also very different on wax and on asphaltene precipitation. An increase in concentration of light hydrocarbons such as C_3 and nC_5 and nonhydrocarbons such as CO_2 decreases the CPT (see Fig. 5.24, to be discussed later). On the other hand, an increase in the amount of these species can significantly enhance asphaltene precipitation. The composition of the precipitate is also very different in wax and asphaltene precipitation. The precipitated wax does not contain asphaltenes, resins or other aromatics. Asphaltenes and resins are the principal components of the precipitate in asphaltene precipitation. The mixing of a crude oil with some polar species in small amounts may have a strong effect on asphaltene precipitation, whereas it may have very little effect on wax precipitation. These substantial differences between wax and asphaltene precipitation are the main reasons that we have opted to present a thermodynamic micellization model (Victorov and Firoozabadi, 1996) for asphaltene precipitation. In the following, we will first present the standard Gibbs free energy of micellization, then briefly review the state of the precipitated phase followed by Gibbs free energy expressions for the liquid and the precipitated phase. The solution of the equilibrium problem and results will complete the presentation on asphaltene precipitation.

Standard Gibbs free energy of micellization

The micelle sketched in Fig. 5.10 contains an asphaltene core with n_1 asphaltene molecules; n_2 resin molecules are adsorbed onto the surface of the core. In addition to resins that are part of the solvation shell surrounding the core, asphalt-free oil species are also present in the shell. The formation of the solvation shell around the asphaltene core lowers the Gibbs free energy. The standard Gibbs free energy of micellization, ΔG_m^{00}, represents the standard Gibbs free energy difference between (1) n_1 asphaltene molecules in the core, n_2 resin molecules in the shell, and (2) those n_1 and n_2 molecules in an infinite-dilution petroleum mixture:

$$\Delta G_m^{00} = \mu_m^* - n_1 \mu_{1a}^* - n_2 \mu_{1r}^*, \tag{5.37}$$

where μ_m^* is the standard-state chemical potential of the micelle and μ_{1a}^* and μ_{1r}^* are the standard-state chemical potentials of the monomeric asphaltenes and resins, respectively. The standard state is the infinite-dilution state with respect to micelles, monomeric asphaltenes, and resins (see Chapter 1). The magnitude of ΔG_m^{00} reflects many complex physiochemical factors, such as the lipophobic effect, the interaction between asphaltene molecules in the micellar core, the micellar core–bulk crude interfacial effect, and the adsorption of resin molecules onto the surface of the micellar core. Here we will explain the term lipophobic. Because of the synonymity with the term hydrophobic, we will explain the terms hydrophobicity and hydrophilicity first. Water is composed of tiny H_2O dipoles and, therefore, polar substances dissolved in water are called hydrophilic substances, that is, water-loving substances. On the other hand, hydrocarbons such as paraffins, which are nonpolar and have a nonpolar bond between carbon and hydrogen, will not dissolve in water and are called hydrophobic or water-fearing substances. A group of substances that can be dissolved in lipids are called lipophilic, and those that cannot dissolve in lipids are called lipophobic. Since the asphaltene molecules are polar and the asphalt-free oil is nonpolar, the asphaltene molecules are lipophobic and transfer from the crude to form the micellar core. It should be mentioned that adjectives hydrophobic, hydrophilic, lipophobic, and lipophilic denote only general trends. Pan and Firoozabadi (1998b) suggest a reversible micelle formation process sketched in Fig. 5.12 for the evaluation of ΔG_m^{00}. There are four steps in the process.

1a Free energy transfer of asphaltene molecules. Part of the first step is the change in free energy as a result of the transfer of n_1 asphaltene molecules from the infinite-dilution solution to the aggregated state (see Fig. 5.12). The aggregate state is assumed to be a pure-liquid asphaltene state; $(\Delta G_a^0)_{tr}$ expresses the free energy change due to this transfer:

$$(\Delta G_a^0)_{tr} = n_1(\mu_a^m - \mu_{1a}^*), \tag{5.38}$$

where μ_a^m is the liquid chemical potential per asphaltene molecule in the micellar core. In order to evaluate μ_a^m, we assume that the Gibbs free

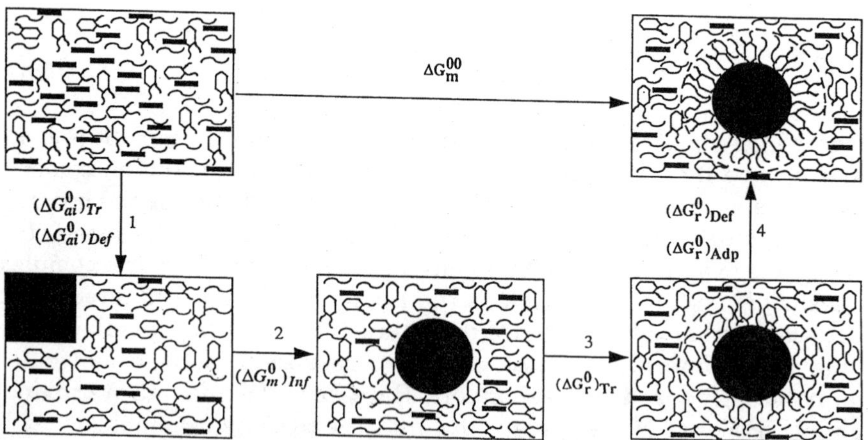

Figure 5.12 Schematic representation of the reversible process of micellar formation (adapted from Pan and Firoozabadi, 1998b).

energy of an asphaltene molecule in the micellar core consists of: (1) association and (2) standard state:

$$\mu_a^m = (\Delta g_a^0)_{as} + \mu_a^0(T, P), \tag{5.39}$$

where $(\Delta g_a^0)_{as}$ is the association free energy and μ_a^0 is the chemical potential of an asphaltene molecule in the pure-liquid state at temperature T and pressure P. The expression for μ_{1a}, the chemical potential of the monomeric asphaltenes is

$$\mu_{1a} = \mu_a^0(T, P) + kT \ln x_{1a} + kT \ln \gamma_{1a}, \tag{5.40}$$

where x_{1a} is the molecular fraction of the monomeric asphaltenes in the solution and γ_{1a} is the activity coefficient. In Eq. (5.40), k is the Boltzmann constant which replaces the gas constant R when molecules rather than moles are used as a basis. [The relation between the number of molecules n, and the number of moles N, can be readily obtained from $PV = NRT$ where R is the gas constant, and $PV = nkT$; $n = (k/R)N = N_A N$, where N_A is the Avogadro number.] When infinite dilution is used as the standard state, the expression for μ_{1a} is

$$\mu_{1a} = \mu_{1a}^*(T, P) + kT \ln x_{1a} + kT \ln \gamma_{1a}^*, \tag{5.41}$$

where γ_{1a}^* is the activity coefficient of the asphaltene molecules using infinite dilution as the standard state. Since the left side of Eqs. (5.40) and (5.41) are the same, then

$$\mu_a^0 - \mu_{1a}^* = -kT \ln \frac{\gamma_{1a}}{\gamma_{1a}^*}, \tag{5.42}$$

which is also valid at infinite dilution (that is, $x_{1a} \to 0$),

$$\mu_a^0 - \mu_{1a}^* = -kT \ln \frac{\gamma_{1a}^\infty}{\gamma_{1a}^{*,\infty}}. \tag{5.43}$$

At infinite dilution, $\gamma_{1a}^{*,\infty} = 1$ (see Eq. (1.140) of Chapter 1), and Eq. (5.43) simplifies to

$$\mu_a^0 - \mu_{1a}^* = -kT \ln \gamma_{1a}^\infty. \tag{5.44}$$

The activity coefficient γ_{1a}^∞ can be estimated from

$$\gamma_{1a}^\infty = \frac{\varphi_a^\infty}{\varphi_a^L(T, P)}, \tag{5.45}$$

where φ_a^∞ and $\varphi_a^L\,(T, P)$ are the asphaltene fugacity coefficients at infinite dilution and the pure-liquid state, respectively, both at temperature

T and pressure P (see Problem 1.14, Chapter 1). The fugacity coefficients φ_a^∞ and φ_a^L can be estimated from an EOS (in the estimation of φ_a^∞, one may use Eq. (3.32) of Chapter 3 and set $x_a = 0$). Combining Eqs. (5.38), (5.39), (5.44), and (5.45),

$$(\Delta G_a^0)_{tr} = n_1 \left[(\Delta g_a^0)_{as} - kT \ln \frac{\varphi_a^\infty}{\varphi_a^L} \right]. \tag{5.46}$$

The first and second terms in Eq. (5.46) are negative. The association results in lowering the Gibbs free energy. As a result, the Gibbs free energy of transfer in Eq. (5.46) lowers the Gibbs free energy and, there-fore, micellar formation and growth are favored. The transfer of asphal-tenes also results in the deformation of asphaltene molecules, which is described next.

1b Free energy of deformation of asphaltene molecules. The Gibbs free energy of deformation of asphaltene molecules can be obtained from the expression (Flory, 1953; Nagarajan and Ruckenstein, 1991; Nagara-jan and Ganesh, 1989a, 1989b),

$$(\Delta G_a)_{def} = \frac{n_1 kT}{2} \left[\frac{R^2}{m_a L_{oil}^2} + 2\frac{m_a^{1/2} L_{oil}}{R} - 3 \right], \tag{5.47}$$

where R is the radius of the asphaltene core, m_a is the ratio of the mole-cular volumes of asphaltene and asphalt-free oil species, V_a/V_{oil}, and L_{oil} is the characteristic length of the asphalt-free oil molecules (see Example 5.5 for the calculation of molecular volumes). The characteris-tic length is estimated from $(V_{oil})^{1/3}$. In order to deform asphaltene mole-cules, one needs to perform work; therefore, $(\Delta G_a^0)_{def}$ is positive, which prevents the micelle from infinite growth.

2 Interfacial free energy of asphaltene-core formation. In order to form a bubble or a droplet from a bulk phase, as we have seen in Chapter 2, work must be performed. Therefore, the contribution of the interfacial effect $(\Delta G_m^0)_{int}$ to the standard Gibbs free energy of micelle formation is positive and prevents the micelle from infinite growth. The Gibbs free energy of formation of a micelle is given by

$$(\Delta G_m^0)_{int} = \sigma \mathcal{A}, \tag{5.48}$$

where σ is the asphaltene core–bulk crude oil interfacial tension and $\mathcal{A}$ is the total surface area of the micellar core, $\mathcal{A} = 4\pi R^2$. There are three implicit assumptions in the above equation: (1) the pressure is assumed to be the same in the surrounding liquid and in the asphaltene core, (2) the interfacial tension is assumed to be independent of the core radius, and (3) the osmotic pressure effect is neglected. Equation (5.48) can be written in an alternative form by dividing $\mathcal{A}$ into n_2 equal surface areas and using the symbol a to represent the surface area of one unit, $a = 4\pi R^2/n_2$,

$$(\Delta G_m^0)_{int} = n_2 \sigma a. \tag{5.49}$$

3 Transfer free energy of resin molecules.　The expression for the transfer free energy of n_2 resin molecules from an infinite-dilution solution to the shell is given by

$$(\Delta G_r^0)_{tr} = n_2(\mu_r - \mu_{1r}^*),$$

where μ_r is the chemical potential of a resin molecule in the shell. The expression for μ_r is

$$\mu_r = \mu_r^0(T, P) + kT \ln x_{r,sh} + kT \ln \gamma_{r,sh}. \tag{5.50}$$

In the above equation, μ_r^0 (T,P) is the chemical potential of a resin molecule in the pure liquid state, $x_{r,sh}$ is the molecular fraction of resin in the shell, and $\gamma_{r,sh}$ is the activity coefficient of the resin in the shell. Similarly to Eq. 5.44,

$$\mu_r^0 - \mu_{1r}^* = -kT \ln \gamma_{1r}^\infty, \tag{5.51}$$

where γ_{1r}^∞ is the activity coefficient of the resin in an infinite-dilution solution with the standard state being a pure liquid state. Combining Eqs. (5.49) to (5.51),

$$(\Delta G_r^0)_{tr} = n_2 kT[\ln \gamma_{r,sh} x_{r,sh} - \ln \gamma_{1r}^\infty]. \tag{5.52}$$

Both γ_{1r}^∞ and $\gamma_{r,sh}$ can be evaluated by an EOS:

$$\gamma_{1r}^\infty = \frac{\varphi_r^\infty}{\varphi_r^L} \tag{5.53}$$

$$\gamma_{r,sh} = \frac{\varphi_{r,sh}^L}{\varphi_r^L}, \tag{5.54}$$

where $\varphi_{r,sh}^L$ is the fugacity coefficient of the resin in the shell and φ_r^∞ is the fugacity coefficient of the resin at infinite dilution. Combining the above three equations, one obtains

$$(\Delta G_r^0)_{tr} = n_2 kT \ln\left[\frac{\varphi_{r,sh}^L x_{r,sh}}{\varphi_r^\infty}\right]. \tag{5.55}$$

The expression for $x_{r,sh}$ can be derived as follows:

$$V_{sh} = n_2 V_r + n_{oil} V_{oil},$$ (5.56)

where V_{sh} is the volume of the micellar shell and n_{oil} is the molecular number of the asphalt-free oil species in the shell. Equation (5.56) can be written as

$$n_{oil} = \frac{(V_{sh} - n_2 V_r)}{V_{oil}}.$$ (5.57)

Therefore,

$$x_{r,sh} = \frac{n_2}{n_2 + n_{oil}} = \frac{n_2}{[n_2 + (V_{sh} - n_2 V_r)/V_{oil}]}.$$ (5.58)

The shell volume is given by

$$V_{sh} = \frac{4\pi}{3}[(R + D)^3 - R^3],$$ (5.59)

where D denotes the shell thickness. The final expression for $(\Delta G_r^0)_{tr}$ is given by

$$(\Delta G_r^0)_{tr} = n_2 kT \ln \varphi_{r,sh}^L \left\{ \frac{n_2}{n_2 + \frac{4\pi}{3}[(R + D)^3 - R^3 - n_2 V_r]/V_{oil}} \right\} / \varphi_r^\infty.$$

(5.60)

The above term is also negative, implying that the transfer of resin molecules from the bulk phase to the solvation shell is the preferred state.

4a Adsorption free energy of resins. When n_2 resin molecules adsorb onto the surface of the micellar core at constant temperature and pressure, without consideration of the interfacial effect, the Gibbs free energy change is given by $n_2[(\Delta h_r^0)_{ads} - T(\Delta s_r)_{ads}]$. The adsorption of resins, however, decreases the contact surface between the asphaltene core and the asphalt-free oil species by $n_2 \sigma a_0$ where a_0 is the contact area of the resin molecule onto the surface of the micellar core. This decrease of interfacial free energy is called the interfacial screening effect in micellar science. The screening free energy for each resin molecule adsorbed onto the micellar core is σa_0. Therefore, the adsorption free energy for each resin molecule consists of three parts: (1) adsorption enthalpy, $(\Delta h_r^0)_{ads}$, (2) adsorption entropy, $(\Delta s_r^0)_{ads}$, and (3) interfacial screening energy:

$$(\Delta G_r^0)_{ads} = n_2[(\Delta h_r^0)_{ads} - T(\Delta s_r)_{ads} - \sigma a_0].$$ (5.61)

In the above equation, $(\Delta h_r^0)_{ads}$ is a measure of heat interaction between asphaltene and resin molecules. The adsorbed resin molecules generate steric repulsion among the polar heads, which results in a loss of entropy compared with the unadsorbed molecule; the entropy change is shown by $(\Delta s_r)_{ads}$. While (Δh_r^0) can be measured directly or estimated as we will see later, $(\Delta s_r)_{ads}$ can be calculated from (Nagarajan and Ruckenstein, 1991)

$$(\Delta s_r)_{ads} = k \ln\left[1 - \frac{a_p}{a}\right], \tag{5.62}$$

where a_p is the effective cross-sectional area of a resin molecule attached to the polar head. The adsorption enthalpy is negative, whereas the adsorption entropy term is positive. The interfacial screening energy is also negative. Once we add the last term of Eq. (5.61) to Eq. (5.49), the term $n_2\sigma(a - a_0)$ also represents the interfacial effect, which is positive.

4b Free energy of resin deformation. The deformation free energy of n_2 resin molecules is also expressed by an equation similar to that of asphaltene deformation,

$$(\Delta G_r^0)_{def} = \frac{n_2 k T}{2}\left[\frac{D^2}{m_r L_{oil}^2} + 2\frac{m_r^2}{D} - 3\right], \tag{5.63}$$

where m_r is the ratio of the molecular volumes of the resin to the asphalt-free oil species.

Now we can sum up all the terms to provide the expression for ΔG_m^{00}:

$$\Delta G_m^{00} = [(\Delta G_a^0)_{tr} + (\Delta G_a^0)_{def}] + (\Delta G_m^0)_{int} + [(\Delta G_r^0)_{tr}] + [(\Delta G_r^0)_{ads} + (\Delta G_r^0)_{def}]. \tag{5.64a}$$

Substituting from Eqs. (5.46), (5.47), (5.49), (5.60–5.62), and (5.63) into Eq. (5.64a),

$$\frac{\Delta G_m^{00}}{kT} = n_1\left[\frac{(\Delta g_a^0)_{as}}{kT} + \ln\frac{\varphi_a^L}{\varphi_{1a}^\infty} + \frac{1}{2}\left(\frac{R^2}{m_a L_{oil}^2} + 2\frac{m_a^{1/2}L_{oil}}{R} - 3\right)\right]$$

$$+ n_2\frac{\sigma(a - a_0)}{kT} + n_2\left[\frac{(\Delta h_r^0)_{ads}}{kT} - \ln\left(1 - \frac{a_p}{a}\right) + \ln\frac{\varphi_{r,sh}^L x_{r,sh}}{\varphi_r^\infty}\right.$$

$$\left.+ \frac{1}{2}\left(\frac{D^2}{m_r L_{oil}^2} + 2\frac{m_r^{1/2}}{D} - 3\right)\right]. \tag{5.64b}$$

The first bracket on the left represents the Gibbs free energy transfer of n_1 asphaltene molecules and their deformation. The second term represents the interfacial Gibbs free energy, and the second bracket represents various terms for resin contribution to the micellization process. In Example 5.6, an alternative form of Eq. (5.64a) will be presented.

Next we will derive the expression for the Gibbs free energy of the liquid phase, making the assumption that all the micelles are of the same size (that is, they are monodispersed). The monodispersity assumption, although supported by experimental data (Storm *et al.*, 1995) can be relaxed at the expense of computational complexity.

Gibbs free energy of the liquid phase

As the sketch in Fig. 5.10 shows, the bulk liquid phase consists of micelles, monomeric asphalts and resins, and asphalt-free oil monomers in the bulk phase and in the shell. Let us denote asphaltenes and resins as the solute and the rest of the species (that is, asphalt-free oil species) as the solvent. Then the Gibbs free energy of the liquid phase, G^{L_1}, can be written as

$$G^{L_1} = G_{solvent} + G_{solute}.$$ (5.65)

Later we will represent the precipitated phase by L_2. We assume a total of c components/pseudocomponents in the crude oil. The component indices for the asphaltenes and resins are c and $(c-1)$, respectively. Therefore, the number of species in the asphalt-free oil is $(c-2)$. The expression for $G_{solvent}$ is

$$G_{solvent} = \sum_{i=1}^{c-2} n_i^{L_1} \mu_i^{L_1},$$ (5.66)

where $n_i^{L_1}$ is the molecular number and $\mu_i^{L_1}$ is the chemical potential of the solvent species i (that is, the asphalt-free oil species). The expression for $\mu_i^{L_1}$ is

$$\mu_i^{L_1} = \mu_i^0(T) + kT \ln \hat{f}_i^{L_1}(T, P, n_1^{L_1}, \ldots, n_c^{L_1}) \qquad i = 1, \ldots, (c-2), \quad (5.67)$$

where $\mu_i^0(T)$ is the chemical potential per molecule of the asphalt-free oil species in the reference state at temperature T and $\hat{f}_i$ is the fugacity of solvent-species i in the liquid phase using the mean-field approximation. In mean-field approximation, one can picture the molecules immersed in a continuum having the fluid's average properties. The symbol " ^ " denotes the mean-field approximation. Note that in Eq. (5.67) the fugacity of the reference state for component i is assumed to be unity. An EOS can be used to calculate $\hat{f}_i$. Combining Eqs. (5.66) and (5.67) gives

$$G_{solvent} = \sum_{i=1}^{c-2} n_i^{L_1} \mu_i^0(T) + kT \sum_{i=1}^{c-2} n_i^{L_1} \ln \hat{f}_i^{L_1}(T, P, n_1^{L_1}, \ldots, n_c^{L_1}). \qquad (5.68)$$

The calculation of the Gibbs free energy of the solute components is more complex (see Pan and Firoozabadi, 1997a). We can divide G_{solute} into three parts,

$$G_{solute} = G^0 + G_{mix} + G_{inter}, \qquad (5.69)$$

where G^0, G_{mix}, and G_{inter} are the standard state, the mixing, and the interaction Gibbs free energies. The standard-state Gibbs free energy is given by

$$G^0 = n_{1a}^{L_1} \mu_{1a}^* + n_{1r}^{L_1} \mu_{1r}^* + n_m \mu_m^*. \qquad (5.70)$$

The material balance equations in the liquid phase are

$$n_a^{L_1} = n_{1a}^{L_1} + n_1 n_m \qquad (5.71)$$

$$n_r^{L_1} = n_{1r}^{L_1} + n_2 n_m, \qquad (5.72)$$

where $n_a^{L_1}$ and $n_r^{L_1}$ are the total number of asphaltene and resin molecules and n_m is the total number of micelles in the liquid phase L_1. Combining Eqs. (5.37) and (5.70) with (5.72), one obtains,

$$G^0 = n_a^{L_1} \mu_{1a}^* + n_r^{L_1} \mu_{1r}^* + n_m \Delta G_m^{00}. \qquad (5.73)$$

The mixing Gibbs free energy is given by (see Example 1.4, Chapter 1)

$$G_{mix} = kT[n_{1a}^{L_1} \ln x_{1a} + n_{1r}^{L_1} \ln x_{1r} + n_m \ln x_m], \qquad (5.74)$$

where x_{1a} and x_{1r} are the species fractions of the monomeric asphaltenes and resins, respectively, and x_m is the molecular fraction of the micelles.

The interaction Gibbs free energy of solute species, G_{inter}, which represents the interactions between the micelles, the monomeric asphaltenes and resins, and the asphalt-free oil species in the liquid solution L_1, can be estimated using the mean-field approximation,

$$G_{inter} = kT[n_a^{L_1} \ln \hat{\gamma}_a^* + n_r^{L_1} \ln \hat{\gamma}_r^*], \qquad (5.75)$$

where $\hat{\gamma}^*$ is the activity coefficient; it is a function of the gross concentration of all species in the liquid solution. The chemical potentials of the asphaltene and resin molecules using the mean-field approximation can be expressed as

$$\mu_i^{L_1} = \mu_i^0(T) + kT \ln x_i^{gross} + kT \ln \hat{\varphi}_i(T, P, n_1^{L_1}, \ldots, n_c^{L_1})P$$
$$i = (c-1) \text{ and } c \quad (5.76)$$

$$\mu_i^{L_1} = \mu_i^*(T, P) + kT \ln x_i^{gross} + kT \ln \hat{\gamma}_i^*(T, P, n_1^{L_1}, \ldots, n_c^{L_1})$$
$$i = (c-1) \text{ and } c, \quad (5.77)$$

where x_i^{gross} is the gross molecular fraction of asphaltenes or resins in liquid phase L_1. From Eqs. (5.76) and (5.77),

$$\mu_i^0(T) + kT \ln \hat{\varphi}_i P = \mu_i^*(T, P) + kT \ln \hat{\gamma}_i^* \qquad i = (c-1) \text{ and } c. \quad (5.78)$$

Combining Eqs. (5.75) and (5.78) gives

$$G_{inter} = n_a^{L_1} \mu_a^0(T) + n_r^{L_1} \mu_r^0(T) + kT[n_a^{L_1} \ln \hat{\varphi}_a P + n_r^{L_1} \ln \hat{\varphi}_r P]$$
$$- [n_a^{L_1} \mu_a^*(T, P) + n_r^{L_1} \mu_r^*(T, P)]. \quad (5.79)$$

Substituting Eqs. (5.73), (5.74), and (5.79) into Eq. (5.69) provides the Gibbs free energy of the solute species,

$$G_{solute} = kT[n_{1a}^{L_1} \ln x_{1a} + n_{1r}^{L_1} \ln x_{1r} + n_m \ln x_m] + n_m \Delta G_m^{00}$$
$$+ n_a^{L_1}[\mu_a^0(T) + kT \ln \hat{\varphi}_a P] + n_r^{L_1}[\mu_r^0(T) + kT \ln \hat{\varphi}_r P]. \quad (5.80)$$

The Gibbs free energy of the liquid phase, L_1, is obtained by adding Eqs. (5.68) and (5.80)

$$\boxed{\begin{aligned} G^{L_1} = {} & \sum_{i=1}^{c} n_i^{L_1} \mu_i^0(T) + kT \sum_{i=1}^{c-2} n_i^{L_1} \hat{f}_i^{L_1} \\ & + kT[n_{1a}^{L_1} \ln x_{1a} + n_{1r}^{L_1} \ln x_{1r} + n_m \ln x_m] \\ & + n_m \Delta G_m^{00} + kT[n_a^{L_1} \ln \hat{\varphi}_a P + n_r^{L_1} \ln \hat{\varphi}_r P] \end{aligned}}. \quad (5.81)$$

To repeat again, at constant T and P the fugacities and fugacity coefficients in Eq. (5.81) are a function only of gross concentrations of all the species:

$$\hat{f}_i = \hat{f}_i(n_1^{L_1}, n_2^{L_2}, \ldots, n_{c-2}^{L_1}, n_{c-1}^{L_1}, n_c^{L_1}) \qquad i = 1, \ldots, (c-2) \qquad (5.82a)$$

$$\hat{\varphi}_i = \hat{\varphi}_i(n_1^{L_1}, n_2^{L_2}, \ldots, n_{c-2}^{L_1}, n_{c-1}^{L_1}, n_c^{L_1}) \qquad i = (c-1) \text{ and } c, \qquad (5.82b)$$

according to the mean-field approximation.

Gibbs free energy of the precipitated phase.   In order to evaluate the Gibbs free energy of the precipitated phase, first the state of this phase should be ascertained. Most of the asphaltene precipitation calculations are based on the assumption that the precipitate is a solid phase. This assumption is most probably true at room temperature but may not be valid at reservoir temperatures which are often higher than 330 K. Experimental studies have shown that the precipitated phase is in the form of dark solid particles when a crude oil is diluted with propane at room temperature (Kokal *et al.*, 1992). When the same crude oil is diluted with *n*-pentane or a heavier normal alkane at room temperature, the precipitated phase has a crystalline state (Kokal *et al.*, 1992). Chung *et al.* (1991) also observed the precipitate to be in a solid state when a crude oil was mixed with *n*-pentane at room temperature.

At high temperatures, Kokal *et al.* (1992), Hirschberg *et al.* (1984) and Godbole *et al.* (1995) observed the precipitation of a black-liquid mixture. A transition from solid to liquid state was observed by Storm *et al.* (1996) at above 330 K. Here we will formulate the Gibbs free energy of the precipitated phase as a liquid. The expression for the Gibbs free energy of the precipitated solid phase will be presented in Example 5.7.

In the following, we will assume that the precipitated phase only consists of asphaltenes and resins. We will also assume that the precipitated phase does not form a micellar solution; there will be no association either between asphaltene molecules or between asphaltene and resin molecules in the precipitate. The Gibbs free energy expression for the precipitated liquid phase L_2 will be simply

$$G^{L_2} = n_a^{L_2}[\mu_a^0(T) + kT \ln f_a^{L_2}] + n_r^{L_2}[\mu_r^0(T) + kT \ln f_r^{L_2}], \qquad (5.83)$$

where $f_a^{L_2}$ and $f_r^{L_2}$ are the fugacities of the asphaltenes and resins in phase L_2 (see Fig. 5.9). Note that Eq. (5.83) does not assume an ideal liquid solution. An EOS can be used to readily calculate fugacities in the above equation.

Gibbs free energy of the liquid and the precipitate system and equilibrium

The total Gibbs free energy of the light-liquid phase L_1 and the precipitated phase L_2 is given by $G = G^{L_1} + G^{L_2}$. Using Eqs. (5.81) and (5.83),

$$
\begin{aligned}
G = {} & \sum_{i=1}^{c} n_i \mu_i^0(T) + kT \sum_{i=1}^{c-2} n_i^{L_1} \ln \hat{f}_i^{L_1} \\
& + kT(n_{1a}^{L_1} \ln x_{1a}^{L_1} + n_{1r}^{L_1} \ln x_{1r}^{L_1} + n_m \ln x_m) \\
& + n_m \Delta G_m^{00} + kT(n_a^{L_1} \ln \hat{\varphi}_a^{L_1} P + n_r^{L_1} \ln \hat{\varphi}_r^{L_1} P) \\
& + kT(n_a^{L_2} \ln f_a^{L_2} + n_r^{L_2} \ln f_r^{L_2}).
\end{aligned}
\tag{5.84}
$$

The material balance expressions for asphaltene and resin molecules are

$$
n_a = n_{1a}^{L_1} + n_1 n_m + n_a^{L_2}
\tag{5.85}
$$

$$
n_r = n_{1r}^{L_1} + n_2 n_m + n_r^{L_2}.
\tag{5.86}
$$

In the above two equations, n_a and n_r are the total number of molecules of asphaltenes and resins in the system, which are fixed quantities.

The minimization of Gibbs free energy of the total system with respect to the independent variables subject to the constraints of Eqs. (5.85) and (5.86) provides the amount and composition of each phase. In general there are eight variables for the calculation of an equilibrium state at constant T and P. Those are $n_1, n_2, D, n_{1a}^{L_1}, n_{1r}^{L_1}, n_m, n_a^{L_2}$, and $n_r^{L_2}$. Once these variables are obtained, the micellar size can be readily determined. Pan and Firoozabadi (1997a) have used the feasible sequential quadratic programming (FSQP) to minimize G (Zhou, Tits, and Lawrence, 1996).

Parameters of the micellization model

A number of parameters are introduced in the standard Gibbs free energy of micellization (that is, Eq. (5.64b)) and the total Gibbs free energy expression (that is, Eq. (5.84)). These parameters can be estimated from the chemical structure of the species and other measurements. In the following, numerical values for these parameters will be discussed.

Adsorption enthalpy $(\Delta h_r^0)_{ads}$. The nature of the resin adsorption onto the asphaltene core is a weak acid–base interaction, similar to a hydrogen bond. Therefore, the formation enthalpy of a hydrogen bond can be used, $(\Delta h_r^0)_{ads} = -11.2 \times 10^3$ J/gmole (see Prausnitz et al., 1986).

Interfacial tension between liquid asphaltene and bulk crude oil.   There is no reported measurement of the interfacial tension between the asphaltene liquid and the asphalt-free oil species in the literature. A value of about 37 dynes/cm may be reasonable at room temperature (Firoozabadi and Ramey, 1998; Firoozabadi, *et al.*, 1998). The interfacial tension will change as the oil composition changes because of dilution with different amounts of a diluent. It will also change with temperature. Based on the work of Nagarangan and Ganesh (1989a, 1989b), Pan and Firoozabadi (1997a) proposed the following expression for the interfacial tension to be used in Eq. (5.49) and Eq. (5.64):

$$\sigma = \sigma_0 \left(\frac{T}{T_0}\right)^{0.35} \left(\frac{L_{oil}}{L_{oil_0}}\right)^2 \left(\frac{\delta_a - \delta_{oil}}{\delta_{a0} - \delta_{oil_0}}\right)^{1.3} \tag{5.87}$$

where $\sigma_0 = 37$ dynes/cm is the interfacial tension between the asphaltene liquid and the asphalt-free crude oil at stock-tank conditions. Other symbols in Eq. (5.87) are δ_{oil}, the solubility parameter of the asphalt-free oil species at temperature T (that is, asphalt-free oil); δ_a, the asphaltene solubility parameter at temperature T; δ_{a0}, and δ_{oil_0}, the solubility parameters of the asphaltene and asphalt-free oil species at temperature T_0, respectively, and L_{oil} and L_{oil_0}, characteristic lengths of asphalt-free oil species at T and T_0, respectively. The estimation of solubility parameters is discussed in Problem 3.10 of Chapter 3 using an EOS.

Gibbs free energy of association, $(\Delta g_a{}^0)_{as}$.   Based on the experimental data of Storm and Sheu (1995), the Gibbs free energy change due to association of asphaltene molecules is estimated to be around $(\Delta g_a^0)_{as} = -10kT$. This term is an important parameter of the micellization model and results in a substantial reduction of the system Gibbs free energy.

Asphaltene and resin molecular parameters.   The molecular weights of asphaltenes and resins are $M_a = 1000 = $ g/gmole (Storm and Sheu, 1995) and $M_r = 850$ g/mole, based on the work of Lian, Lin, and Yen (1994).

The contact area of a resin on the surface of a micellar core is $a_0 = 21$ Å^2 (assuming that the polar head of the resin being the OH group), and the effective cross-sectional area of a resin attached to the polar head is $a_p = 40$ Å^2, using the cross-sectional area of one benzene ring (Nagarajan and Ruckenstein, 1991).

The estimated molar volume of asphaltene molecules is $V_a = 0.9$ m^3/kgmole (see Example 5.5), and that of resin, $V_r = 0.7$ m^3/kgmole. The molar volume of asphalt-free oil can be readily estimated from an EOS.

Results from the micellization model

Pan and Firoozabadi (1997a, 1998b, 1998c) have used the model described above to study asphaltene precipitation from several crude oils. Some results from their work will be presented here.

Figure 5.13 shows the effect of pressure on precipitation from the crude 2 in Hirschberg *et al.* (1984) with propane at a temperature of 367 K (94°C). The composition of crude oil is provided in Table 5.6. Note that the measured asphaltene and resin contents in the stock-tank oil are 0.6 and 5.4 percent (weight), respectively; the ratio of resin to asphaltene is very high. The crude oil of Table 5.6 is mixed with propane at a weight ratio of 1 to 7. The resulting crude/propane mixture has a bubblepoint pressure of about 40 bar at 367 K. According to Fig. 5.13, at pressures above 1250 bar, there is no precipitation. At about 1250 bar, precipitation begins. The amount of the precipitation increases slowly until a pressure of 300 bar and then increases rapidly to the saturation pressure of 40 bar. Below the saturation pressure, as a result of rapid change of composition of propane in the light liquid phase, the amount of precipitate decreases very fast. At about 34 bar, there is no precipitation. Figure 5.14 provides the precipitation amount of the asphaltenes and resins. Note that the precipitation is expressed in weight percent of tank oil. At the precipitation onset, both asphaltenes and resins precipitate. As the pressure decreases

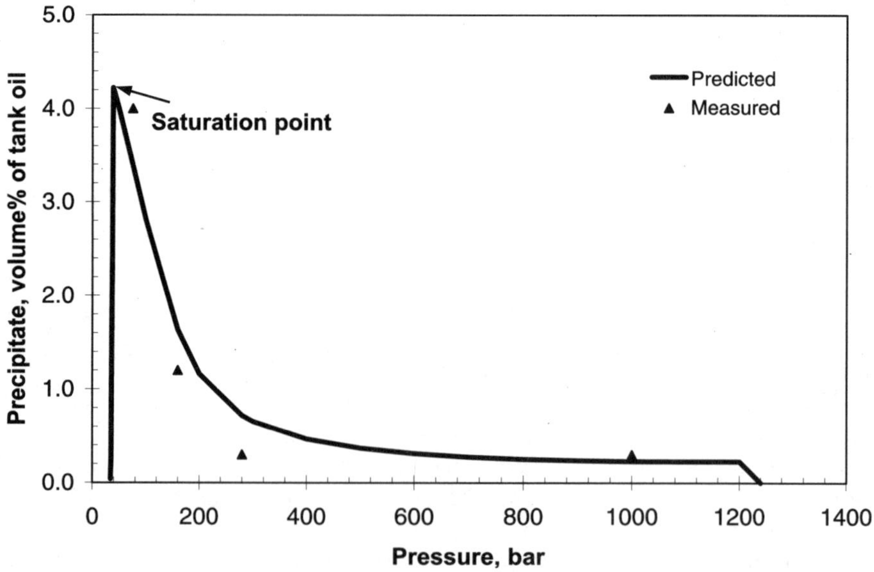

Figure 5.13 Pressure dependence of precipitation for the mixture of crude oil of Table 5.6 with propane at 367 K (adapted from Pan and Firoozabadi, 1997a).

TABLE 5.6   Composition of Hirshberg Stock-Tank Oil 2 (from Victorov and Firoozabadi, 1996)

Comp.	Mole%	M, g/mole
CO_2	0.0	
C_1	0.07	
C_2	0.07	
C_3	0.87	
iC_4	0.53	
nC_4	2.44	
iC_5	1.71	
nC_5	2.36	
C_6	4.32	
C_{7+}	87.63	227
CP1	30.737	140
CP2	20.785	191
CP3	18.951	239
CP4	12.288	309
CP5	3.443	601
r	1.302	850
a	0.123	1000

Weight% of asphaltene and resin in tank oil is 0.6 and 5.4, respectively.

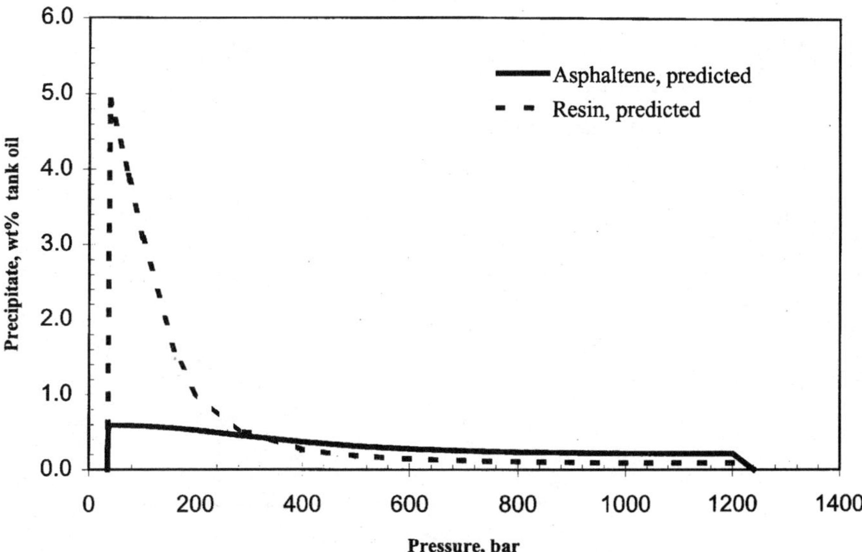

Figure 5.14   Predicted pressure dependence of asphaltene and resin precipitation from the mixture of crude oil in Table 5.6 and propane at 367 K (adapted from Pan and Firoozabadi, 1997a).

below 300 bar, the amount of resin increases much faster than that of asphaltenes. At the saturation pressure, all asphaltenes and nearly all resins precipitate (see Table 5.6).

In asphaltene precipitation calculations, T_c, P_c, ω and interaction coefficients should be provided for the EOS. The calculations in Figs. 5.13 and 5.14 are based on the PR-EOS. The critical pressures of resins and asphaltenes were set to 8.8 and 8.2 bar, respectively. The acentric factors of resins and asphaltenes were set to 1.8 and 2.0, respectively. The Cavett (1964) correlation was used to estimated T_c and P_c except for asphaltenes and resins. The binary interaction coefficient between asphaltene and resin is set to zero, and between methane and asphaltene, and methane and resin are set to 0.14 and 0.1, respectively. Other binary interaction coefficients are set according to Table 3.3 of Chapter 3.

When CO_2 is mixed with some crude oils, asphaltene may precipitate above a certain concentration. Figure 5.15 depicts the amount of the precipitation vs. CO_2 concentration for the Weyburn reservoir fluid at 160 bar and 332 K (Kokal *et al.*, 1992). The composition of the Weyburn oil is given in Table 5.7. Note that the amounts of the asphaltenes and resins in the stock-tank oil are 4.9 and 8.9 weight percent, respectively (see Table 5.7). In Table 5.7, the C_{28+} composition is 13.75 mole percent which comprises fractions CP1, CP2, resins and asphaltenes. Figure 5.15 shows that when CO_2 content in the Weyburn oil is less than 50

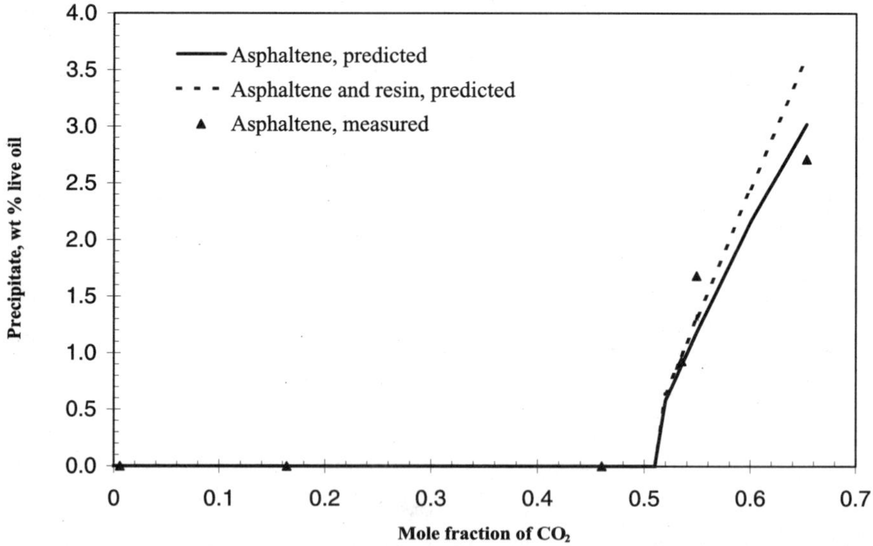

Figure 5.15   Amount of precipitated asphaltene and resin from the Weyburn reservoir fluid (see Table 5.7) and CO_2 mixture at 160 bar and 332 K (from Pan and Firoozabadi, 1997a).

TABLE 5.7   Composition of Weyburn Reservoir Fluid (from Pan and Firoozabadi, 1997b)

Comp.	Mole%	M^*, g/mole
N_2	0.96	
CO_2	0.58	
H_2S	0.30	
C_1	4.49	
C_2	2.99	
C_3	4.75	
iC_4	0.81	
nC_4	1.92	
iC_5	1.27	
nC_5	2.19	
C_{6-9}	25.73	105
C_{10-17}	26.98	179
C_{18-27}	13.28	312
C_{28+}	13.75	576
CP1	6.69	442
CP2	4.359	582
r	1.839	850
a	0.862	1000

Weight% of asphaltene and resin in tank oil is 4.9 and 8.9, respectively.
* Estimated from TBP data.

mole%, there is no precipitation. Above 50 mole%, precipitation begins and increases sharply with CO_2 concentration. Note that the resin precipitation is a small part of the precipitant. The experimental data are also presented in Fig. 5.15. The effect of resin concentration in precipitation is shown in Fig. 5.16. Note that there is a modest effect of resin concentration in precipitation inhibition. Figure 5.17 provides the effect of CO_2 concentration and resin concentration on the micellar core radius in Å. The predicted micellar-size changes due to the resin effect are in agreement with experimental data (Espinat and Ravey, 1993). The predicted concentration effect is also in line with experimental data (Ferworn, Svrcek, and Mehrotra, 1993). Figure 5.18 provides the asphaltene precipitation onset vs. mole fraction of the separator gas in a North Sea separator crude oil. The compositions of the separator oil and separator gas are provided in Table 5.8. The amounts of the asphaltenes and resins in the stock tank oil are 0.9 and 2.9 wt%, respectively. The results in Fig. 5.18 show that temperature variation of 32°C has a small effect on the onset of precipitation (Fotland et al., 1997). Different trends have been reported concerning the effect of temperature on asphaltene precipitation from crude oils. The amount of asphaltene precipitation may increase for a mixture of crude oil and propane as temperature increases

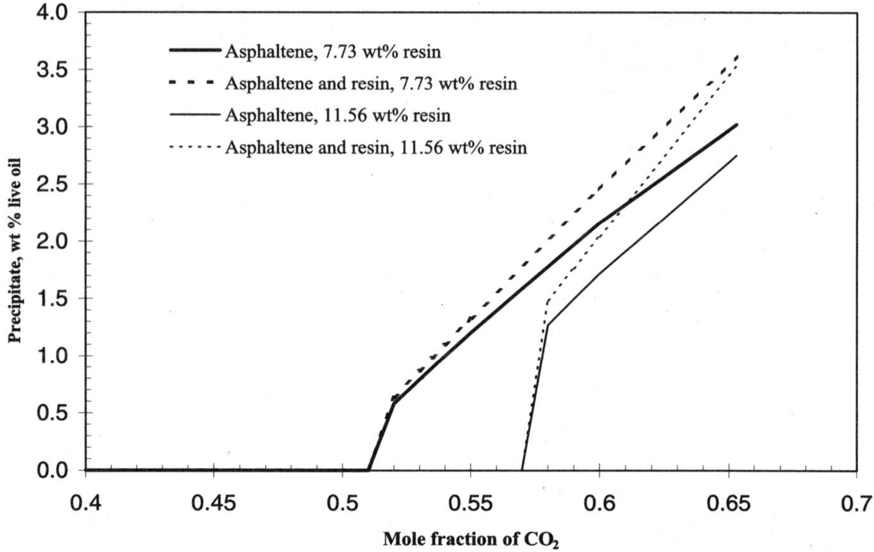

Figure 5.16   Effect of resin concentration on asphaltene and resin precipitation from the Weyburn reservoir fluid and CO_2 mixture at 160 bar and 332 K (adapted from Pan and Firoozbadi, 1997a).

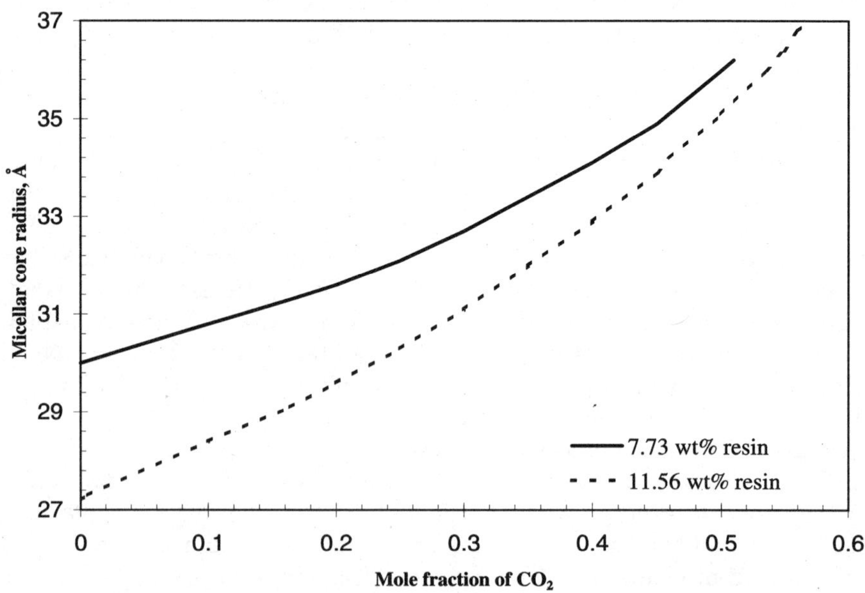

Figure 5.17   Effect of resin concentration on micellar size of the Weyburn reservoir fluid and CO_2 mixture at 160 bar and 332 K (adapted from Pan and Firoozabadi, 1997a).

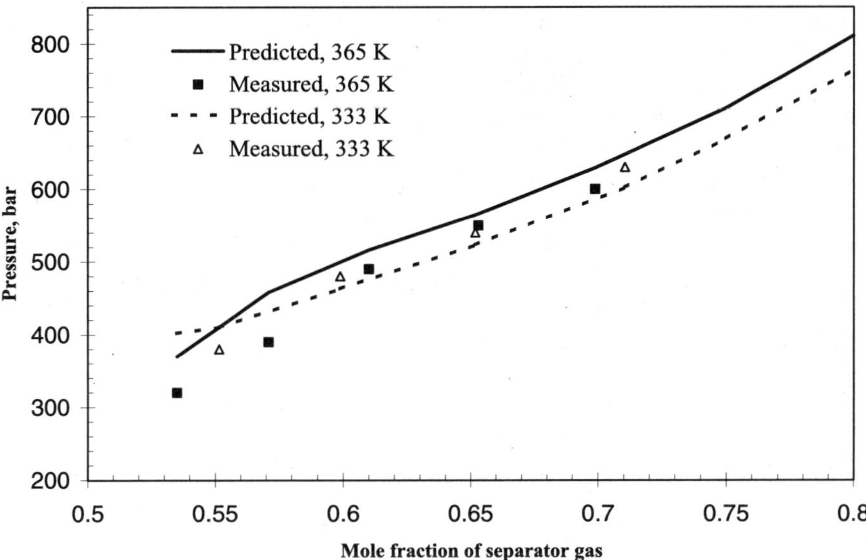

Figure 5.18   Onset pressure for asphaltene precipitation of the mixture of separator oil and separator gas of a North Sea reservoir fluid (adapted from Pan and Firoozabadi, 1997a).

TABLE 5.8   Composition of separator oil and separator gas of the North Sea reservoir crude (from Pan and Firoozabadi, 1997a)

Comp.	Sep. oil mole%	Sep. gas mole%	M, g/mole
N_2	0.0	2.357	
CO_2	0.0	0.426	
C_1	8.365	67.496	
C_2	4.848	12.793	
C_3	7.888	9.733	
iC_4	1.486	1.215	
nC_4	5.586	3.027	
iC_5	2.201	0.815	
nC_5	3.263	0.761	
C_6	4.530	0.619	86
C_7	7.620	0.517	93
C_8	8.348	0.201	107
C_9	6.515	0.040	121
C_{10+}	39.348	0.0	282
CP1	15.648		189
CP2	11.786		261
CP3	7.955		338
CP4	3.370		548
r	0.459		850
a	0.130		1000

Molecular Weight of separator oil is 190 g/mol.
wt% of asphaltene and resin are 0.9 and 2.8 in tank oil, respectively.

(Andersen, 1994). For normal-alkane diluents with carbon number above 5, the precipitated amount may fall with increasing temperature (Fuhr *et al.*, 1991; Ali and Al-Ghannam, 1981). Temperature may also have a negligible effect as the data in Fig. 5.18 reveal.

In conclusion to this chapter, we would like to point out that the term precipitation, whether used for wax or for asphaltene, is synonymous with thermodynamic equilibrium. Wax and asphaltene deposition, which are more complex, include kinetics and require detailed knowledge of the deposition process and the associated transport phenomena. The topic of deposition is in a very early stage of development and, therefore, is not covered.

Examples and theory extension

Example 5.1   Derive Eq. (5.36) of the text.

Solution   At equilibrium at the wax precipitation onset,

$$f_i^S(T,(P) = f_i^L(T, P, \boldsymbol{x}^L),$$

where the index i represents the precipitating component. From Eqs. (1.42) and (1.94) of Chapter 1

$$\left[d\ln f_i^S(T,(P) = \left(\frac{v_i^S(T,P)}{RT} \right) dP \right]_T,$$

where v_i^S is the molar volume of the precipitating component. One can write

$$f_i^S(T,(P) \approx f_i^{0,S}(T, P^0) \exp\left[\frac{v_i^S}{RT}(P - P^0) \right],$$

where $f_i^{0,S}$ is the fugacity of solid component at pressure P^0. In the above equation, it is assumed that v_i^S is independent of pressure. The vapor pressure of solid component (which is a heavy normal-alkane) at temperature T, is often very small in comparison to P (that is, $P^0 \ll P$); therefore one may make a further approximation,

$$f_i^S(T, P) \approx f_i^{0,S}(T, P^0) \exp\left[\frac{Pv_i^S}{RT} \right].$$

For an ideal liquid solution (see Eq. 1.122, Chapter 1) $f_i^L(T, P, \boldsymbol{x}^L) = x_i^L f_i^L(T, P)$; one can also write the expression for $f_i^L(T, P)$ similar to $f_i^S(T, P)$,

$$f_i^L(T, P) \approx f_i^{0,L}(T, P^0) \exp\left[\frac{Pv_i^L}{RT} \right].$$

Combining the above equations,

$$x_i^L \approx \frac{f_i^{0,S}(T, P^0)}{f_i^{0,L}(T, P^0)} \exp\left[-\frac{P(v_i^L - v_i^S)}{RT}\right]$$

where the $f_i^{0,S}/f_i^{0,L}$ ratio is a function of only temperature, and therefore Eq. (5.36) is established.

Example 5.2 *Vapor–solid equilibrium model* Consider the vapor-multisolid equilibrium system sketched in Fig. 5.19. Write down the system of equations that can be used to solve the problem.

Solution   Similarly to the liquid–multisolid equilibrium at constant temperature and pressure, one can write

$$f_i^V(P, T, y) = f_{pure\ i}^S(P, T) \qquad i = (c - c_S + 1), \ldots, c.$$

The material balances for the nonprecipitating components are

$$z_i - y_i\left[1 - \sum_{j=(c-c_S+1)}^{c} n_j^S/F\right] = 0 \qquad i = 1, \ldots, (c - c_S),$$

where the symbols were defined earlier. For the precipitating components, where all the solid phases are pure,

$$z_i - y_i\left[1 - \sum_{j=(c-c_S+1)}^{c} n_j^S/F\right] - n_i^S/F = 0 \qquad i = (c - c_S + 1), \ldots, c - 1.$$

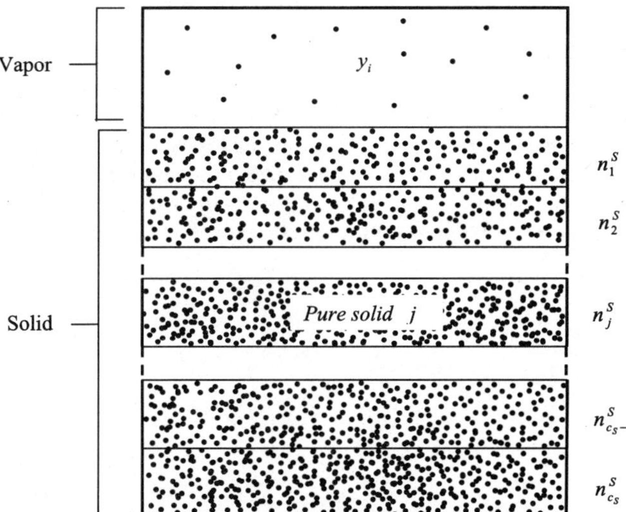

Figure 5.19   Vapor–multisolid-phase model for wax precipitation.

The constraint equation for component i in the gas phase is given by

$$\sum_{i=1}^{c} y_i = 1.$$

The above $(c + c_S)$ equations define the $(c + c_S)$ unknowns: c_S unknowns in n_i^S and c unknowns in y_i. The criterion for the formation of solid phases is given by Eq. (5.22).

Example 5.3 *Wax precipitation from natural gases* Consider a natural-gas fluid of composition shown in Table 5.9 at $P = 563$ bar and $T = 338$ K, where the fluid is in the gaseous state (Ungerer *et al.*, 1995). Use the multisolid-wax model to calculate the temperature at which the solid precipitates at 563 bar. Ungerer *et al.*, measured the phase transition from gas to (gas+solid) at a temperature of 62.0°C at constant pressure of 563 bar. Also calculate the amount of the precipitated solid vs. temperature at $P = 563$ bar.

Solution The model described in Example 5.2 can be used to calculate the temperature at which the solid phase precipitates. The composition and the properties of the synthetic fluid are given in Table 5.9. This mixture is denoted by SHF4 in the work of Ungerer *et al.*

Note that Δh^f values in Table 9 are different than values calculated from Eq. 5.20. Some n-alkanes undergo a solid-solid transition below their fusion-point temperature. As an example, nC_{36} undergoes two transition solid states; from melting temperature of 348.9 K, the first transition temperature is 346.8 K, and the second transition temperature is 341.5 K. The melting point enthalpy at the melting point is 21,230 cal/gmole, and the sum of enthalpy changes at the two transition temperatures is 9670 cal/gmole. Therefore, the total enthalpy change from the liquid state to the solid state after the second transition state is about 30,900 cal/gmole. Finke, *et al.* (1954) and Shaerer, *et al.* (1955) provide the data in solid transitions for nC_{16} and nC_{36}, respectively. Also note that in this problem, there is no need to obtain the enthalpy difference between the gas and the solid state at T^f.

We can use the PR-EOS (Peng and Robinson, 1976) to calculate fugacity and density. The interaction coefficients from Chapter 3 are used in the PR-EOS, we set $\delta_{C_1-C_{36}} = 0.12$ and $\delta_{C_1-toluene} = 0.05$. Figure 5.20 plots the amount of the

TABLE 5.9 The composition and physical properties of synthetic fluid mixture SHF4 from Ungerer *et al.* (1995)

Comp.	z mol. fraction	M, g/gmole	T_c, K	P_c, bar	ω	T^f, K	Δh^f, cal/gmole
C_1	0.756	16	190.6	46.0	0.008	90	—
C_2	0.113	30	305.3	48.8	0.098	90	—
nC_4	0.049	58	425.3	37.9	0.193	138	—
Toluene	0.020	92	591.7	41.1	0.253	113	—
nC_8	0.027	114	568.8	24.8	0.394	143.4	—
nC_{16}	0.023	226	717.0	14.2	0.742	291.3	12,050
nC_{36}	0.012	506	872.7	7.1	1.3	347.9	30,900

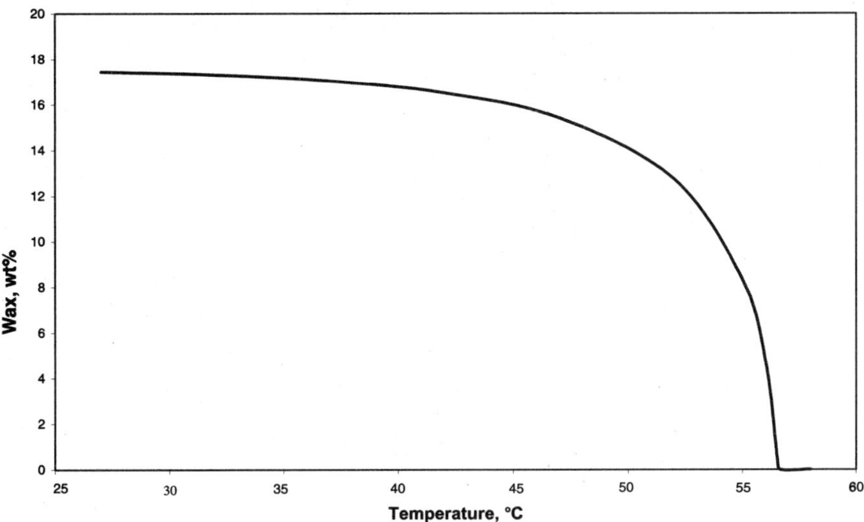

Figure 5.20   Wax precipitation from the synthetic gas: $P = 563$ bar.

precipitated solid phase (wt% of solid with respect to the weight of the original synthetic fluid) vs. temperature. Note that the predicted solid-phase formation temperature is around 56.5°C, which is in fair agreement with the measured value of 62.8°C. The multisolid model predicts the precipitation of only nC_{36} in the whole temperature range. In fact, below 45°C very little nC_{36} remains in the gas phase.

This simple example indicates that in certain cases, wax may precipitate from a natural gas before the dewpoint pressure is reached.

Example 5.4   *Vapor–liquid–multisolid equilibrium formulation.* Set up the equations that can be used to solve the vapor–liquid–multisolid system sketched in Fig. 5.3.

Solution   We will use two approaches to solve the vapor–liquid–multisolid phase equlibrium. The first approach will be based on the flash-type calculations, and a second approach from direct minimization of the Gibbs free energy.

(1) Flash approach. The fugacity equalities, material balance expressions, and contraints define the problem.

(a) The equilibrium between vapor and liquid phases provides

$$f_i^V(P, T, y_1, \ldots, y_{c-1}) = f_i^L(P, T, x_1^L, \ldots, x_{c-1}^L) \quad i = 1, \ldots, c.$$

(b) The equilibrium between liquid and solid phases provides

$$f_i^L(P, T, x_i^L, \ldots, x_{c-1}^L) = f_{\text{pure } i}^S(P, T) \quad i = (c - c_S + 1), \ldots, c.$$

(c) Material balance for the nonprecipitating components is

$$z_i - x_i^L \left[1 - \sum_{j=(c-c_S+1)}^{c} n_j^S/F \right] - K_i^{VL} x_i^L (V/F) = 0 \quad i = 1, \ldots, (c - c_S)$$

(d) Material balance for the precipitating components is

$$z_i - x_i^L \left[1 - \sum_{j=(c-c_S+1)}^{c} n_j^S/F - (V/F) \right] - n_i^S/F - K_i^{VL} x_i^L (V/F) = 0$$

$$i = (c - c_S + 1), \ldots, c - 1.$$

(e) The constraint equations are

$$\sum_{i=1}^{c} y_i = 1$$

$$\sum_{i=1}^{c} x_i^L = 1.$$

The above $(2c + c_S + 1)$ equations provide the $(2c + c_S + 1)$ unknowns, which are y_i, x_i^L, n_i^S, and V. Newton's method can be used to solve the above system of equations. The K_i^{VL}s are given by $\varphi_i^L(P, T, x^L)/\varphi_i^V(P, T, y)$. The Wilson correlation discussed in Chapter 4 can be used for the first estimate of K_i^{VL} to initiate calculations.

(2) Direct minimization of Gibbs free energy. The Gibbs free energy of the system in Fig. 5.3 is given by

$$G = G^V + G^L + G^S.$$

In terms of chemical potentials,

$$G = \sum_{i=1}^{c} n_i^V \mu_i^V(T, P, y) + \sum_{i=1}^{c} n_i^L \mu_i^L(T, P, x^L) + \sum_{i=1}^{c} n_i^S \mu_i^S(T, P).$$

The chemical potentials μ_i^V, μ_i^L and μ_i^S are given by

$$\mu_i^V(T, P, y) = \mu_i^0(T) + RT \ln \frac{f_i^V(T, P, y)}{f_i^0(T)}, \qquad i = 1, \ldots, c,$$

$$\mu_i^L(T, P, x^L) = \mu_i^0(T) + RT \ln \frac{f_i^L(T, P, x^L)}{f_i^0(T)}, \qquad i = 1, \ldots c,$$

$$\mu_i^S(T, P) = \mu_i^0(T) + RT \ln \frac{f_i^S(T, P)}{f_i^0(T)}, \qquad i = 1, \ldots, c.$$

Note that we do not know how many solid phases will precipitate. Therefore, we assume that all the components can precipitate as a solid phase. The material balance equation for the nonprecipitating and precipitating components is

$$n_i = n_i^V + n_i^L + n_i^S \qquad i = 1, \ldots, c.$$

The above equations can be combined to obtain the following expression for the total Gibbs free energy of the system:

$$G = \left\{ \sum_{i=1}^{c} n_i \mu_i^0(T) - RT \sum_{i=1}^{c} n_i \ln f_i^0(T) \right\}$$

$$+ RT \left[\sum_{i=1}^{c} n_i^V \ln f_i^V(T, P, \underline{n}^V) + \sum_{i=1}^{c} n_i^L \ln f_i^L(T, P, \underline{n}^L) \right.$$

$$\left. + \sum_{i=1}^{c} (n_i - n_i^V - n_i^L) \ln f_i(T, P) \right]$$

At constant temperature T and pressure P, for given overall moles n_i $(i = 1, \ldots, c)$, the terms inside the braces { }, in the above expression are constant and do not vary with n_i^V and n_i^L $(i = 1, \ldots, c)$. Therefore, at equilibrium, in order to obtain the $2c$ unknowns, n_i^V and n_i^L, one can minimize the function given by

$$\bar{G} = \sum_{i=1}^{c} n_i^V \ln f_i^V(T, P, \underline{n}^V) + \sum_{i=1}^{c} n_i^L \ln f_i^L(T, P, \underline{n}^L)$$

$$+ \sum_{i=1}^{c} (n_i - n_i^V - n_i^L) \ln f_i(T, P)$$

with respect to those $2c$ unknowns. An algorithm such as simulated annealing discussed in Chapter 4 can be used to perform the minimization. Once n_i^V, and n_i^L are obtained, the amount of the vapor, liquid, and solid phases can be calculated.

Example 5.5 *Molecular volume of asphaltene and other substances, and asphaltene-micellar core radius* Calculate the molecular volume in Å^3 from mass density and molecular weight. Calculate also the number of asphaltene molecules in the asphaltene core from the asphaltene micellar core radius of Fig. 5.17.

Solution Avogadro's number gives the number of molecules in a mole: $N_A = 6.02 \times 10^{23}$ molecules/gmole. Therefore, the relation between mass density, ρ, molecular weight, M, and molecular volume, V, is given by

$$\frac{\rho}{M} = N_A V.$$

Let us use the units of g/cm^3, g/gmole, and Å^3 for ρ, M and V, respectively. Then

$$V = \left(\frac{10}{6.028} \right) \frac{M}{\rho}.$$

For nC_{10}, $M = 142$ g/gmole, and $\rho \approx 0.724$ g/cm^3 at 20°C; then the molecular volume is

$$V = (10/6.028)/(142/0.724) = 35 \text{ Å}^3.$$

For asphaltenes, if we asume $M = 1000$ g/gmole and $\rho \approx 1.1$ g/cm^3 at 25°C (molar volume ≈ 0.9 m^3/kgmole), then the molecular volume is $V \approx 1500$ Å^3.

From Fig. 5.17 the micellar core radius of the Weyburn reservoir fluid before mixing with CO_2 is about 30 Å. The number of asphaltene molecules in the core is given by

$$n_1 V = \frac{4}{3} \pi R^3.$$

For $R = 30$ Å, $V \approx 1500$ Å^3, one obtains $n_1 \approx 75$.

Example 5.6 Show that the standard Gibbs free energy of micellization can be expressed by

$$\Delta G_m^{00} = [(\Delta G_a^0)_{tr} + (\Delta G_a^0)_{def}] + [(\Delta G_r^0)_{tr} + (\Delta G_r^0)_{ads} + (\Delta G_r^0)_{def}]$$
$$+ (\Delta G_m^0)_{int} + (\Delta G_r^0)_{str},$$

where $(\Delta G_r^0)_{str}$ is the steric repulsion energy of resins in the shell. What is the advantage of the above alternative expression for ΔG_m^{00}?

Solution Consider Eq. (5.64a), and let us write the expression for $(\Delta G_r^0)_{ads}$ from Eq. (5.61),

$$(\Delta G_r^0)_{ads} = n_2[(\Delta h_r^0)_{ads} - T(\Delta s_r)_{ads} - \sigma a_0].$$

We can designate the steric repulsion term $-n_2 T(\Delta s_r)_{ads}$ by $(\Delta G_r^0)_{str}$ and exclude it from $(\Delta G_r^0)_{ads}$. We can also exclude $-n_2 \sigma a_0$ from the $(\Delta G_r^0)_{ads}$ term and add it to $(\Delta G_m^0)_{int}$ to obtain $(\Delta G_m^0)_{int} = n_2 \sigma(a - a_0)$ as was stated earlier. The advantage of this alternative expression for ΔG_m^{00} is that the formulation for co-adsorption of the amphiphiles onto the micellar core can be facilitated, as we will see in Example 5.9.

Example 5.7 Derive the Gibbs free energy expression for the precipitated asphaltene phase when it is in a solid state.

Solution The expression for the Gibbs free energy of the precipitated solid phase comprised of asphaltene and resin molecules is given by

$$G^S = n_a^S \mu_a^S + n_r^S \mu_r^S,$$

assuming that only asphaltenes and resins constitute the solid-solution phase. For the ideal solid solution

$$G^S = n_a^S[\mu_a^0(T) + kT \ln x_a^S f_a^S] + n_r^S[\mu_r^0(T) + kT \ln x_r^S f_r^S].$$

In the above two equations, n is the number of molecules, f_a^S and f_r^S are the fugacities of pure asphaltene and resin solids at temperature T and pressure P, respectively, and x is the molecular fraction in the precipitated-solid phase. The fugacity of pure solid-component i can be obtained from Eq. (5.16). Acevedo *et al.* (1995) suggest the enthalpy of fusion and the melting-point temperature for asphaltenes to be 7300 cal/gmole and 583 K, respectively. Pan and Firooza-badi (1998a) suggest the enthalpy of fusion and the melting-point temperature for resins to be 5000 cal/gmole, and 483 K, respectively. The last two terms of the exponent in Eq. (5.16) may be neglected in asphaltene precipitation calculations when the state of the precipitate is solid.

Example 5.8   *Effect of aromatics on asphaltene precipitation*   Some aromatic solvents have been used to inhibit the asphaltene precipitation in crudes (Cimino *et al.*, 1995), although aromatics such as benzene and toluene in general may not be very effective for asphaltene-precipitation inhibition because high concentrations may be required. When an aromatic such as benzene or toluene is mixed with a crude, the micellization model described in this chapter should be modified. The modification is necessary because as a result of crude/aromatic mixing, the aromatics mainly appear in the solvation shell and reduce the interfacial tension between the asphaltene-liquid core and liquid solution outside the core. Modify the formulation presented earlier for this purpose.

Solution   Figure 5.11 provides the sketch for a micelle when an aromatic solvent is added to the crude oil. The effect of aromatic solvent in asphaltene precipitation is mainly because of the decrease of the interfacial tension. In other words, when aromatics are added to a crude oil, the term $(\Delta G_m^0)_{int} = n_2 \sigma a$ in Eq. (5.49) or the term $n_2 \sigma (a - a_0)$ reduces. The first step is, therefore, the calculation of σ between the pure asphaltene liquid and the liquid mixture of asphalt-free crude oil and the aromatic. Pan and Firoozabadi (1998b) provide the following expression for the interfacial tension between the asphaltene liquid phase and the surrounding liquid:

$$\sigma = \frac{kT}{L_{shell}^2} \left[\frac{V_a(\delta_{shell} - \delta_a)^2}{6kT} \right]^{0.65} ,$$

where δ_a and δ_{shell} are solubility parameters of the asphaltene liquid and the asphalt-free oil/aromatic mixture, respectively. The solubility parameter δ_{shell} is given by (see Eq. (1.162) of Chapter 1),

$$\delta_{shell} = \delta_{oil}\Phi_{oil} + \delta_{aromatic}\Phi_{aromatic},$$

where Φ_{oil} and $\Phi_{aromatic}$ are volume fractions of the asphalt-free oil, and the aromatic solvent, respectively, and δ_{oil} and $\delta_{aromatic}$ are the solubility parameters of the asphalt-free oil and the aromatic solvent, respectively. Because $\delta_{aromatic}$ is closer to δ_a than δ_{oil}, (see Problem 5.10), σ decreases as a result of high concentration of aromatics in the shell.

The expressions for ΔG^{00} and G^{L_1} (see Eqs. (5.64) and (5.81)) are, therefore, modified for the purpose of studying the effect of aromatics on asphaltene precipitation. The indices for the asphalt-free oil species are from 1 to $(c - 3)$, the index for the aromatic species is $(c - 2)$; resin and asphaltene indices are $(c - 1)$ and c, respectively. The main modification for G^{L_1} is due to $G_{solvent}$ (see

Eqs. (5.66) to (5.68)). The expression for the Gibbs free energy for the solute species remains the same (see Eq. (5.80)). The final expression for Gibbs free energy of liquid phase L_1 is

$$G^{L_1} = \sum_{i=1}^{c} n_i^{L_1} \mu_i^0(T) + kT \sum_{i=1}^{c} n_i^{L_1} \ln \hat{\varphi}_i(\hat{n}_1^{L_1}, \ldots, \hat{n}_c^{L_1}) P + n_m \Delta G_m^{00}$$

$$+ kT \left(\sum_{i=1}^{c-3} (n_i^{L_1} - n_{i,sh}^{L_1}) \ln x_{i,b} + \sum_{i=1}^{c-3} n_{i,sh}^{L_1} \ln x_{i,sh} + (n_{aro}^{L_1} - n_{aro,sh}^{L_1}) \ln x_{aro,b} \right.$$

$$\left. + n_{aro,sh} \ln x_{aro,sh} + [n_{1a}^{L_1} \ln x_{1a} + n_{1r}^{L_1} \ln x_{1r} + n_m \ln x_m \right).$$

In the above equation the subscript *aro*, and *b* represent the aromatics, and the bulk phase, respectively; $x_{i,b}$ and $x_{i,sh}$ represent the mole fractions of component i in the bulk liquid phase and in the shell, respectively. Pan and Firoozabadi (1998c) have used the above modified formulation to study the effect of toluene on a crude oil/CO_2 system. Figure 5.21 shows the effect of toluene on asphaltene precipitation from the mixture of the Weyburn crude oil and CO_2. Note that as the concentration of toluene increases, the precipitate decreases.

Example 5.9 *Effect of amphiphiles on asphaltene-precipitation inhibition* Chang and Fogler (1994a,b) demonstrate that amphiphiles can be used to inhibit asphaltene precipitation. Formulate the expression for the standard Gibbs free energy of micellization and the Gibbs free energy of phase L_1, when polar substance such as amphiphiles are mixed with a crude oil. Amphiphiles have a polar head and a nonpolar tail.

Solution   Since amphiphiles have a polar head, they are expected to co-adsorb with resins onto the micellar-core surface. Therefore, one needs to modify both ΔG_m^{00} and the Gibbs free energy of the liquid phase L_1. In the following, we will briefly present the modifications to ΔG_m^{00}. The standard Gibbs free energy of micellization, including the co-adsorption of the amphiphiles, can be expressed as,

$$\Delta G_m^{00} = \mu_m^* - n_1 \mu_{1a}^* - n_2 \mu_{1r}^* - n_f \mu_{1f}^*,$$

where n_f is the molecular number of the amphiphiles in the solvation shell and μ_{1f}^* is the chemical potential of the monomeric amphiphile molecule at infinite dilution. ΔG_m^{00} can also be expressed by

$$\Delta G_m^{00} = [(\Delta G_a^0)_{tr} + (\Delta G_a^0)_{def}] + [(\Delta G_r^0)_{ad} + (\Delta G_r^0)_{tr} + (\Delta G_r^0)_{def}]$$

$$+ [(\Delta G_f^0)_{ads} + (\Delta G_f^0)_{tr} + (\Delta G_f^0)_{def}] + (\Delta G_m^0)_{int} + (\Delta G_m^0)_{ste},$$

where the terms in the first two brackets and the last two terms were discussed in Example 5.6. The terms in the third bracket represent the contribution for the amphiphile. The terms representing the interfacial effect $(\Delta G_m^0)_{int}$ and the steric effect $(\Delta G_m^0)_{ste}$ for two polar species (that is, resin and amphiphile) differ from the case of a single polar species (that is, resin) which was presented earlier. Figure 5.22 shows a schematic of a micelle in a mixture of crude oil and amphiphiles.

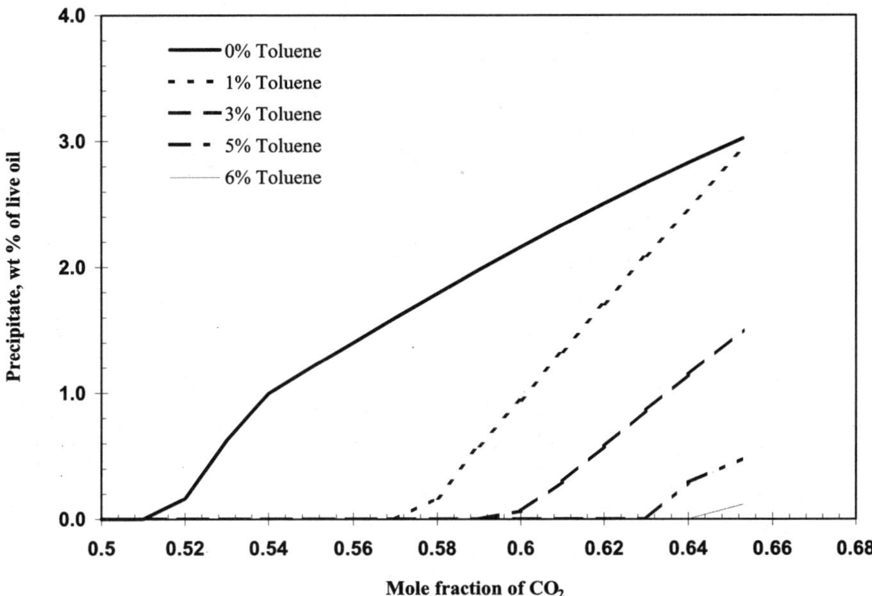

Figure 5.21   Effect of toluene concentration (mole%) on asphaltene precipitation: Weyburn reservoir fluid and CO_2 mixture at 160 bar and 332 K (adapted from Pan and Firoozabadi, 1998c).

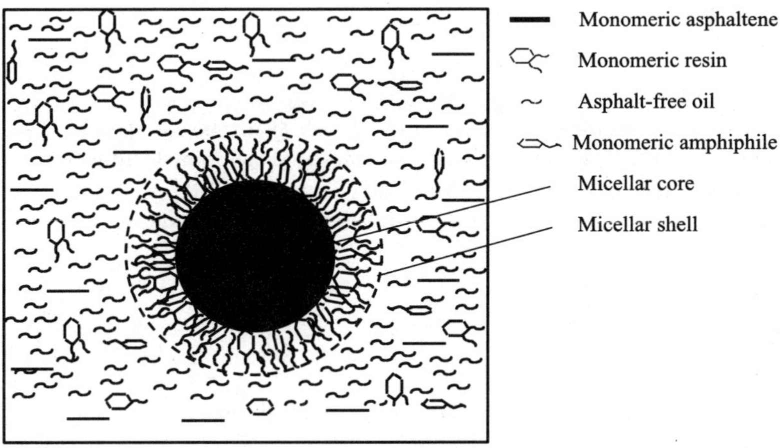

Figure 5.22   Schematic representation of a micelle in the crude/amphiphile mixture (adapted from Pan and Firoozabadi, 1998c).

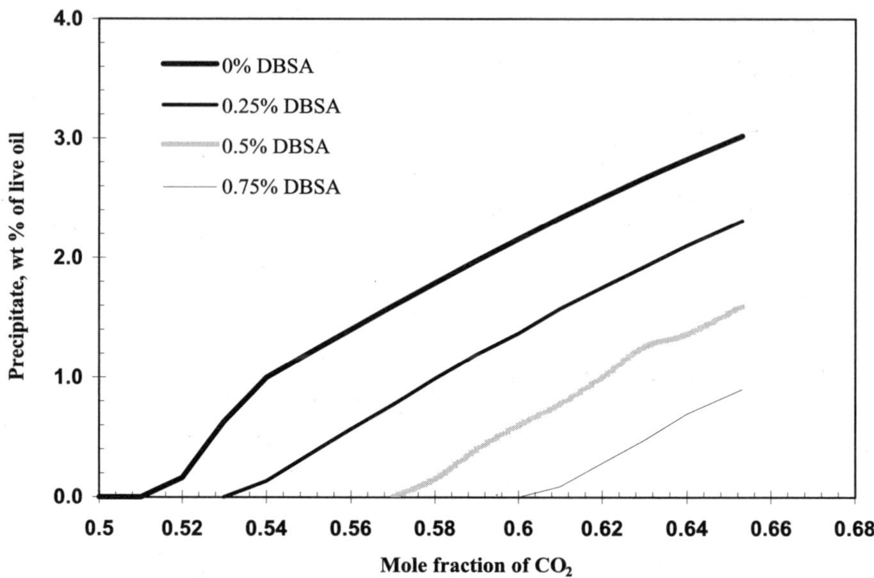

Figure 5.23   Effect of DBSA concentration (mole%) on asphaltene precipitation: Weyburn reservoir fluid and CO_2 mixture at 160 bar and 332 K (adapted from Pan and Firoozabadi, 1998c).

TABLE 5.10   Contribution of various terms to ΔG_m^{00} (unit, kT) for the mixture of Weyburn Oil with 60 mole% CO_2 and 0.5 mole% DBSA (from Pan and Firoozabadi, 1998c)

Asphaltene: $(\Delta g_a^0)_{as} = -4.6$; $(\Delta g_a^0)_{tr} = -6.8$; $(\Delta g_a^0)_{def} = 0.6$

Resin: $(\Delta g_r^0)_{ste} = 2.6$; $(\Delta h_r^0)_{ads} = -4.1$; $(\Delta g_r^0)_{int} = 4.6$; $(\Delta g_r^0)_{tr} = -4.9$; $(\Delta g_r^0)_{def} = 0.07$

DBSA: $(\Delta g_f^0)_{ste} = 0.5$; $(\Delta h_f^0)_{ads} = -20.3$; $(\Delta g_f^0)_{int} = 5.4$; $(\Delta g_f^0)_{tr} = -2.7$; $(\Delta g_f^0)_{def} = 0.74$

Figure 5.23 shows the effect of amphiphile p-(n-dodecyl)benzenesulfonic acid (DBSA) when mixed with the Weyburn crude oil/CO_2 system. Note that the amphiphile has a strong effect on asphaltene-precipitation inhibition. The reason for the strong effect of DBSA is a large negative $(\Delta h_f^0)_{ads}$. The forms of various terms for ΔG_m^{00} are provided in Problem 5.11. Table 5.10 shows the contribution of various terms to ΔG_{00}^m.

Problems

5.1.   Derive Eq. (5.9) of the text.

Hint: You may use Eq. (3.114) of Chapter 3 in your derivation.

5.2.   Show that a higher entropy of fusion for the first precipitating component can lead to a higher CPT.

5.3.   How would you reason that an ideal solid solution is a good assumption for the precipitated wax phase?

5.4.   Derive the Gibbs Phase Rule for the system sketched in Fig. 5.3.

5.5.   Consider Fig. 5.24 which shows the calculated CPT of a crude oil (stock-tank oil), as well as the results when the crude oil is mixed with CO_2, C_1, C_3, and nC_5. Explain the effects of pressure and composition on CPT from that figure.

5.6.   Consider a mixture of nC_{15} and nC_{20} at temperature T and pressure P. Suppose the mixture forms a liquid solution and two solid solutions; one solid solution is rich in nC_{15} and the other solid solution is rich in nC_{20}. Set up the equations that can be used to solve the liquid–solid–solid equilibrium.

5.7.   *Effect of pressure on CPT for gases*   Unlike liquids where a pressure increase often enhances the wax precipitation at constant temperature, for gases pressure has often the opposite effect: pressure increase may inhibit the precipitation at constant temperature. The solubility of nC_{28} in CO_2 at 308 K, and the solubility of nC_{20} in C_1 are provided in Tables 5.11 and 5.12 below. First derive the expression

$$y_1 = P_1^{sat} \exp\left[\frac{v_1^{sat}(P - P_1^{sat})}{RT}\right]/(\varphi_1 P)$$

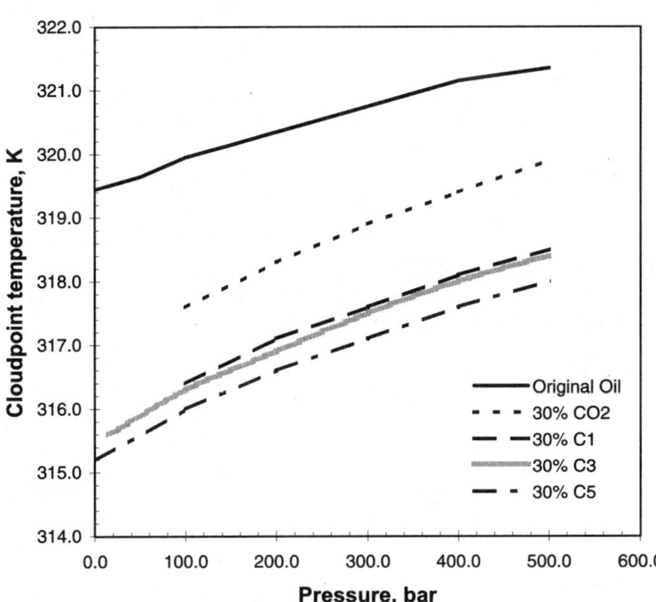

Figure 5.24   Calculated cloudpoint temperature vs. pressure for a stock-tank oil diluted by various light components (adapted from Pan *et al.*, 1997b).

TABLE 5.11   Solubility of nC_{28} in CO_2 at 308 K (data from Smith *et al.*, 1996)

Pressure, bar	Solubility in CO_2, mole%
10.0	0.0039
14.98	0.00532
17.51	0.00647
19.98	0.00689
21.51	0.00724
34.03	0.00858

TABLE 5.12   Solubility of nC_{20} in C_1 (data from van der Kooi, Flöter, and de Loos, 1995)

Temperature, K	Pressure, bar	Solubility in C_1, mole%
305.66	73.5	0.80
305.75	76.5	1.0
305.73	89.30	2.0

where v_1^{sat} is the solid-solute molar volume at temperature T, φ_1 is the fugacity coefficient of the solute component in the gas phase, P_1^{sat} is the sublimation pressure (at which a solid vaporizes), and y_1 is the mole fraction of solute component in the gas phase (that is, solubility). Note that v_1^{sat} and P_1^{sat} are functions of temperature T. The equation above shows that the solubility of the heavy normal-alkane solids in CO_2 and C_1 gases increases with the pressure increase. Spell out the assumptions that are made to derive the above expression and then relate the solubility to CPT. Finally use this expression to predict the solubilities of nC_{20} in C_1 and nC_{28} in CO_2, and compare the results with the data.

5.8.   A student has simplified the understanding of the effect of pressure on wax precipitating from natural gases and crude oils through a P vs. T plot shown in Fig. 5.25. Can you extend the plot for the effect of pressure on wax precipitation from a mixture of equilibrium gas and liquid phases?

Would there be any change in Fig. 5.25 when a near-critical gas or a near-critical liquid mixture is considered for wax precipitation?

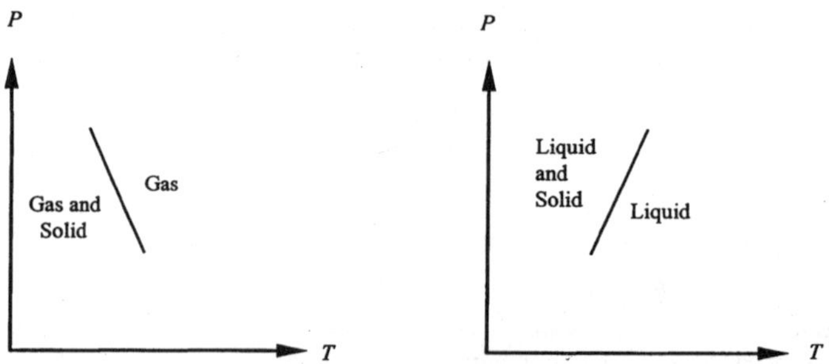

Figure 5.25 P–T plot for the multicomponent gases and liquids showing pressure effect on wax precipitation.

5.9. Show that the thickness of the shell in Fig. 5.11 can be obtained from

$$\frac{4\pi}{3}[(R+D)^3 - R^3] = (n_2 + n_{aromatic} + n_{oil})V_{sh},$$

where V_{sh} is the volume of the mixture in the shell comprised of asphalt-free oil, resins, and the aromatic solvent.

5.10. Use the expression for the solubility parameter (see Problem 3.10, Chapter 3) to estimate solubility parameters of nC_5 to nC_{10}, and toluene at 20°C and compare your results with the values below. If the solubility parameter of asphaltenes at 20°C is $\delta_a = 9.5$ $(cal/cm^3)^{0.5}$ and the other values are: δ_{nC_5} to $\delta_{nC_{10}} = 7.12, 7.33, 7.46, 7.53, 7.62, 7.66$ and $\delta_{toluene} = 8.87$ $(cal/cm^3)^{0.5}$, then plot the precipitation data of Fig. 5.8 vs. the difference between the solubility parameter of the asphaltene and the solubility parameter of the above diluents. (The solubility parameters for nC_5 to nC_{10} and toluene above were obtained from the PR-EOS. However, the densities were measured values.)

5.11. Show that the interfacial effect for the monolayer co-adsorption of resin and amphiphile onto the asphaltene micellar core can be represented by

$$(\Delta G^0)_{int} = (\Delta G_r^0)_{int} + (\Delta G_f^0)_{int}$$

where
$$(\Delta G_r^0)_{int} = n_2\sigma(a - a_{o,r})$$
$$(\Delta G_f^0)_{int} = n_f\sigma(a - a_{o,r}).$$

In the above equations, $a = 4\pi R^2/(n_2 + n_f)$, and $a_{o,r}$ and $a_{o,f}$ are the cross-sectional areas of the polar head of resin molecule and the polar head of an amphiphile molecule, respectively. Note that in the above formulation, the interfacial screening energy is included in the expression for $(\Delta G^0)_{int}$.

5.12. In general, in the expression for ΔG_m^{00} for an amphiphile–crude oil mixture, the following terms appear: (1) $n_1 kT \ln \varphi_a^{L_1}/\varphi_{al}^\infty$, (2) $n_1(\Delta h_r^0)_{ads}$, (3) $n_2 kT \ln \varphi_{r,sh}^{L_1} x_{r,sh}/\varphi_r^\infty$, (4) $n_f(\Delta h_f^0)_{ads}$, and (5) $n_f \ln \varphi_{f,sh}^L/\varphi_f^\infty$. Explain what each term represents.

5.13. For a crude oil–amphiphile mixture $(\Delta h_f^0)_{ads} = 5(\Delta h_r^0)_{ads}$. Would you suggest the amphiphile is much more effective than the resin in precipitation inhibition and why?

5.14. Experimental data indicate that for some amphiphiles, once their concentration increases beyond a certain value, their effectiveness for precipitation inhibition decreases. In other words, there is an optimum concentration for an amphiphile. What type of micellization model can describe the optimum concentration?

References

1. Acevedo, S., M. A. Ranaudo, G. Escobar, L.B. Gutierrez, and X. Gutierrez: "A Unified View of the Colloidal Nature of Asphaltenes," *Asphaltenes: Fundamentals and Applications,* E.Y. Sheu and O.C. Mullins (eds.), Plenum Press, NY, Chap. 4, 1995.
2. Ali, L.H., and K.A. Al-Ghannam: "Investigations into Asphaltenes in Heavy Crude Oils: 1. Effect of Temperature on Precipitation by Alkane Solvents," *Fuel,* vol. 60, p. 1045, 1981.
3. Andersen, S.I: "Effect of Precipitation Temperature on the Composition of n-Heptane Asphaltene," *Fuel Sci. Tech. Int.,* vol. 12, p. 51, 1994.
4. Burke, N.E., R.E. Hobbs, and S.F. Kashou: "Measurement and Modeling of Asphaltene Precipitation," *J. of Pet. Tech.,* p. 1440, Nov. 1990.
5. Cavett, R.H.: "Physical Data for Distillation Calculations–Vapor-Liquid Equilibrium," Proc., 27th. API Meeting, San Francisco, CA, vol. 52, p. 351, 1964.
6. Chang, C.-L. and H.S. Fogler: "Stabilization of Asphaltenes in Aliphatic Solvents Using Alkylbenzene-Derived Amphiphiles. 1. Effect of the Chemical Structure of Amphiphiles on Asphaltene Stabilization," *Langmuir,* vol. 10, p. 1749, 1994a.
7. Chang, C.-L. and H.S. Fogler: "Stabilization of Asphaltenes in Aliphatic Solvents Using Alkylbenzene-Derived Amphiphiles. 2. Study of the Asphaltene-Amphiphile Interactions and Structures Using Fourier Transform Infrared Spectroscopy and Small-Angle X-ray Scattering Techniques," *Langmuir,* vol. 10, p. 1758, 1994b.
8. Chung, F., P. Sarathi, and R. Jones: "Modeling of Asphaltene and Wax Precipitation," Topical Report NIPER-498, U.S., D.O.E., Jan. 1991.
9. Cimino, R., S. Correra, A.D. Bianco, and T.P. Lockhat: "Solubility and Phase Behavior of Asphaltenes in Hydrocarbon Media," *Asphaltenes: Fundamentals and Applications,* E.Y. Sheu and O.C. Mullins (eds.), Plenum Press, NY, Chap. 3, 1995.
10. Countinho, J.A.P., and V. Ruffier-Meray: "Experimental Measurements and Thermodynamic Modeling of Paraffinic Wax Formation in Undercooled Solutions," *Ind. Eng. Chem. Res.,* vol. 36, no. 11, p. 4977, 1997.
11. Dorset, D.L.: "Chain Length and the Cosolubility of n-Paraffins in the Solid State," *Macromolecules,* vol. 23, p. 623, 1990.
12. Espinat, D. and J.C. Ravey: "Colloidal Structure of Asphaltene Solutions and Heavy-Oil Fractions Studied by Small-Angle Neutron and X-Ray Scattering," paper SPE 25187 presented at the 1993 *SPE International Symposium on Oilfield Chemistry,* New Orleans, March 2–5, 1993.
13. Ferworn, K.A., W.Y. Svrcek, and A.K. Mehrotra: "Measurement of Asphaltene Particle Size Distributions in Crude Oils Diluted with n-Heptane," *Ind. Eng. Chem. Res.,* vol. 32, p. 955, 1993.
14. Finke, H. L., M. E. Gross, G. Waddington, and H. M. Huffman. "Low-Temperature Thermal Data for the Nine Normal Paraffin Hydrocarbons from Octane to Hexadecane," *J. Amer. Chem. Soc.,* vol. 76, p. 333, 1954.
15. Firoozabadi, A., D.L. Katz, H. Soroosh, V.A. Sajjadian: "Surface Tension of Reservoir Crude-Oil/Gas Systems Recognizing the Asphalt in the Heavy Fraction," *SPE Res. Eng.,* p. 265, Feb. 1988.
16. Firoozabadi, A., and H. J. Ramey: "Surface Tension of Water-Hydrocarbon Systems at Reservoir Conditions," *J. of Can. Pet. Tech.,* vol. 27, No. 3, p. 41, 1988.
17. Flory, P.J., *Principles of Polymer Chemistry,* Cornell Univ. Press., Ithaca, NY, 1953.
18. Fotland, P., H. Anfinsen, H. Foerdedal, and H.P. Hjermstad: "The Phase Diagrams of Asphaltenes: Experimental Technique, Results and Modeling on Some North Sea Crude Oils," paper presented at *the Symposium on the Chemistry of the Asphaltene and Related Substances,* Cancun, Mexico, Nov. 12–15, 1997.
19. Fuhr, B.J., C. Cathrea, L. Coates, H. Kalya, and A.I. Majeed: "Properties of Asphaltenes from a Waxy Crude," *Fuel,* vol. 70, p. 1293, 1991.
20. Godbole, S.P., K. J. Thole, and E.W. Reinbold: "EOS Modeling and Experimental Observations of Three-Hydrocarbon-Phase Equilibria," *SPE Res. Eng.,* p. 101, May 1995.
21. Gonzalez, G. and A. Middea: "Peptization of Asphaltene by Various Oil Soluble Amphiphiles," *Colloids and Surfaces,* vol. 52, p. 207, 1991.
22. Hansen, A.B., E. Larsen, W.B. Pedersen, A.B. Nielsen, and H.P. Rønningsen: "Wax

Precipitation from North Sea Crude Oils. 3. Precipitation and Dissolution of Wax Studied by Differential Scanning Calorimetry," Energy & Fuels, vol. 5, p. 914, 1991.

23. Hirschberg, A., L.N.J. de Jong, B.A. Schipper, and J.G. Meijers: "Influence of Temperature and Pressure on Asphaltene Flocculation," *Soc. of Pet. Eng. J.*, p. 283, June 1984.

24. Kawanaka, S., S.J. Park, and G.A. Mansoori: "Organic Deposition from Reservoir Fluids: A Thermodynamic Predictive Technique," *SPE Res. Eng.*, vol. 6, p. 185, 1991.

25. Kokal, S.L., J. Najman, S.G. Sayegh, and A.E. George: "Measurement and Correlation of Asphaltene Precipitation from Heavy Oils by Gas Injection," *J. Can. Pet. Tech.*, vol. 31, p. 24, 1992.

26. Leontaritis, K.J. and G.A. Mansoori: "Asphaltene Flocculation During Oil Production and Processing: A Thermodynamic Colloidal Model," paper SPE 16258 presented at 1987 *SPE Intl. Symposium on Oilfield Chemistry*, San Antonio, TX, Feb. 4–6, 1987.

27. Lian, H., J.-R. Lin, and T.F. Yen: "Peptization Studies of Asphaltene and Solubility Parameter Spectra," *Fuel*, vol. 73, p. 423, 1994.

28. Lin, J.R., H. Lian, K.M. Sadeghi, and T.F Yen.: "Asphalt Colloidal Types Differentiated by Korcak Distribution," *Fuel*, vol. 70, pp. 1439, 1991.

29. Lira-Galeana, C., A. Firoozabadi, and J.M. Prausnitz: "Thermodynamics of Wax Precipitation in Petroleum Mixtures," *AIChE J.*, vol. 42, p. 239, 1995.

30. MacMillan, D.J., J.E. Tackett Jr., M.A. Jessee, and T.G. Monger-McClure: "A Unified Approach to Asphaltene Precipitation: Laboratory Measurement and Modeling," *J. Pet. Tech.*, p. 788, Sept. 1995.

31. Madsen, H.E.L., and R., Boistelle: "Solubility of Long-Chain *n*-Paraffins in Pentane and Heptane," *J. Chem. Soc., Faraday Trans.*, vol. 72, p. 1078, 1976.

32. Madsen, H.E.L., and R. Boistelle: "Solubility of Octacosane and Hexatriacontane in Different n-Alkane Solvents," *J. Chem. Soc., Faraday Trans.*, vol. 75, p. 1254, 1979.

33. Mitchell, D.L., and J.G. Speight: The Solubility of Asphaltenes in Hydrocarbon Solvents, *Fuel*, vol. 52, p. 149, April 1973.

34. Nagarajan, R. and K. Ganesh: "Block Copolymer Self-Assembly in Selective Solvents: Theory of Solubilization in Spherical Micelles," *Macromolecules*, vol. 22, p. 4312, 1989a.

35. Nagarajan, R. and K. Ganesh: "Block Copolymer Self-Assembly in Selective Solvents: Spherical Micelles with Segregated Cores," *J. Chem. Phys.*, vol. 90, p. 5843, May, 1989b.

36. Nagarajan, R. and Ruckenstein, E.: "Theory of Surfactant Self-Assembly: A Predictive Molecular Thermodynamic Approach," *Langmuir*, vol. 7, p. 2934, 1991.

37. Nghiem, L.X., M.S. Hassam, R. Nutakki, and A.E.D. George: "Efficient Modelling of Asphaltene Precipitation," SPE 26642 presented at 1993 SPE *Annual Technical Conference and Exhibition*, Houston, TX, Oct. 3–6, 1993.

38. Nielsen, B.B., W.Y. Svrcek, and A.K. Mehrotra: "Effect of Temperature and Pressure on Asphaltene Particle Size Distributions in Crude Oils Diluted with *n*-Pentane," *Ind. Eng. Chem. Res.*, vol. 33, p. 1324, 1994.

39. Pan, H. and A. Firoozabadi: "Thermodynamic Micellization Model for Asphaltene Precipitation from Reservoir Crudes at High Pressures and Temperatures," paper *SPE* 38857 presented at the 1997 *SPE Annual Technical Conference and Exhibition, San Antonio, Texas, Oct. 5–8, SPE Prod. & Fac.* (in review) 1997a.

40. Pan, H., A. Firoozabadi, and P. Fotland: "Pressure and Composition Effect on Wax Precipitation: Experimental Data and Model Results", *SPE Prod. and Fac.*, p. 250, Nov. 1997b.

41. Pan, H., and A. Firoozabadi, "Complex Multiphase Equilibrium Calculations by Direct Minimization of Gibbs Free Energy Using Simulated Annealing" *SPE Res. Eng.*, p. 36, Feb. 1988a.

42. Pan, H. and A. Firoozabadi, "Thermodynamic Micellization Model for Asphaltene Aggregation and Precipitation in Petroleum Fluids," *SPE Prod. & Fac.*, p. 118, May 1998b.

43. Pan, H. and A. Firoozabadi: "Thermodynamic Micellization Model for Asphaltene Precipitation Inhibition," 1998c (to be submitted).

44. Pederson, K.S., P. Skovborg, and H.P., Rønningsen: "Wax Precipitation from North Sea Crude Oils. 4. Thermodynamic Modelling," *Energy and Fuels*, vol. 5, No. 6, p. 924, 1991.

45. Peng, D.-Y. and D.B. Robinson: "A New Two-Constant Equation of State," *Ind. Eng. Chem. Fund.*, vol. 15, p. 59, 1976.

46. Philp, R.P.: "High Temperature Gas Chromatography for the Analysis of Fossil Fuels: A Review," *J. High Resolution Chromatography*, vol. 17, p. 398, 1994.

47. Prausnitz, J.M., R.N. Lichtenthaler, , and E.G. Azevedo: *Molecular Thermodynamics of Fluid Phase Equilibria,* Prentice-Hall Inc., Englewood Cliffs, NJ, Chap. 9., 1986

48. Riazi, M.R., T.A. Al-Sahhaf: "Physical Properties of n-Alkanes and n-Alkylhydrocarbons: Application to Petroleum Mixtures," *Ind. Eng. Chem. Res.,* vol. 34, p. 4145, 1995.

49. Rønningsen, H.P., B. Bjørndal, A.B. Hansen, and W.B. Pedersen: "Wax Precipitation from North Sea Crude Oils. 1. Crystallization and Dissolution Temperatures, and Newtonian and Non-Newtonian Flow Properties," *Energy & Fuels,* vol. 5, 895, 1991.

50. Shaerer, A.A., C.J. Busso, A.E. Smith, and L.B. Skinner: "Properties of Pure Normal Alkanes in the C_{17} to C_{36} Range," *J. Amer. Chem. Soc.,* vol. 77, p. 2017, 1955.

51. Sheu, E.Y. and D.A. Storm: "Colloidal Properties of Asphaltenes in Organic Solvents," in *Asphaltenes: Fundamentals and Application,* E.Y. Sheu and O.C. Mullins (eds.), Plenum Press, NY, Chap. 1, 1995.

52. Smith, V.S, P.O. Campbell, V. Vandana, and A.S. Teja: "Solubilities of Long-Chain Hydrocarbons in Carbon Dioxide," *Int. J. Thermophysics,* vol. 17, no. 1, p. 23, 1996.

53. Snyder, R.G, G. Conti, H.L. Strauss, and D.L. Dorset: "Thermally-Induced Mixing in Partially Microphase Segregated Binary n-Alkane Crystals," *J. Phys. Chem.,* vol. 97, p. 7342, 1993.

54. Snyder, R.G., M.C. Goh, V.J.P. Srivatsavoy, H.L. Strauss, and D.L. Dorset: "Measurement of the Growth Kinetics of Microdomains Binary n-Alkane Solid Solutions by Infrared Spectroscopy," *J. Phys. Chem.,* vol. 96, p. 10008, 1992.

55. Snyder, R.G., V.J.P. Srivatsavoy, D.A. Cates, H.L. Strauss, J.W. White, and D.L. Dorset: "Hydrogen/Deuterium Isotope Effects on Microphase Separation in Unstable Crystalline Mixtures of Binary n-Alkanes," *J. Phys. Chem.,* vol. 98, p. 674, 1994.

56. Storm D.A. and E.Y. Sheu: "Characterization of the Colloidal Asphaltenic Particles in Heavy Oil", *Fuel,* vol. 74, p. 1140, 1995.

57. Storm D.A., R.J. Barresi, and E.Y. Sheu: "Development of Solid Properties and Thermochemistry of Asphalt Binders in the 26–65°C Temperature Range", *Energy & Fuels,* vol. 10, p. 855, 1996.

58. Storm, D.A., R.J. Barresi, and E.Y. Sheu: "Evidence for the Micellization of Asphaltenic Molecules in Vacuum Residue," PREPR, ACS, Division of Petroleum Chemistry, vol. 40, p. 776, 1995.

59. Ungerer, P., B. Faissat, C. Leibovici, H. Zhou, E. Behar, G. Moracchini, and J.P. Courcy: "High Pressure-High Temperature Reservoir Fluids: Investigation of Synthetic Condensate Gases Containing a Solid Hydrocarbon," *Fluid Phase Equilibrium,* p. 287, 1995.

60. Van der Kooi, H.J., E. Flöter, and Th.W de Loos: "High-Pressure Phase Equilibria of $\{(1-x)CH_4 + xCH_3(CH_2)_{18}CH_3\}$," *J, Chem. Thermodynamics,* vol. 27, p. 847, 1995.

61. Victorov, A.I. and A. Firoozabadi: "Thermodynamics of Asphaltene Precipitation in Petroleum Fluids by Micellization Model," *AIChE J.,* vol. 42, p. 1753, 1996.

62. Wiehe, L.A. and K.S. Liang: "Asphaltenes, Resins, and Other Petroleum Macromolecules," *Fluid Phase Equilibria,* vol. 117, p. 201, 1996.

63. Won, K.W.: "Thermodynamics for Solid Solution-Liquid-Vapor Equilibria: Wax Phase Formation from Heavy Hydrocarbon Mixtures," *Fluid Phase Equilibria,* vol. 30, p. 265, 1986.

64. Wu, J.: "Molecular Thermodynamics of Some Highly Asymmetric Liquid Mixtures," PhD Dissertation, Chemical Engineering Department, University of California-Berkeley, 1998.

65. Zhou, J.L., A.L. Tits, and C.T. Lawrence: "Users Guide for FFSQP Version 3.6: A FORTRAN Code for Solving Constrained Nonlinear (Minimax) Optimization Problems, Generating Iterates Satisfying All Inequality and Linear Constraints," Electrical Engineering Department, University of Maryland, 1996.

Index

About the Author

Abbas Firoozabadi is an internationally known expert in petroleum reservoir engineering. The director of Reservoir Engineering Research Institute (RERI) in Palo Alto, California, he is also a professor at Imperial College in London, U.K. He has taught at Stanford University and the University of Texas, Austin and has published more than 90 papers